Lithic Raw Material Economies in Late Glacial and Early Postglacial Europe

Edited by

Lynn E. Fisher
Berit Valentin Eriksen

BAR International Series 1093
2002

Published in 2016 by
BAR Publishing, Oxford

BAR International Series 1093

Lithic Raw Material Economies in Late Glacial and Early Postglacial Europe

ISBN 978 1 84171 471 4

BAR Publishing is the trading name of British Archaeological Reports (Oxford) Ltd. British Archaeological Reports was first incorporated in 1974 to publish the BAR Series, International and British. In 1992 Hadrian Books Ltd became part of the BAR group. This volume was originally published by Archaeopress in conjunction with British Archaeological Reports (Oxford) Ltd / Hadrian Books Ltd, the Series principal publisher, in 2002. This present volume is published by BAR Publishing, 2016.

Printed in England

BAR
PUBLISHING

BAR titles are available from:

BAR Publishing
122 Banbury Rd, Oxford, OX2 7BP, UK
EMAIL info@barpublishing.com
PHONE +44 (0)1865 310431
FAX +44 (0)1865 316916
www.barpublishing.com

Table of Contents

Preface

During several hundred thousand years of human prehistory siliceous rocks such as flint and chert probably were the most important raw materials used for tool production. Admittedly, this observation is to some extent a reflection of the fact that lithic artifacts are largely unaffected by factors severely limiting the survival of organic remains. However, the fact remains that lithic inventories or assemblages, and the context they belong to, generally represent the best or even the only source of information on prehistoric Stone Age societies.

Evidently, lithic analyses of all kinds have a long history in archaeological studies. For many years, morphological studies of tool-types and elaborate discussions of relative chronologies were a primary focus, but this has changed and the past few decades have witnessed a steadily growing interest in many other aspects of lithic studies. Detailed analyses of inventories from the Paleolithic through the Neolithic document characteristic changes in lithic raw material economy through time. Changes express themselves in variations in the technologically and culturally determined patterns of procurement, use and discard applied to any flint nodule during its entire 'life cycle' (the so called '*châine opératoire*', or operational sequence). Different cultural traditions are thus not only characterized by the well known typological succession of artifact types, but also to a very high degree by different technological and socio-economic processes pertaining to raw material procurement and use, as well as the technical or functional mode of exploitation (knapping and further modification) of nodules and blanks.

The papers in the present volume provide a variety of perspectives on lithic exploitation patterns in late glacial and early postglacial Western Europe. It began as a symposium held at the 59th Annual Meeting of the Society for American Archaeology in Anaheim, April 1994. The intention of the Anaheim symposium was to bring together scholars working in several different areas of Europe to present comparative data and evaluate theoretical perspectives on the use of lithic raw materials by late glacial and early postglacial hunter-gatherers. The symposium was highly successful in this respect and initial versions of the chapters by Bicho, Eriksen, Fisher, Floss, Olive and Taborin, Thacker, and Valentin *et al.* were first presented at this symposium. In order to secure a geographically and chronologically more balanced coverage of the topic, we decided to solicit additional contributions from researchers working in neighboring parts of Europe. Along with revised versions of the papers presented in Anaheim, we are accordingly pleased to include additional chapters by Sulgostowska, Hahn, Otte *et al.*, and Straus in the current volume.

Following a brief introduction by the editors, the volume presents eleven chapters addressing various aspects of lithic raw material exploitation patterns in regions spanning much of continental Europe, from Poland to Portugal. It is our hope that this collection of papers will stimulate more inter-regional discussions of data and ideas as well as general interpretive problems regarding lithic raw material economy in late glacial and early postglacial.

Acknowledgements

The editors would like to thank all the contributors for their interest in the project and for their patience throughout the protracted process of bringing this volume to completion. Thanks are also due to Robert L. Kelly and Randall White, discussants at the Anaheim symposium, for their thoughtful comments on the original set of papers.

1. Lithic Raw Material Economy in Late Glacial and Early Postglacial Western Europe: Introduction

Berit Valentin Eriksen and Lynn E. Fisher

Strategies of lithic raw material acquisition, transport and use are an important source of information about regional organization of hunter-gatherer land use. A growing literature in hunter-gatherer archaeology has developed around discussions of how lithic raw material use can shed light on prehistoric territorial organization, mobility and communication, as well as the overall character of regional and inter-regional connections between social groups. A fundamental premise throughout the present volume thus is that regional and inter-regional patterns in the use and procurement of lithic raw materials from well-defined sources provide reliable information on hunter-gatherer mobility, territorial organization, and inter-group communication. To a comparative study of the highly mobile hunter-gatherers of late glacial and early postglacial Western Europe, such an approach brings major advantages.

Large parts of Western Europe were covered by ice in the Pleniglacial and only (re-) colonized during late glacial and early postglacial times. Mobility strategies, territorial organization and communication patterns must have been particularly vital aspects of the contemporary socioeconomic systems. Moreover, past hunter-gatherer mobility and settlement patterns are likely to have changed significantly during this period as a reflection of major environmental changes. Within a few millennia, landscapes throughout Europe changed markedly from open tundra, steppe, taiga or steppe-woodland to more densely wooded habitats, and it is widely accepted that the increased density of Boreal forests must have led to a wide array of changes in resource distribution and in hunter-gatherer settlement and subsistence systems.

Evidently, the late glacial and early postglacial comprise a period of major socioeconomic and cultural changes. These changes are also reflected in the relationships between the management of lithic raw material and other resources, subsistence-settlement systems, and social organization. Given the presumably great environmental and cultural variability within an area the size of Western Europe during the period in question it is reasonable to expect some spatial and temporal differences with respect to these issues on a macro-regional scale. By encouraging discussion of analogous or contrasting patterns in lithic raw material economy among geographically and geo-morphologically different areas (Poland, Belgium, Paris Basin, Central Germany, Southern Germany, Southern France, Northern Spain, Central Portugal), this volume aims at a detailed inter- as well as intra-regional comparison and contrast of approaches and results.

The following eleven chapters address various aspects of late glacial and early postglacial raw material economy in regions spanning much of continental Europe, from Poland to Portugal (Figure 1). The papers provide case studies and/or exhaustive reviews of currently available data on the use and procurement of lithic raw materials in relation to subsistence-settlement systems and socioeconomic organization of prehistoric hunter-gatherer groups.

To assure a certain degree of coherence and consistency throughout the volume authors were asked to address aspects of the following questions:

1. How is the acquisition of lithic raw materials used for tool production organized around other activities in a subsistence-settlement system and in a social system?
2. What is the relation of lithic raw material use to scheduling of activities, social organization (control of resources, specialization of production), and management of resources (quarrying, heat treatment, etc.)?
3. How much can raw material distributions on an archaeological landscape tell us about the range and organization of hunter-gatherer mobility?
4. Over what distances are lithic raw materials transported in the late glacial and early postglacial (compared to other materials)?
5. Can we distinguish between exchange and direct acquisition through mobility?

Authors were very cooperative in orienting their contributions toward these questions, which might explain the high degree of uniformity among approaches. We find, for instance, that a recurrent theme throughout the volume is the scheduling of activities, i.e., the question of planning ahead in anticipation of future lithic raw material needs. Virtually all chapters address this issue more or less explicitly.

On the other hand, the high degree of uniformity of approaches may also largely be a reflection of methodology. Although the contributions are derived from several different research traditions, several broadly shared analytical and methodological approaches unite the contributions.

The contributions to this volume share a strong central focus on conceptualizing technology as a cultural system. For example, although the French concept of '*châine opératoire*' or operational sequence (Geneste 1991) and the Anglophone notion of technological organization (Nelson 1991) diverge in their overarching theoretical goals (Perlès 1993), they tend to converge in analytical approaches that emphasize 1) modeling technological processes in terms of a series of steps including raw material acquisition, processing, use and discard, 2) considering the various alternatives or perceived costs and benefits that might have influenced or determined choices made by prehistoric

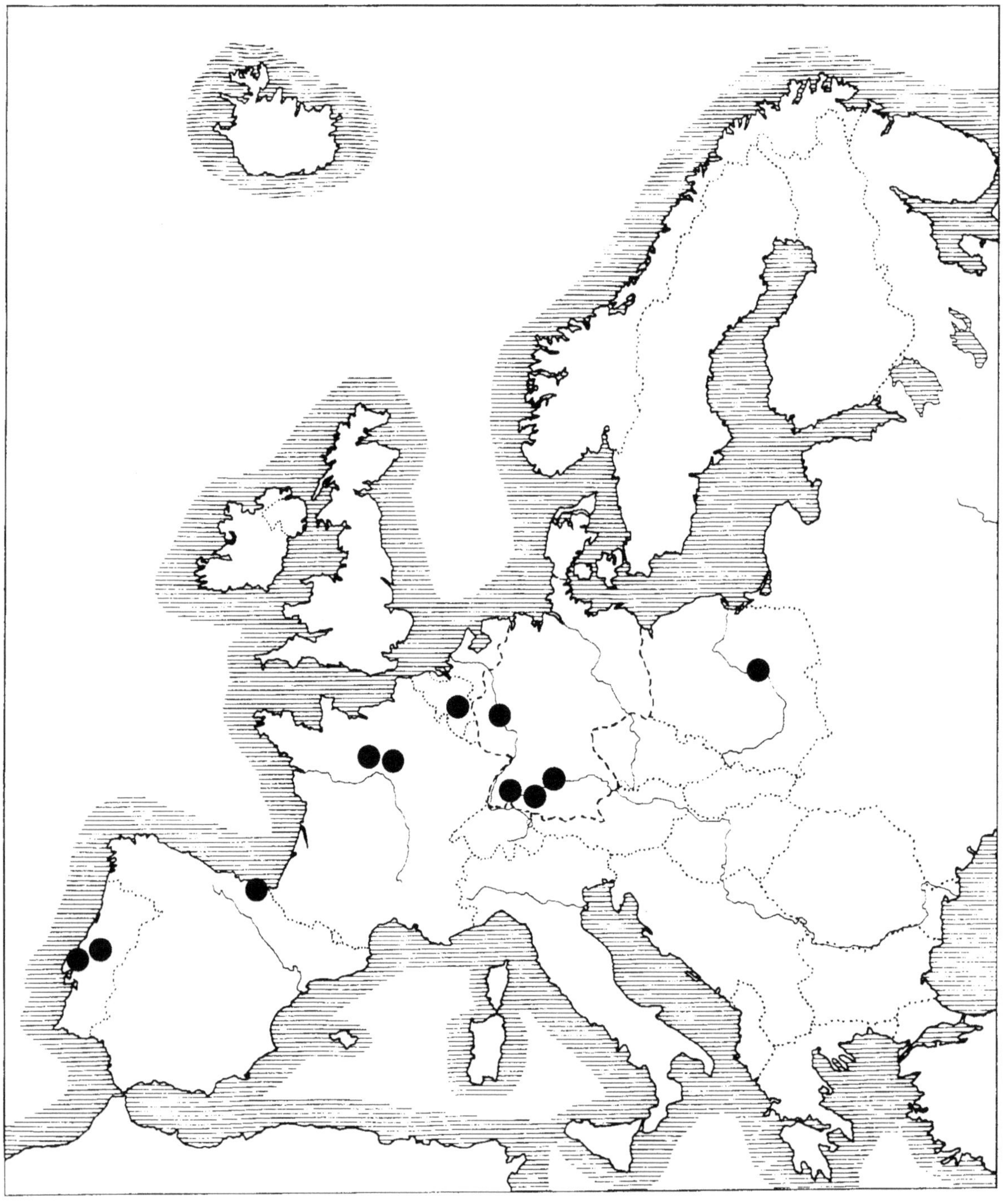

Figure 1: Map showing the location of the case studies in Europe.

flintknappers at each of these steps, and 3) exploring variability in these choices in different seasonal and/or socioeconomic contexts. Work within these three shared areas constitutes a growing body of specifically archaeological theory focused on technological choices. This body of theory is rooted in empirical observations drawn from ethnoarchaeology (Binford 1979; Yellen 1977). Accordingly, most of the authors make use of the collector-forager model of mobility organization in hunter-gatherer societies (Binford 1979, 1980; Kelly 1983), and a set of technological constructs such as embedded procurement, curation/expediency, and "gearing up". Archaeological tests of these models have led to some revisions and criticism (e.g., Odell, ed. 1996; Thacker, this volume), but it cannot be denied that these ideas have created a lively and growing area of debate that has profoundly affected our thinking on prehistoric technology.

In addition, most of the papers share methodological limitations in the identification of raw materials. Virtually all observations and discussions with respect to sourcing are based on macroscopic examinations of the lithic raw materials. This methodological bias was not intended from an editorial point of view. It is rather a reflection of the widespread use of more inexpensive methods (macroscopic examination) by comparison to costly laboratory methods (microscopic analysis, trace element analysis, emission spectrometry, neutron activation analysis, etc.). Laboratory methods generally are expensive and time-consuming and results are accordingly limited. Only very few researchers have applied such methods in the study of late glacial and early postglacial lithic inventories from Western Europe (for references see for instance Affolter 1991 and Grünberg 1988). A very productive direction in lithic studies that can be seen in several contributions in this volume, though, is the use of systematic survey for locating raw material sources. The papers by Floss, Olive and Taborin, Otte *et al.*, and Thacker all draw on results of major field research aimed at locating and describing raw material sources.

Another similarity of approach in many chapters is the focus on exploring relationships between variations in core technology (e.g., reduction sequences, core preparation) and variations in raw material quality or availability (Hahn, Thacker, Bicho), site function (Valentin *et al.*), or settlement mobility (Hahn). The papers in nearly all cases present detailed information on the uses of various kinds of raw materials under varying circumstances. The result is a large body of data that can be used to assess different sources of variation in core technology across regions. The methodological homogeneity of the chapters allows for almost direct comparison of both data and results concerning lithic raw material procurement strategies among prehistoric hunter-gatherer groups. The volume thus confirms the existence of common tendencies as well as regional diversity in lithic raw material economy in late glacial and early postglacial Western Europe.

It seems widely accepted that lithic raw materials in the late glacial and early postglacial were obtained primarily through embedded procurement, that is, raw materials were gathered and transported by hunter-gatherers in the course of their other activities, without significant travel motivated purely by the search for raw material. However, contributions to this volume also provide evidence for late glacial quarry workshop sites from Poland (Sulgostowska), Belgium (Otte *et al.*), the Paris Basin (Olive and Taborin, Valentin *et al.*), and Portugal (Thacker). The model of Magdalenian raw material economy in the Middle Rhine region presented by Floss also implies a logistic pattern of long distance raw material transport which is more than just a question of embedded procurement.

It is also generally assumed that lithic raw materials were typically not traded during the period in question. At the same time, papers in this volume clearly show that the relationship between distance to raw material sources and hunter-gatherer seasonal or annual ranges is not a simple one. Several interesting questions are raised.

First is the question of the temporal and spatial scale of mobility reflected in raw material assemblages. Since stone raw materials may be rapidly exhausted and "drop out" of a mobile technological system, transport of lithics may represent only relatively short-term mobility (Eriksen). In areas with abundant raw materials, an overall focus on local raw materials may obscure differences in mobility between time periods (Straus). A second area of interest is the scheduling of tool production in a mobile activity pattern. Are stone raw material acquisition and lithic production distributed relatively evenly across space and time, or are these activities (or particular stages of work) clustered at sites where flintknapping is a major activity (Olive and Taborin, Otte *et al.*, Valentin *et al.*)? Third, several papers tackle the problem of understanding superimposed patterns of long- and short-distance transport of materials in a region, as a function of the technical requirements of making different tools (Otte *et al.*, Thacker), or in a context of "macro-" and "micromoves" (Hahn). Finally, Sulgostowska notes the possibility that changes in raw material distribution in the early postglacial may reflect the presence of territorial boundaries or other social barriers to movement of materials.

Overall, discussions of raw material acquisition strategies in this volume move well beyond the simple dichotomy of embedded vs. direct procurement. The distance over which raw materials are transported by mobile hunter-gatherers should be affected by a number of factors, including the seasonal and annual range of hunter-gatherer groups, the distribution of raw materials on the landscape, stone tool needs, and tool use lives. The complexity of these possible contributing factors suggests a need for careful consideration of modes of transport.

One very productive direction for research indicated by several authors (Eriksen, Olive and Taborin, Sulgostowska) is in comparison of differences in distance and mode of distribution of stone raw materials versus other materials and objects such as mollusks, jet, and ochre. This comparative perspective provides an empirically based approach to understanding the role of various transport mechanisms in prehistory. Sulgostowska's contribution is interesting in making a link between lithics and other transported materials, in the special case of very wide distribution of chocolate flint, which might be symbolically associated with extraction and distribution of ochre. This suggests a context for exploring variations in distance of lithic transport in terms of regional social networks.

Modes of transport of raw materials are a major focus of work in most papers in the volume. This approach should shed light on the significance of data on raw material distributions in terms of mobility and possibly in terms of social networks. However, due to general limitations of available data, the question initially posed concerning the social organization of lithic production was taken up for consideration by fewer authors. Indeed, one of the major challenges facing the stone age archaeologist is—how to extract information about past sociocultural and socioeconomic relationships from find assemblages consisting almost exclusively of lithic artifacts. The extraordinary record of behavior provided by intrasite spatial analysis and refitting on some Paris Basin Magdalenian sites provides an unusual perspective on use of stone raw materials by interacting residential groups (Olive and Taborin). Analysis of raw material use by different residential groups suggests variation in skill, productivity of knapping, and in the quality of raw materials used. At the same time, it is clear that lithic production among the highly mobile Magdalenian hunter-gatherers varied in response to seasonal changes in activities (Valentin *et al.*).

The papers as a group document considerable regional and temporal variation in raw material procurement and use. Most papers deal with long-term chronological aspects or the question of change or continuity in raw material procurement strategies during a period of several thousand years, i.e. either the Pleniglacial (Floss, Otte *et al.*, Valentin *et al.*, Thacker) or the late glacial and early post-glacial (Sulgostowska, Fisher, Eriksen, Otte *et al.*, Straus, Bicho). Interestingly we find that many authors do observe an increase in the opportunistic use of locally available raw materials in the early postglacial. Several authors also note marked changes in the scheduling of lithic production in the early Holocene. These observations are independent of geographical region and might perhaps be related to the marked climatic and paleo-environmental changes during the period in question and the accompanying adaptive changes in weapon and tool design (Fisher, Otte *et al.*), as well as mobility and land-use patterns (Sulgostowska, Floss, Eriksen).

Given the extent of the region covered in this volume, regional diversity understandably is marked, but this is at least in part due to geological and geomorphological differences. That is, differences in raw material types, quality, and not least availability, account for a major part of the observed regional variation in raw material use and procurement strategies. The studies in this volume also show differences in the distances over which raw materials were transported in the southern (tens of kilometers) versus northern continental regions (hundreds of kilometers) of Europe.

As previously mentioned, papers in this volume also allow for a detailed intra-regional comparison and contrast of approaches and results. The Paris Basin is the object of two papers presented by prominent exponents of the French school of lithic research. The single site case study of the Magdalenian site Etiolles (Olive and Taborin) is complemented by a more general view on lithic raw material economy through time in the late glacial Paris Basin (Valentin *et al.*).

Southwestern German data, on the other hand, has been looked at through three different types of glasses: a German presentation of the Magdalenian site Buttentalhöhle (Hahn) is here complemented by two more general views from abroad—an Anglo-American study of raw material choices in the context of changing tool design goals over time (Fisher) and a continental, North-European study on the use and procurement of non-local lithics and materials used for personal adornment (Eriksen). The professor (Hahn) and his two foreign students (Fisher and Eriksen) here look at largely the same group of sites from both a common perspective (the Tübingen school of lithic research) and from three quite independent research traditions.

In the case of Southwestern Germany we find this possibility for intra-regional comparison and contrast of different research traditions highly rewarding in the way it opens up new directions for joint research on past hunter-gatherer mobility. This would also seem to be the case with respect to the Portuguese Estremadura where Bicho's suggestion that increased logistic mobility can be attributed generally to the period 12,000 - 9,000 BP contrasts with Thacker's proposition that increased logistical mobility associated with longer occupation of camps and diversified resource use occurs only in the early postglacial. This difference in interpretation invites further examination of intersite variation in raw material use in the Late Magdalenian in this region.

The intra-regional comparisons reveal that some of the variation between case studies may be attributed partly to past sociocultural and socio-economic differences and partly to differing scholarly traditions. The evidence of differential use of materials by different residential groups in the late glacial Paris Basin is thus first and foremost the result of endless hours of meticulous refitting of inventories. Analogous observations could very likely be made in other regions if equally exhaustive analytical techniques were applied.

At least one fundamental conclusion may be drawn from the contributions in this volume: the fact that raw material studies cannot stand alone. They need to be combined with other analytical approaches providing information on, for example, season, duration, extent and function of the occupation as reflected in the archaeological remains. The use and procurement of lithic raw materials for tool manufacture must be seen within the broader context of hunter-gatherer subsistence and settlement systems.

References Cited

Affolter, J.
1991 Analyse pétrographique du silex: origine des matières premières. *Cahiers d'archéologie jurasienne* 2:81-89.

Binford, L. R.
1979 Organization and formation processes: Looking at curated technologies. *Journal of Anthropological Research* 35:255-273.
1980 Willow smoke and dogs' tails: Hunter-gatherer settlement systems and archaeological site formation. *American Antiquity* 45:4-20.

Geneste, J.-M.
1991 Systèmes techniques de production lithique: variations techno-économiques dans les processus de réalisation des outillages paléolithiques. *Techniques et culture* 17-18:1-35.

Grünberg, J. M.
1988 *Das Rohmaterial der Steinartefakte von Andernach: Ein Beitrag zur Anwendung naturwissenschaftlichen Verfahren in der Archäologie*. British Archaeological Reports, International Series 448. BAR Publishing, Oxford.

Kelly, R. L.
1983 Hunter-gatherer mobility strategies. *Journal of Anthropological Research* 39:277-306.

Nelson, M. C.
1991 The study of technological organization. In *Archaeological Method and Theory, Volume 3,* edited by M. Schiffer, pp. 57-100. University of Arizona Press, Tucson.

Odell, G., editor
1996 *Stone Tools: Theoretical Insights into Human Prehistory.* Plenum Press, New York.

Perlès, C.
1993 Ecological determinism, group strategies, and individual decisions in the conception of prehistoric stone assemblages. In *The Use of Tools by Human and Non-Human Primates*, edited by A. Berthelet and J. Chavaillon, pp. 266-277. Clarendon Press, Oxford.

Yellen, J. E.
1977 *Archaeological Approaches to the Present.* Academic Press, New York.

2. Flint Raw Material Economy During the Late Glacial and Early Postglacial in the Oder-Daugava-Prypet Basin

Zofia Sulgostowska

In Poland, studies of lithic raw material sources have a long tradition of advanced research. This chapter presents the current state of research in the Late Glacial and early Postglacial of the Oder-Daugava-Prypet Basin. This area is particularly interesting because of evidence for Late Glacial pit quarrying of high-quality Jurassic cherts including the well-known chocolate flint of the Holy Cross region. Late Glacial systems of flint processing and transport in Poland include multi-stage processing and transport of materials between workshops, habitation sites, and caches. The distribution of flint and distance and mode of raw material transport is compared in Late Paleolithic (Magdalenian and Tanged Point) and Mesolithic contexts, and between regions with varying flint sources. Interesting differences exist between the distribution of Jurassic cherts in the Cracow and Warta regions, where local cherts make up the majority of all assemblages and long-distance imports are relatively rare, and the wide-spread Late Glacial transport of blades and blade tools of chocolate flint, which comprises up to 90% of some assemblages as distant as 200 kilometers from the source, with lower frequencies common up to 350 kilometers. Current evidence suggests no specialization in lithic production, and no barriers to access in the Late Glacial, though a system of social control over resources may have existed at the ochre processing site of Rydno. Finally it is suggested that a general pattern of increased reliance on local rather than imported raw materials in the early Postglacial may reflect a reduced range of mobility due to the decline of migratory fauna, and possibly also the first appearance of territorial boundaries that limit access to regional raw material sources.–*Editors.*

Introduction

Flint raw material economy is a fascinating subject of study, which sheds light on the development of technology, the environment and its exploitation, the economy and the structure of society. In the territory under discussion, several sources of macroscopically identifiable flints make it possible to trace a wide distribution of flint imports and a system of contacts between human group that changes during the late glacial and early postglacial. It is not easy to gain an overview of this subject, because of the long history of research by at least three generations of archaeologists, inaugurated by the eminent prehistorian Stefan Krukowski (Krukowski 1920, 1922, 1939-48, 1961) and reflected in a large literature on the subject. This paper is an attempt to identify some problems and to present the current state of research. Less information is available from the eastern area, accounting for the restriction I have imposed on the territory under consideration. The archaeological investigations are not sufficiently supported by petrographic evidence, although petrographic studies began early in the region (Samsonowicz 1923); as result of this, the identification of flint is based mostly on macroscopic analysis.

This paper will address the following points:

1) flint material supply,
2) methods of exploitation of flint sources,
3) flint processing in specialized workshops,
4) access to flint and the flint-using economy,
5) distribution of different kinds of flint, and
6) the flint economy as a reflection of social structure.

Flint Material Supply

An uneven distribution of flint sources in the Oder-Daugava-Prypet basin (Figure 1) is due to varying underlying geology. The northern part of the area belongs to the European Plain, while the southern part is characterized by uplands with sediments rich in flint outcrops. In the part of Europe under discussion the main lithic raw material is flint, while others, such as radiolarite or obsidian, are rare imports.

Cretaceous flint is available over almost the whole territory, except the southern mountain region and the lower Neman and Daugava basins in the north. So-called "Baltic flint" occurs in post-glacial sediments as erratic nodules from several to a dozen or more centimeters in diameter. Traces of aeolian polish, wear and cracking due to transport and frost on the surface of these nodules indicate that they were collected from the surface. Outcrops of good quality Cretaceous flint are located on the rocks near the rivers Bug, Narew and Pripat, and provided nodules up to 30 centimeters in diameter. The macro- and microscopic similarity of Cretaceous flints from different regions excludes them from analysis of distribution.

There are also several flint outcrops in the southern part of the territory. In the Kraków-Wieluń upland, there are outcrops of Upper Jurassic flint from limestones of Oxfordian age on the Upper Warta river and in the Cracow region. The Turonian flint (referred to in Polish literature as Swieciechów or grey dotted) outcrops are located in the Vistula valley near the mouth of the San river. The Late Oxfordian or Early Kimmeridgian "chocolate flint" comes from the Holy Cross mountain region, where several macroscopically different groups can be recognized (Schild 1971).

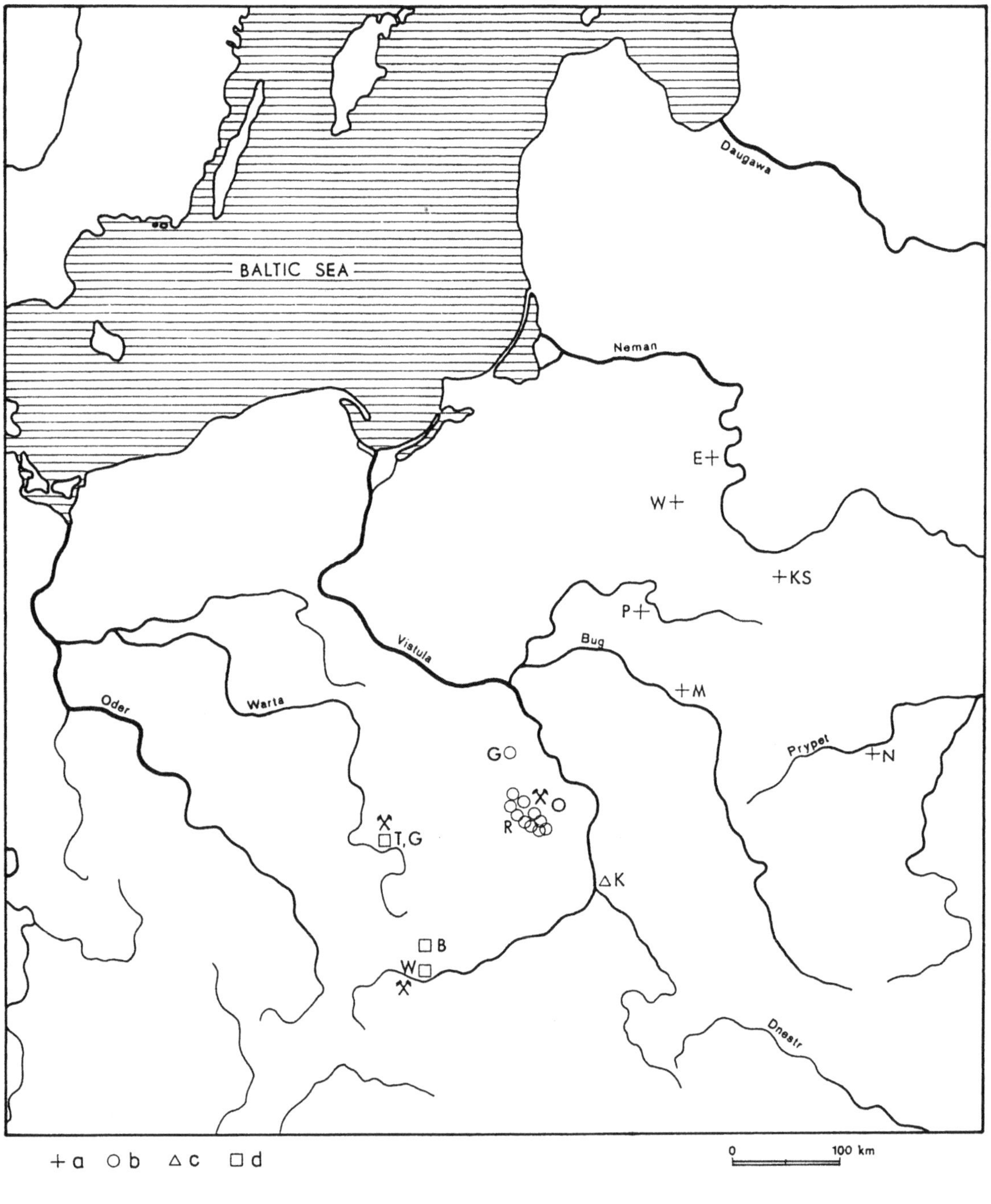

Figure 1. Distribution of flint sources, extraction points and flint workshops during the Late Glacial. Cretaceous flint (a): Eżerynai (E), Krasne Selo (KS), Mielnik (M), Nobel (N), Płonka (P), Wołkusz (W); Chocolate flint (b): Gulin (G), Orońsko (O), Rydno (R); Turonian flint (c): Kopiec (K); Jurassic flint (d): Brzoskwinia (B), Gojść (G), Trzebca (T), Wołowice (W). Sources: Krukowski 1939-48; Rimantiene 1971; Schild 1971, 1975; Ginter 1974, 1993; Gieysztor-Szymczak 1981; Sulgostowska 1989, 1990; Szymczak 1992; Sobczyk 1993; Florek and Libera 1994; Kudriashov and Lipnitsckaia 1994.

In connection with the lithic raw material economy it is necessary to recall the requirements of producers, which were determined by the technology used and by the final product. The most complex flint processing, connected with multi-stage preparation, can be seen in the Late Paleolithic Tanged Point Technocomplex (Mazovian cycle, i.e., the Sviderian culture and Ahrensburgian culture). Here, blade cores with two opposing platforms were worked with an organic punch and soft hammer. These complexes show evidence of extraction activities, specialized flint processing and wide distribution of imports. During the Mesolithic Vistulian cycle, i.e., the Janisławice culture, such specialized flint processing based on single-platform blade cores is known, though on a minor scale.

Methods of Source Exploitation

Evidence of flint quarrying from the late glacial is known for chocolate flint from Orońsko (Krukowski 1939-48; Schild 1971), where pits 1.2 to 2 meters in diameter and up to 3.2 meters deep were dug in residual karstic clays. In the sediments filling one of those pits, an arched backed bladelet typical of assemblages dated to Allerød or Early Younger Dryas was found.

There is also evidence of Jurassic flint quarrying in the Kraków-Wieluń upland. At Wołowice on the Magdalenian site I, six pits up to 1.5 meters in diameter and 1.7 meters deep were dug in sand and clay sediments (Dagnan-Ginter 1976; Sobczyk 1993). The C^{14} age is 10,920 ± 200 BP (Gd 4654). The second center of Jurassic flint exploitation, ascribed to the Tanged Point Technocomplex (Mazovian cycle, Sviderian culture), is located north of the Warta valley near Gojść (Ginter 1974, 1993). There are several pits, most of them 1.5 to 2 meters in diameter and slightly smaller in depth, though an example 6 meters in diameter and 2.5 meters deep has also been recorded. These were probably dug using axe preforms which constitute typical elements of the workshop tool inventory (Figure 2). On the surface, this activity can be observed over an area of approximately 100 by 20 meters.

The same method of exploitation (pits) is assumed for chocolate flint on the Late Mesolithic site II at Tomaszów, though the relevant parts of the site were destroyed (Schild *et al.* 1985). Also, concentrations of Mesolithic sites are associated with the accumulation of Baltic erratic flint sources in the Great Poland region, though no pits were recorded (Kobusiewicz 1989).

There was probably a common exploitation method for all kinds of flint outcrops, including Turonian and Cretaceous, but the majority of flint nodules seem to be collected from the surface.

Flint Processing in Specialized Workshops

The type of site observed by S. Krukowski (1920, 1922, 1939-48) constitutes specialized workshops for flint processing. They vary in the structure of flint inventories, reflecting the purpose of producers, whether early preparation of nodules, advanced preparation, early exploitation of cores, advanced exploitation, and/or tool production (Schild 1980). Flint inventories indicate the following kinds of workshops: those where preliminary selection of nodules was made (at the quarry); those where core preforms and cores, later exported, were prepared (quarry area workshops); and workshops where blades were processed. There is also a special type of workshop where blades were produced and exported but also made on the spot into tools and used, as shown by the presence of burin spalls, broken tanged points and scrapers with use-wear polish - examples are Rydno, Gulin, Nobel, Krasne Selo, Wołkusz, and Kopiec.

The location of workshops relative to flint sources varies. At the Gojść and Trzebca complexes (Ginter 1974, 1993), Wołowice complex (Sobczyk 1993) and Tomaszów II (Schild *et al.* 1985), workshops for processing core preforms, cores and blades were set up nearby, from a dozen to several hundred meters from the source. Workshops specializing in blade production were located at a distance of one (Nobel) to 15 or more kilometers from outcrops (Rydno 1, Gulin).

The area of workshops, covering from a dozen to tens of flint concentrations, varies from 2 (Gulin) to 2.5 hectares (Gojść, Ginter 1993) up to several square kilometers as at Rydno (Schild and Królik 1981) or Nobel (Sulgostowska 1989). Concentrations are oval in shape, one to ten meters in diameter, and do not usually overlap. The distance between concentrations varies from several dozen centimeters to several meters. The number of items per concentration ranges from hundreds up to 20,000, and density ranges from several tens to more than a thousand objects per square meter (Figure 3).

The amount of flint processed is hard to calculate because the majority of the final product was exported, but the weight of what remains amounts to several tens of kilograms (in blade workshops with habitation elements, e.g., Gulin) or several hundreds of kilograms (Orońsko).

All kinds of flint used during the late glacial were processed in specialized workshops, the distribution of which is shown in Figure 1. Cretaceous flint workshops are known from the territory of northeastern Poland (Mielnik [Szmit 1929], Plonka [Gieysztor-Szymczak 1981], Wołkusz [Szymczak 1992]), southern Lithuania (Eżerynai 8 and 17 [Rimantiene 1971]), western Byelorussia

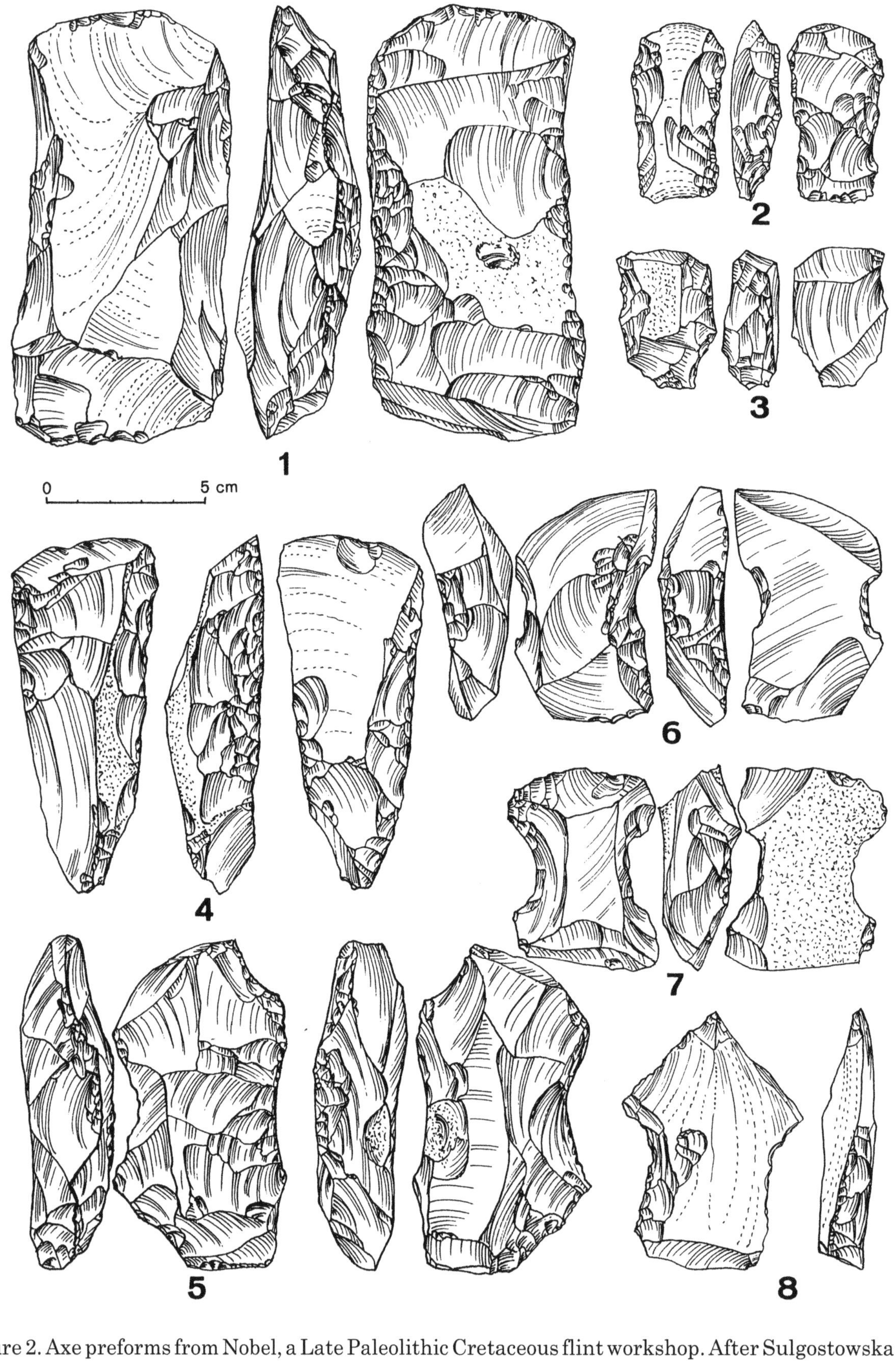

Figure 2. Axe preforms from Nobel, a Late Paleolithic Cretaceous flint workshop. After Sulgostowska 1989.

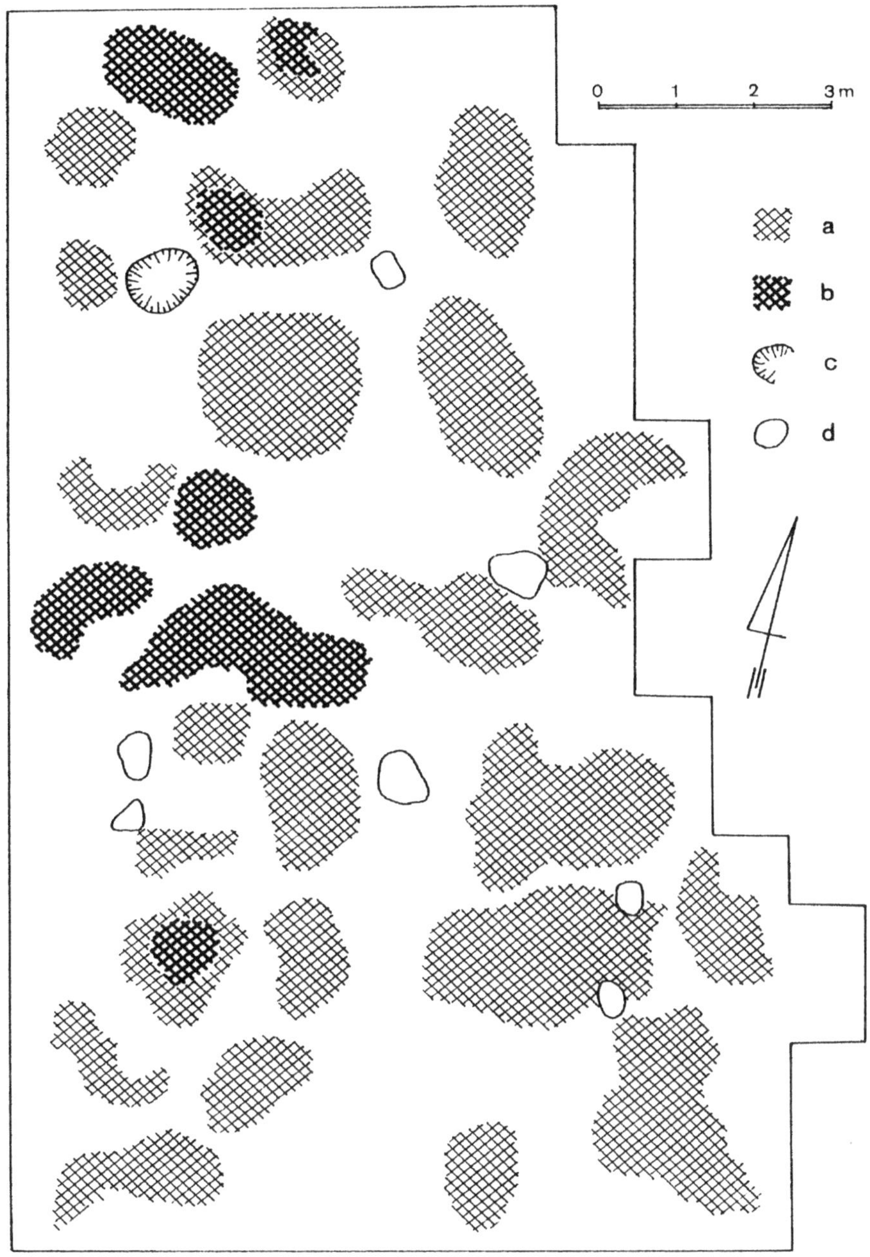

Figure 3. Gojśç 1 (cut III), a Late Paleolithic Jurassic flint workshop. Distribution of flint concentrations ("kshemenitsas") and their density. (a) fewer than 1000 flints per square meter; (b) more than 1000 flints per square meter; (c) disturbed area; (d) stones. After Ginter 1974.

(Krasne Selo [Krukowski 1976b; Kudriashov and Lipnitsckaia 1994]), and northeastern Ukraine (Nobel [Krukowski 1939-48; Sulgostowska 1989]). Jurassic flint was processed in the Cracov region at Wołowice (Dagnan-Ginter 1976; Sobczyk 1993), at Brzoskwinia (Sobczyk 1993) and also in the Upper Warta center at Gojść and Trzebca (Ginter 1974, 1993). Turonian grey dotted flint workshops were recently discovered at Kopiec (Florek and Libera 1994). Chocolate flint was processed at Orońsko (Krukowski 1939-48; Schild 1971), at the Rydno complex (Krukowski 1961; Schild and Królik 1981; Królik and Schild 1993), and at Gulin (Krukowski 1939-48; Sulgostowska 1986, 1991).

The sites mentioned above are associated with several taxonomic units. Wołowice and Brzoskwinia represent the Magdalenian, while the rest belong to the Tanged Point Technocomplex. Among the latter, the Bromme-Lyngby technology with single-platform cores worked by a hard hammer was found only at Eżerynai. Krasne Selo is characterized by abundant Ahrensburgian elements, while the rest of the workshops are associated with the Mazovian cycle (Sviderian culture). The large number of workshops associated with the Mazovian cycle results from the extremely intensive Mazovian technology, in which up to 50% of the nodule is lost during core preparation and rejuvenation, as has been proved by refittings (Fiedorczuk 1992). The second important complex, Arch Backed Piece flint

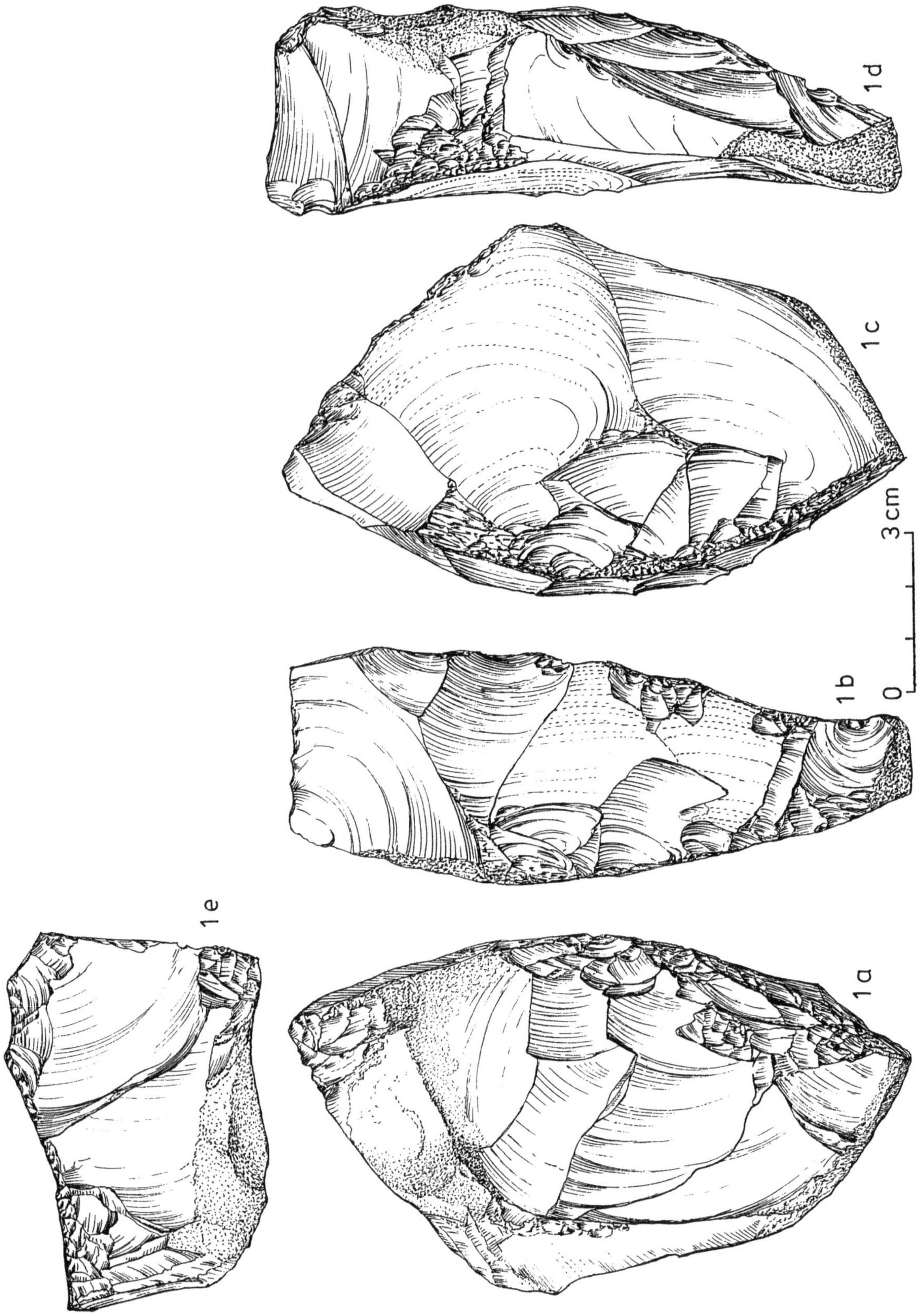

Figure 4. Core preform from Tomaszów II, a Late Mesolithic chocolate flint workshop. After Schild et al. 1985.

technology, was much simpler and did not require such good raw material.

From the Mesolithic, only Tomaszów II has workshops processing core preforms (Figure 4), cores, blades and tools of chocolate flint. These are connected with the Late Mesolithic Vistulian cycle (Janisławice culture) dated to 6555 ± 45 BP (GrN-7051) (Schild *et al.* 1985).

Access to Raw Material and its Utilization

Access to flint is conditioned by distance from sources and assumes a lack of limitation through ownership of sources by certain groups. Differential access to resources is reflected in the quantity, structure and morphology of inventories. Three sites with differing access to Cretaceous flint were chosen, and the structure of inventories was tested on the basis of the model proposed by J. K. Kozłowski (1980). The following sites were considered: the Salaspils Laukskola habitation site in the Daugava river drainage, without local flint (Zagorska 1994); the Augustów-Wójtowskie Włóki habitation site in the Mazurian Lakeland, with local flint; and the Nobel workshop with elements of a habitation site on the Pripat river, with abundant local flint (Sulgostowska 1989).

Results show that in concentrations from Salaspils Laukskola, artifacts were scarce, an absence or minimal number of exploited cores was noted, and the ratio of tools to waste from their production (for methodological reasons burin spalls are always included in the tools) varies from 30% to 56%. At Augustów-Wójtowskie Włóki, cores constitute 0.7% of the assemblage, and tools 9.5%, while among six concentrations from Nobel, cores make up 0.9-4.8% of the total and tools 3.7-16.3% (Sulgostowska 1989:77). In addition, the ratio of unretouched blades to blades made into tools and the proportion of multiple tools shows clearly that on the sites with limited access to flint, more blades were made into tools, flakes were often used as blanks, and multiple tools were produced more frequently (Sulgostowska 1986).

In conclusion, it is clear that raw material was economized in situations of difficult or limited access. How can we explain the fact that artifacts were abandoned on such a site as Salaspils Laukskola, when each piece was imported? Were they lost singly, in which case we must assume tens of settlement episodes for one concentration, or left purposely for some irrational reason? A different situation of prodigal use is observed on sites where stock could be easily renewed. It cannot be excluded that on such sites craftsmanship was improved and a new generation was taught.

Distribution of Flints

The distribution of flint can be calculated on the basis of the number of imports, their structure and scale, and distance from the sources. Some general patterns in flint distribution can be noted, with some exceptions. The general pattern is the following: Paleolithic and Mesolithic inventories within several dozens of kilometers of outcrops are dominated by local flint which constitutes approximately 90% of the assemblages; on workshop sites with multiple concentrations, imports of exotic raw materials are found; and imports are less common during the Mesolithic than the Paleolithic. Only flints from established sources which can be recognized macroscopically were analyzed. Cretaceous flint was excluded because of its similarity in different regions.

Jurassic flint from the Cracow (Sobczyk 1993) and Warta regions (Ginter 1974, 1993; Pelisiak 1987) was used only by local groups, and imports of other flints (i.e., chocolate) by these groups are rare. Occasional imports of Jurassic flint were found in the Paleolithic up to 150 kilometers from their source (Sulgostowska 1989), and in the Mesolithic up to 100 kilometers (Cyrek 1981). Turonian (Swieciechów) flint was also only of local importance, though it is sporadically found up to 200-250 kilometers from the source in the Paleolithic (Sulgostowska 1989) and up to 260 kilometers in the Mesolithic (Cyrek 1981).

The distribution of chocolate flint was exceptional due to the number of imports, their structure and distance from outcrops. This phenomenon has often been analyzed (Krukowski 1920, 1922, 1939-48; Schild 1971, 1975, 1996, n.d.; Cyrek 1981; S. K. Kozłowski 1989; Sulgostowska 1989, 1990), and a different situation in late glacial assemblages connected with the Tanged Point Technocomplex (Figure 5) by comparison with early postglacial assemblages (Figure 6) has been established. The dominant use of chocolate flint, comprising up to 90% of some inventories, was observed up to 200 kilometers from the sources in the late glacial. Lower frequencies at greater distances reflected the amplitude of band seasonal movements (Schild 1976). Almost all recently excavated sites located up to 350 kilometers from outcrops revealed the presence of single chocolate flint artifacts. For the most part, blades without cortex or blade tools such as end-scrapers, burins, tanged points or retouched blades were imported, while chips are rarely found. This pattern is observed at the Salaspils Laukskola site on Daugava river where 17 items (4 chips, blades, end-scrapers, burins and a tanged point) were found in four of the six concentrations.

During the early postglacial, chocolate flint use and distribution (Figure 6) is more sharply limited

(Cyrek 1981; Schild *et al.* 1985; S. K. Kozłowski 1989; Schild n.d.), and only during the Late Mesolithic does it play an important role among societies of the Vistulian cycle (i.e., Janisławice culture).

The Flint Economy seen as a Reflection of Social Structure

A complex system of extraction, processing and distribution of flint raw materials reflects the technical possibilities of exploitation of the environment and also the organization of societies. There are still some problems under discussion, as for example the existence of groups specializing in those activities. It seems that no specialization existed; rather the complex flint economy reflects the seasonal round of activities. Extraction was probably connected with warmer seasons, when digging pits would be easier (Ginter 1993). Another question is whether different groups had free access to outcrops of the material in demand. This was analyzed at the Rydno complex, a prehistoric

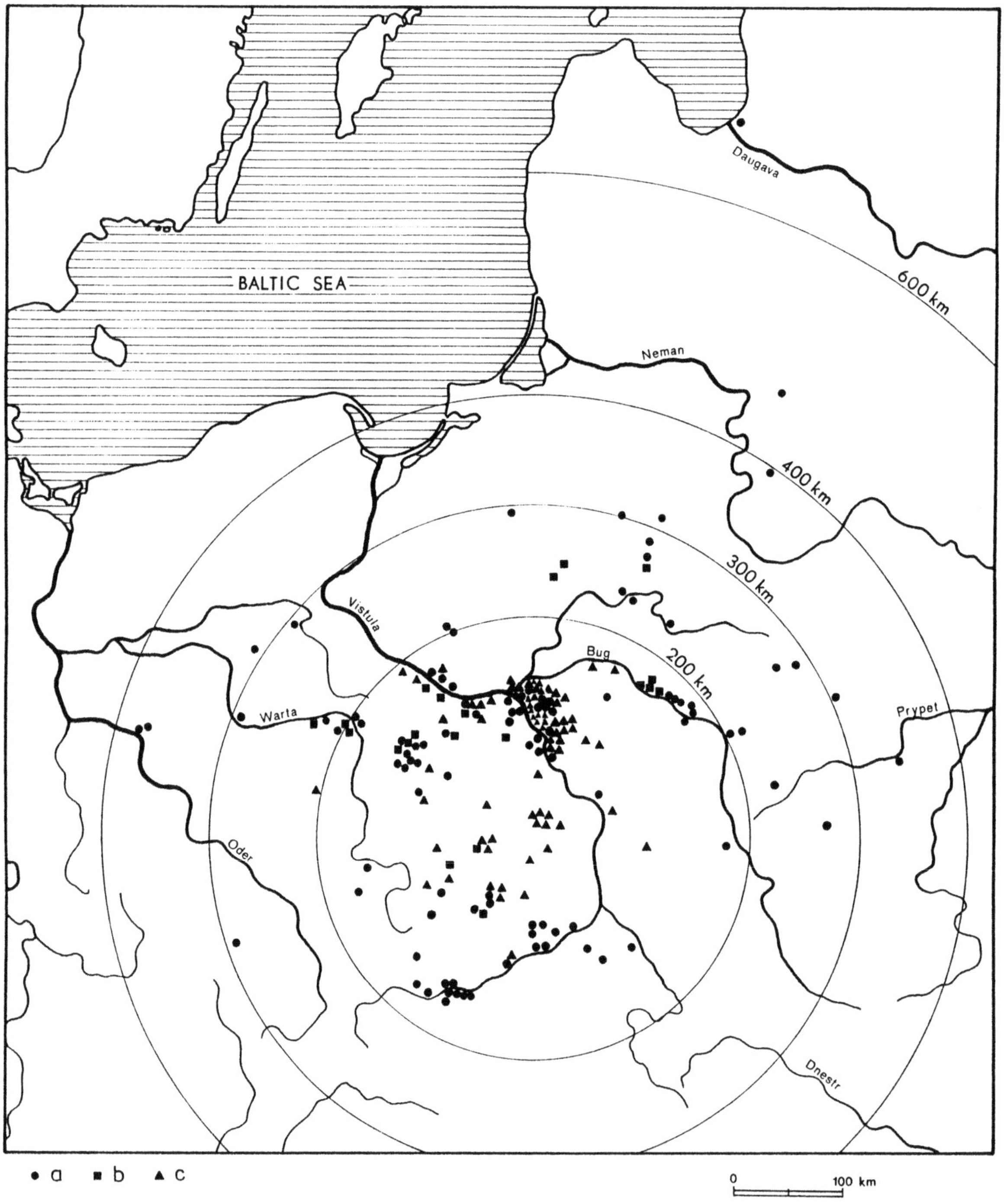

Figure 5. Distribution of chocolate flint artifacts in the Late Glacial inventories. (a) single artifacts; (b) 10-50% of inventory; (c) predominant use. After Schild 1971, 1975; Sulgostowska 1989, 1990.

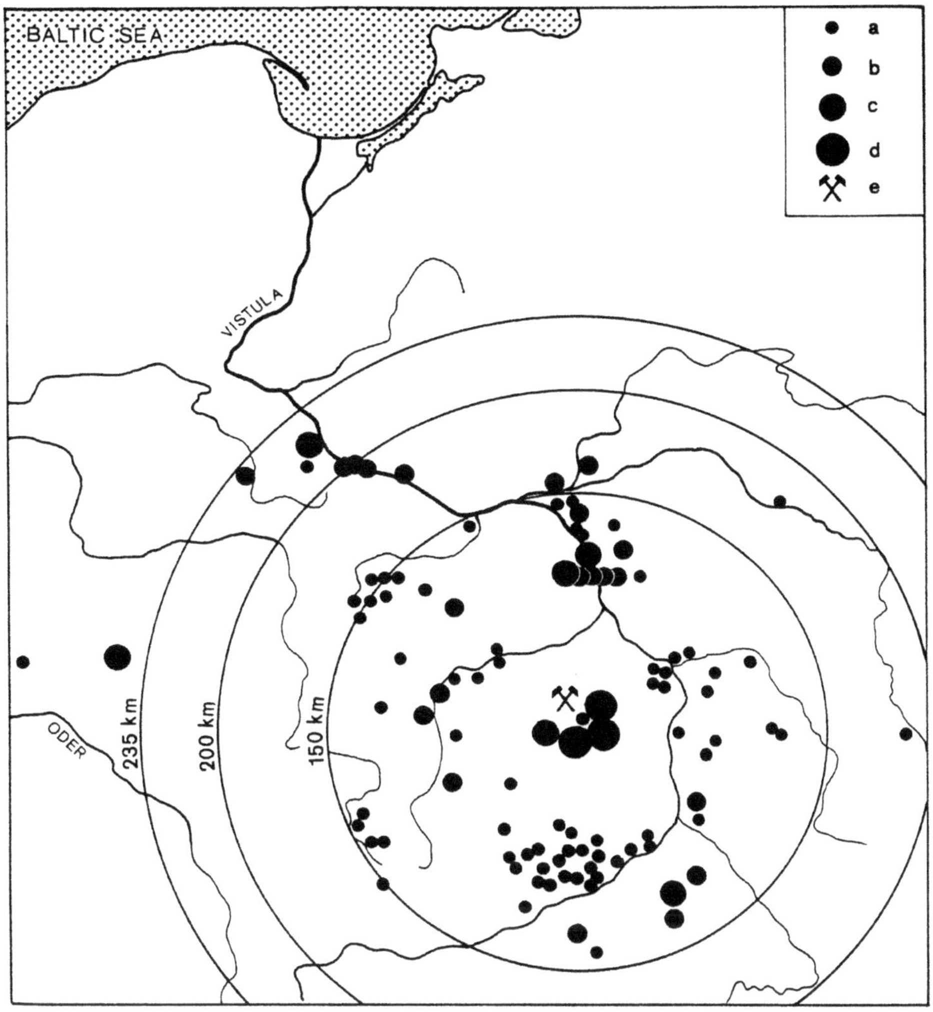

Figure 6. Location of chocolate flint extraction point (e) and distribution of chocolate flint artifacts in Mesolithic inventories. Number of artifacts: (a) 1-10; (b) 11-200; (c) 201-1100; (d) 1101-3400. After Cyrek 1981; S. K. Kozłowski 1989.

center of ochre exploitation and chocolate flint processing with a settlement density indicating almost permanent occupation. Here, a system of control over resources was suggested (Krukowski 1961; Schild and Królik 1981; Królik and Schild 1993). The Rydno complex was functioning from ca. 20,000 C^{14} years ago and was most intensively used between 12,000 and 9,700 C^{14} years BP (Schild 1996).

The formation of flint workshop concentrations was due to individual flintknappers, as is shown by the results of refittings (Fiedorczuk 1992). In the case of workshops where tools in everyday use were also noted (i.e., the Paleolithic workshops Rydno, Gulin, and Nobel, or the Mesolithic workshop Tomaszów II) it is assumed that the everyday, repetitive activity of family groups took place, and not only flint mining and processing for export (Schild *et al.* 1985; Sulgostowska 1991; Ginter 1993).

Distribution of flints seems to indicate a simple, probably family-level system of supply and exchange. Advance planning is reflected in a system of "caches" with hoards of core preforms and cores found on sites which reflect migration routes (Krukowski 1976a; Królik and Schild 1993). Only the wide distribution of chocolate flint during the late glacial suggests a complex, developed network of contacts, which stretches beyond the scope of seasonal reindeer migrations followed by hunting groups. The furthest imports of chocolate flint reached more than 650 kilometers (Salaspils Laukskola), a distance from source similar to the radiolarite artifact from Nobel (Sulgostowska 1989, 1990). Maps indicate strong contacts with north-eastern territories which seem to suggest the route of the first settlement after deglaciation. Such intensive and distant distribution of chocolate flint implies its special character - as an object of social prestige or a symbol of contact with the Rydno ochre mining complex?

The decline in the system of chocolate flint extraction, processing and distribution in the early postglacial and the increasing role of local flint supply can be explained by decreased mobility of human groups hunting non-migratory forest fauna, and probably the initial formation of local territories limiting free access to sources. It is also regarded (Schild 1996) as an indicator of a break in tradition between the Paleolithic and Mesolithic and evidence of the arrival of a new population.

The analysis of lithic raw material economies in this region is far from completed.

References Cited

Cyrek, K.

1981 Uzyskiwanie i użytkowanie surowców krzemiennych w mezolicie w dorzeczu Wisły i górnej Warty (The obtaining and use of flint in the Mesolithic of the Vistula and Upper Warta basins). *Prace i Materiały Muzeum Archeologicznego i Etnograficznego w Łodzi, seria archeologiczna* 28:5-108.

Dagnan-Ginter, A.

1976 Górnopaleolityczna kopalnia krzemienia w Wołowicach, pow. Kraków (Une mine de silex a Wołowice, district de Cracovie, datant du Paleolithique superieur). *Materiały Archeologiczne* 16:133-136.

Fiedorczuk, J.

1992 Późnopaleolityczne zespoły krzemienne ze stanowiska Rydno IV 57 w świetle metody składanek (Late Palaeolithic flint assemblages from Rydno IV 57 site in the light of the refitting method). *Przegląd Archeologiczny* 39:13-65.

Florek, M. and J. Libera

1994 Pierwszy sezon badawczy przykopalnianych pracowni w rejonie wychodni krzemienia świeciechowskiego w Kopcu. *Sprawozdanie z badań terenowych Katedry Archeologii UMCS w roku 1993*, pp. 3-14. Lublin.

Gieysztor-Szymczak, E.

1981 Płonka-Kozły. *Informator Archeologiczny*, Warszawa, pp. 18-19.

Ginter, B.

1974 Wydobywanie, przetwórstwo i dystrybucja surowców i wyrobów krzemiennych w schyłkowym paleolicie północnej części Europy środkowej (The extraction, production and distribution of raw material and flint products at the Late Palaeolithic in the northern part of Central Europe). *Przegląd Archeologiczny* 22:5-122.

1993 Sviderian flint mines and workshops at Gojść on the Upper Warta river. Paper presented at the Symposium on Tanged Point Cultures, Lublin.

Kobusiewicz, M.

1989 Procurement of flint in the Mesolithic of the Polish Plain. In *The Mesolithic in Europe*, edited by C. Bonsall, pp. 442-446. John Donald Publishers, Edinburgh.

Kozłowski, J. K.

1980 Technological and typological differentiation of lithic assemblages in the Upper Palaeolithic. An interpretation attempt. In *Unconventional Archaeology*, edited by R. Schild, pp. 33-35. Ossolineum, Wrocław-Gdańsk.

Kozłowski, S. K.

1989 *Mesolithic in Poland. A New Approach.* Wydawnictwa Uniwersytetu Warszawskiego, Warsaw.

Królik, H. and R. Schild

1993 Rydno, a final Palaeolithic ochre mine and surrounding aggregation grounds. Paper presented at the Symposium on Tanged Point Cultures, Lublin.

Krukowski, S.

1920 Pierwociny krzemieniarskie górnictwa, transportu i handlu w holocenie Polski (Exploitation, transport et commerce de silex dans le holocene de la Pologne), Part I. *Wiadomości Archeologiczne* 5:185-206.

1922 Pierwociny krzemieniarskie górnictwa, transportu i handlu w holocenie Polski (Exploitation, transport et commerce de silex dans le holocene de la Pologne), Part II. *Wiadomości Archeologiczne* 7:37-54.

1939-1948 Paleolit. In *Prehistoria Ziem Polskich*, Encyklopedia Polska PAU 4, pp. 1-117. Kraków.

1961 Rydno. *Przegląd Geologiczny* 9 (4):190-192.

1976a Bejask Jarosławie. In *Skam 1971*, edited by S. Krukowski and A. Nowakowski, pp. 85-102. Ossolineum, Wrocław-Warszawa.

1976b Pierwsze z Krasnego. In *Skam 1971*, edited by S. Krukowski and A. Nowakowski, pp. 113-137. Ossolineum, Wrocław-Warszawa.

Kudriashov, V. E. and O. L. Lipnitsckaia

1994 Kamiennaia industria v raionie Krasnoselskikh kremniedobyvaiushtshikh shakht. Paper presented at the Symposium on Recent Research on the Stone and Early Bronze Ages in the South-Eastern Subbalticum, Supraśl, Poland.

Pelisiak, A.

1987 The flint raw material from the central part of the Polish Jura and its utilization in prehistory. In *International Conference on Prehistoric Flint Mining and Lithic Raw Material Identification in the Carpathian Basin*, pp. 123-127. KMI Rota, Budapest.

Rimantiene, R.
1971 *Paleolit i mezolit Litwy.* Mintis, Vilnius.

Samsonowicz, J.
1923 O złożach krzemieni w utworach jurajskich północno-wschodniego zbocza Gór Swiętokrzyskich (Sur les assises de silex dans le depots jurassiques du versant nord-est des montagnes de święty Krzyż). *Wiadomości Archeologiczne* 8:17-24.

Schild, R.
1971 Location of the so-called chocolate flint extraction sites on the north-eastern footslopes of the Holy Cross Mountains. *Folia Quaternaria* 39:1-81.
1975 Pózny paleolit. In *Prahistoria ziem polskich I, Paleolit i mezolit,* edited by W. Hensel, pp. 159-338. Warszawa-Gdańsk.
1976 Flint mining and trade in Polish prehistory as seen from the perspective of the chocolate flint in the Central Poland. A Second Approach. *Acta Archaeologica Carpathica* 16:147-177.
1980 Introduction to dynamic technological analysis of chipped stone assemblages. In *Unconventional Archaeology,* edited by R. Schild, pp. 57-85. Ossolineum , Wrocław-Gdańsk.
1996 The North European Plain and Eastern Sub-Balticum between 12,700 and 8,000 BP. In *Humans at the End of the Ice Age: The Archaeology of the Pleistocene-Holocene Transition,* edited by L. G. Straus, B. V. Eriksen, J. M. Erlandson, and D. R. Yesner, pp. 129-157. Plenum Press, New York.
n.d. Digging open flint mines and quarries. In *Proceedings of the VIth International Flint Symposium,* Madrid, 1991.

Schild, R. and Królik, H.
1981 Rydno - A final Palaeolithic ochre mining complex. *Przegląd Archeologiczny* 29:53-100.

Schild, R., H. Królik and M. Marczak
1985 *Kopalnia krzemienia czekoladowego w Tomaszowie* (A Chocolate Flint Mine at Tomaszów). Ossolineum, Wrocław-Łódż.

Sobczyk, K.
1993 *The Late Palaeolithic Flint Workshops at Brzoskwinia-Krzemionki near Kraków.* Zeszyty Naukowe Universytetu Jagiellońskiego 177. Prace Archeologiczne 55. Uniwersytet Jagiellonski, Kraków.

Sulgostowska, Z.
1986 The influence of flint raw material on the Final Palaeolithic inventories. In *International Conference on Prehistoric Flint Mining and Lithic Raw Material Identification in the Carpathian Basin,* edited by K. T. Biro, pp. 307-315. KMI Rota, Budapest.
1989 *Prahistoria międzyrzecza Wisły, Niemna i Dniestru u schyłku plejstocenu* (The Prehistory of the Area between the Vistula, Niemen and Dniestr Rivers during the Final Pleistocene). Państwowe Wydawnictwo Naukowe, Państwowe Muzeum Archeologiczne, Warszawa.
1990 Occurrence and utilization of local ochre resources during the Early Holocene in the Oder and Vistula river basins. In *Contribution to the Mesolithic in Europe,* edited by P. M. Vermeersch and P. Van Peer, pp. 317-321. Leuven University Press, Leuven.
1991 A comparative study of Final Palaeolithic flint workshops from the Oder-Vistula river basins. In *Abstracts of the VI International Flint Symposium,* edited by M. A. Bustillo and A. Ramos Milan, pp. 181-184. Instituto Tecnologico GeoMinero de Espana, Madrid.

Szmit, Z.
1929 Badania osadnictwa epoki kamienia na Podlasiu (Recherches des colonisations de l'epoque de pierre en Podlasie). *Wiadomości Archeologiczne* 10:56-117.

Szymczak, K.
1992 *Północno-wschodnia prowincja surowcowa kultury świderskiej* (The North-Eastern Raw Material Province of the Sviderian Culture). Acta Universitatis Lodziensis, Folia archaeologica 15. Wydawnictwa Uniwersytetu Łódzkiego, Łódż.

Zagorska, I.
1994 Salaspils Laukskolas akmens laikmeta apmetne (Die Steinzeitliche Siedlung Laukskola bei Salaspils). *Archeologija un Etnografija* 16:14-28.

3. Mobility and Lithic Economy at the Buttental Site: A Case Study

Joachim Hahn

This chapter presents a case study of acquisition and use of raw materials at a single Magdalenian site, the Buttental cave on the upper Danube. A high proportion of refits at the site allows the author to argue that the assemblage is a relatively homogeneous unit, and to infer reduction sequences and transport of materials and products onto and off of the site. The analytical approach taken relies on macroscopic identification of raw material units (nodules) based on variation in color, zoning, and cortex, confirmed by refitting. Surveys around the site were used to determine possible raw material sources. The resulting detailed view of raw material acquisition, use and transport allows a re-evaluation of current views on mobility and raw material economy in the Swabian Alb. Stone products and raw materials were carried to the Buttental site from local and intermediate areas (within 20 kilometers of the site) and more distant areas (50 - 240 kilometers). Transport from these different sources is attributed to macro- and micromoves. This indicates a pattern of mobility combining relocation of Magdalenian groups on a scale of hundreds of kilometers with shorter-distance moves within the region. The paper finally presents comparative data on use of raw material from distant sources in the Swabian Alb, and identifies regional variations in frequency of imported materials that may be related to site function and seasonality as well as distance to source. *–Editors.*

Late Upper Paleolithic mobility and economy in southwest Germany have for various reasons developed into a favorite topic (Eriksen 1991; Gamble 1978; Hahn 1979; Sturdy 1975; Weniger 1982, 1991). The ephemeral seasonal use of this region has been modeled on subarctic hunter-gatherers. Parallel to these works, a number of sites and assemblages from early excavations have been restudied, including Brillenhöhle, Burkhardtshöhle, Bruckersberg, and Buttental cave.

The Site

In this essay, the assemblage of the late Upper Paleolithic Buttental cave will be used as a test case to examine current concepts of mobility and raw material economy. In 1931, E. Peters excavated this small cave site on the Upper Danube near Beuron in southwestern Germany (Peters 1936). With a single, near-surface archaeological level containing a late Pleistocene fauna and Magdalenian artifacts, it belongs among the many Magdalenian sites in this area (Eriksen 1991). Though little documentation exists for this excavation, analysis of the material was attempted.

Samples of bone from reindeer, horse, and brown bear from this assemblage have been AMS-dated to 13,020 ± 130 BP (OxA-4602), 12,040 ± 120 BP (OxA-4981), and 13,100 ± 140 BP (OxA-4982) (Hahn 1995:27). These dates are somewhat earlier than assumed by conventional typological chronology, because they place this assemblage, characterized by backed points, contemporary with classical Magdalenian assemblages such as Petersfels, Hohle Fels Schelklingen, or Gönnersdorf. A plateau in the late Pleistocene radiocarbon chronology in the range between 12,000 and 13,000 years B.P. has been established (Ammann and Lotter 1989). A calibration for this period is possible, but it has not been used because of the plateau.

The faunal assemblage at Buttental cave consists of horse, reindeer, ibex, brown bear, hare, and fox. Some of the bones, especially the bear, show cut marks indicating human manipulation.

The Current Model

Assemblages from caves are often seen (for example by Peters, as quoted in Mauser 1970:1–4) as a palimpsest of repeated occupations that, due to episodic sedimentation, may appear to represent a homogeneous assemblage. Artifacts and faunal remains therefore may represent a number of relatively independent elements. This assumption is supported by recent or ethnographic use of caves or rock shelters (Gorecki 1991) for shelter during inclement weather or for special purposes. Usually, an archaeological assemblage is treated as a unity, and its typological and quantitative composition is used to establish technocomplexes, industries, stages or facies, or to propose a general settlement pattern (Weniger 1982). The relationship of the archaeological site to the landscape is modeled on the basis of resource scheduling. Raw material provenience and economy are major subjects (Weniger 1991). Weniger presented distance from raw material sources in a three-class hierarchy: to 20 kilometers, to 100 kilometers, and up to 200 kilometers. Generally, it is assumed that raw materials display a predictable distribution on the landscape, being frequent near the source and gradually decreasing farther away. Given the frequency of raw material in southwestern Germany, this type of pattern is thought to represent embedded procurement, in contrast to the Neolithic, where the same spatial distribution is taken for evidence of exchange. Instead of focusing on time-space systematics, lithic analysis attempts to differentiate raw material units, their provenience, reduction sequences, and the function(s) of artifacts and sites. Nodules are macroscopically identified to raw material classes, i.e., chert, radiolarite, quartz, etc. Variants within these raw material classes are distinguished according to color, zoning, texture, and cortex. Within these raw material variants, highly similar pieces appearing to belong to a single nodule are differentiated. These groups are then verified by refitting. This technique attempts to refit broken pieces, cores, and manufacturing or resharpening debris. Refitting can also be used to monitor stratigraphic coherence and spatial dynamics within a site (Lauxmann and Scheer 1986).

Raw Material Provenience

Extensive surveys within a radius of 5–7 kilometers around Buttental cave and around the nearby sites of Probstfels and Jägerhaushöhle have been conducted as part of a field program for all major sites on the Swabian Jura. While a coarse chert was found near the site on the plateau to the south, other sources, especially for a high-quality brown chert, are not certain. Except for one coarse-grained brown chert, these surveys failed to provide a possible source for the major local raw material used on the site. Some of the material is not available locally, such as radiolarite, available within a distance of 50–60 kilometers, and Bavarian tabular chert, probably obtained from sources approximately 240 kilometers from the Buttental site. Radiolarite, though present in morainic deposits near Engen, could also come from the Federsee area, where the Riß and Iller rivers have transported and sorted this Jurassic alpine material into the Danube gravels. In addition, some locally available raw materials were apparently not knapped on the site, but were brought onto it as cores, blanks, or tools.

Considering the high rate of refittings (more than 30%), similar raw materials, and technological and stylistic similarity, the Buttental assemblage can be regarded as a homogeneous unit. It is assumed that a small proportion of artifacts is missing due to the recovery techniques used. The lack of small chips, bladelets, and burin spalls in the assemblage indicates that no dry or water sieving was done. If separated by distance to raw material source, four groups can be established (Table 1): local (2–5 kilometers), neighborhood (20 kilometers), morainic area (50–60 kilometers), and exotic material from Bavaria (240 kilometers). The imported artifacts form three sub-assemblages. The few long-distance transports consist of three backed points, a truncation, and a laterally retouched piece. The radiolarite artifacts from the morainic sources include 5 flakes, 4 blades and two backed points. Local imports, i.e., pieces made on locally available material but apparently not produced on the site, consist of blades, flakes, 11 backed points, laterally retouched pieces, some debris, and a few preparation blanks. The majority of the assemblage (739 pieces) was produced on the site from local raw materials. Here we find a predominance of flakes, blades, many preparation blanks, most of the cores,

Table 1. Raw Material Economy in Buttental Cave

Blank/ Tool	Import far	Import near	Import local	Product local	Export
core	-	-	1	16	2
blade	-	4	33	230	62
flake	-	5	20	337	56
crested	-	-	5	110	39
point	3	2	11	-	-
scraper	-	-	1	-	-
truncation	1	-	2	-	-
burin	-	-	-	5	-
perforator	-	-	1	-	-
other retouched	1	-	10	-	-
backed bladelet	-	-	2	-	-
debris	-	-	5	41	2
TOTAL	5	11	91	739	161

debris, and 5 burins. The estimated number of missing pieces ("exports", see Table 1) from this local production consists of a nearly equal number of flakes and blades, and two cores. The number of missing artifacts from local production is estimated based on the refitted pieces and on removal scars with no match among characteristic individual pieces. The number of exported modified tools cannot be estimated because burin spalls and retouch chips are absent due to the excavation method.

Reduction Sequences

The nearly complete refitting of some nodules provides a means of interpreting raw material economy. Nodules used were small and ovoid in shape, rarely tabular. Many refits apparently reflect raw material flaws such as fissures or generally bad quality, or might be the work of beginning flintknappers. Apparently, three kinds of raw material or nodule shape were selected or modified:

a) long oval nodules with an oval or round cross-section,
b) round flat disks often shaped by fissures or frost cracks, or
c) more rarely, flat slabs with a rhomboid outline.

Reduction sequences are based on the production of a platform and the convex face of the core with a guiding ridge that may be natural. Platform angles on the cores form acute or nearly right angles. On the cores with platform angles near 90 degrees, straight blades were produced, probably by use of a soft limestone hammer. The better, imported blades are more curved and may have been knapped with a soft organic hammer not present in the assemblage. During blade removal, the convex core face becomes flat and has to be reworked, usually with core tablets, crested blades and base preparation that ensures the feathered distal end of the blades. Several cores/nodules display this traditional pattern.

One nodule (Figure 1) allows the reconstruction of stages of work:

1) first, a nodule of brown chert with a natural guiding ridge formed by cortex and a fissure was selected;
2) then, a platform was produced by striking off two long cortex core "tablet" blades forming an acute angle with the ridge;
3) a blade removal stage then followed the natural crested blade;
4) for the second stage of blade production, a crested blade was prepared and the base of the core reshaped; after this, blade removal resumed;
5) the core was abandoned after an unsuccessful attempt to remove a ridge blade.

Compared to the usual reduction sequence of the Magdalenian, with a short core tablet, the first long tablet could be used to the end as a platform without reworking; however, the convex face and the base had to be reshaped. Despite small nodule size, rather long blades could be produced until the core was exhausted. Cortex is present on full production blades until the final phase. Decortication is not advisable due to the small size of the available raw material.

Of the 857 lithic artifacts, 267 (31%) could be refitted, nearly half of which have cortex surfaces. 16% of the refitted artifacts are burned with thermal fractures. Several conjoined blades consist of a burned and an unburned fragment. The high rate of refits points to a rather homogeneous assemblage. The reduction sequences could be established on the basis of conjoined pieces. The refits established the basic raw material units, the nodules, which were used to distinguish the imported, produced and missing artifacts in relation to their known provenience (Table 1).

Raw Material Procurement

Generally, use of the tabular Bavarian chert diminishes in frequency from east to west in a series of Magdalenian sites on the Swabian Jura. The easternmost site considered here, Bärenfelsgrotte, and the nearby sites of Vogelherd and Hohle Fels Hütten, are the only sites to have produced blade cores (Table 2). There seems to be no pattern as to the kind of blank or tool represented.

The pattern of exotic raw materials that diminish to the west is supported by the distribution of radiolarite in Magdalenian assemblages (Figure 2). For Helga Abri at the Hohle Fels Schelklingen, only part of the radiolarite tools have been published. These include 6 backed tools, three endscrapers, one burin and one splintered piece. In the very large assemblage from Felsställe (n=428,485, Kind 1987:114, Table 23), radiolarite is represented by only two backed pieces, though the site is close to radiolarite available in Danube gravels near Ehingen. Similarly, the rich site of Petersfels further west, with access to nearby radiolarite occurrences in Switzerland, has only two pieces of radiolarite (one core, one perforator) among 1773 artifacts, and the nearby site of Kesslerloch produced only one radiolarite cortex flake among 960 artifacts (Pasda, pers. comm.). Hohle Fels Schelklingen, interpreted as a residential site, has the highest percentage of radiolarite (31%), followed by the small site Hohle Fels Hütten (8.1%). The Schussenquelle, located in the middle of the morainic area, has only a modest frequency (3.8%) of radiolarite in the lithic assemblage. It seems from the distribution of radiolarite on Magdalenian sites that the Iller river is mainly responsible for the

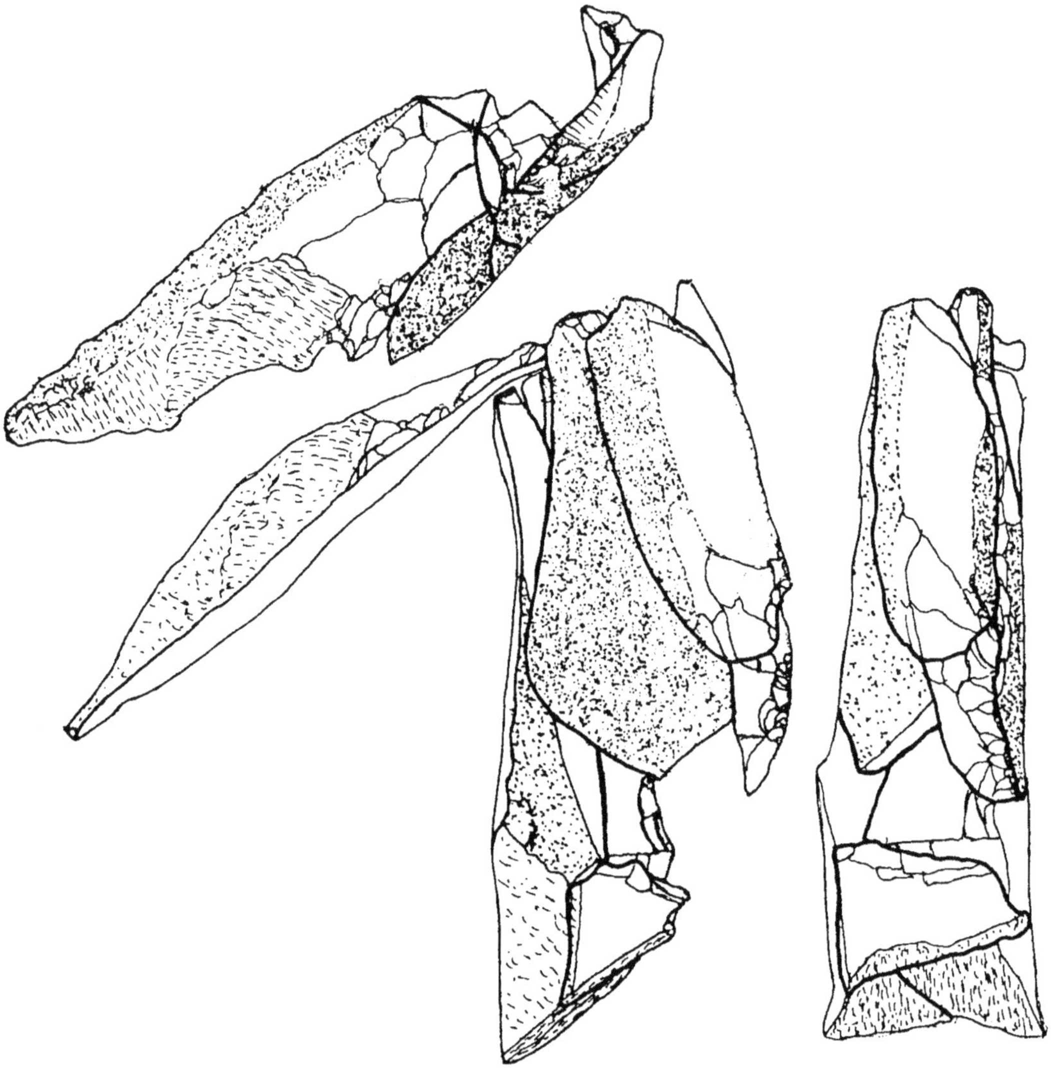

Figure 1. Refitted nodule showing reduction sequence. Buttental cave.

Table 2. Radiolarite (R) and Tabular Chert from Bavaria (B)

Blank/ Tool	BT*		PF**		HF		SI	BU		VO		BA	
	B	R	B	R	B	R	R	B	R	B	R	B	R
core	-	-	-	-	1	1	-	-	-	1	-	1	-
blade	-	4	6	13	-	18	1	17	7	1	1	-	4
flake	-	5	5	1	-	8	-	-	1	-	-	-	14
crested	-	-	1	-	-	4	-	1	-	2	-	-	1
point	3	2	-	4	2	-	-	3	-	-	-	3	2
truncation	1	-	-	1	-	1	-	3	-	-	-	4	-
burin	-	-	-	1	1	1	-	2	1	-	1	-	2
perforator	-	-	-	-	-	-	-	1	2	-	-	3	-
other retouched	1	-	-	-	-	-	-	3	-	-	-	-	-
debris	-	-	1	1	-	-	-	-	-	-	-	-	10
TOTAL	5	11	13	21	4	33	1	30	11	4	2	11	33

*BT, Buttental cave (n=857); PF, Probstfels (n=738); HF, Hohle Fels Hütten (n=409); SI, Sirgenstein-Süd (n=87); BU, Burkhardtshöhle (n=825); VO, Vogelherd II (n=118); BA, Bärenfelsgrotte II-III (n=424).

**Source for data on Probstfels, Pasda 1990.

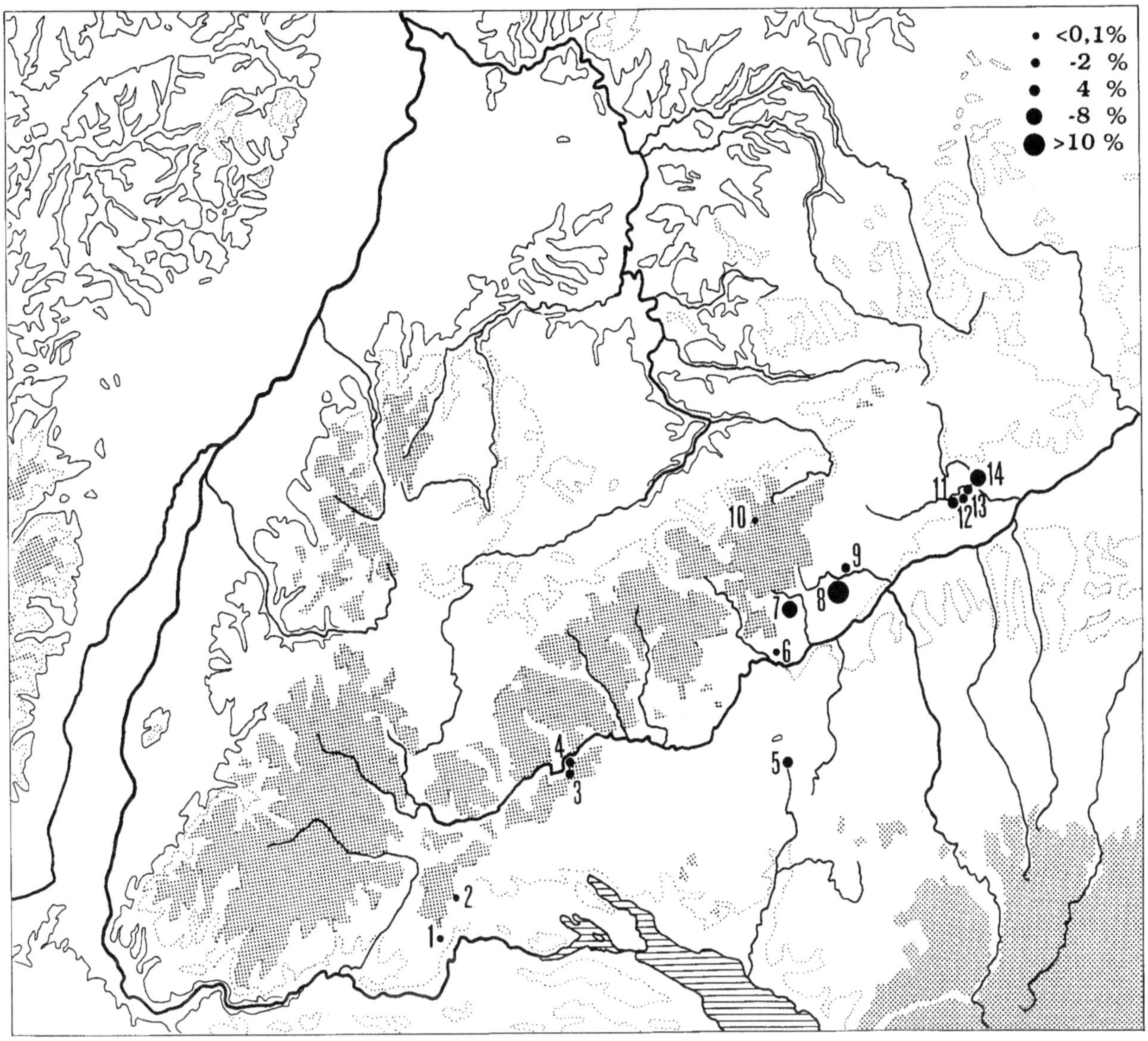

Figure 2. Frequency of radiolarite in Magdalenian assemblages. 1, Kesslerloch; 2, Petersfels; 3, Buttental Cave; 4, Probstfels; 5, Schussenquelle; 6, Felsställe; 7, Hohle Fels Hütten; 8, Hohle Fels Schelklingen; 9, Sirgenstein; 10, Burkhardtshöhle; 11, Hohlenstein Bärenhöhle I; 12, Hohlenstein Stadel III; 13, Vogelherd II; 14, Bärenfelsgrotte.

transport of this raw material.

Except for Hohle Fels Schelklingen, the large, intensively occupied sites have little radiolarite. The small Hohlenstein assemblages (Stadel III and Bärenhöhle I) have a higher percentage of radiolarite, between 4 and 13.5%. Perhaps the frequency of radiolarite is correlated with site seasonality, since the large sites show evidence of fall or winter occupation (Felsställe: winter; Petersfels and Schussenquelle: fall), while the small sites are generally earlier in season (spring to summer) (Weniger 1982). Only Burkhardtshöhle among the small sites is also attributed to a winter occupation. Radiolarite seems to be used most intensively in the Ach valley, with a second minor peak in the Lone valley. Farther west, radiolarite decreases in frequency.

To the east, an increase in use of the Bavarian chert, and decrease or parity in use of radiolarite from Danube gravels would be predicted. At Kaufertsberg (Kaulich 1983), located immediately east of the Swabian Jura in the Nördlinger Ries, both raw materials are represented by only a few pieces, which is the opposite of the expected frequency. This difference may again be related to site seasonality and/or function, since Kaufertsberg is another rather large assemblage.

Summary and Conclusions

Some additional assemblages, such as the important Magdalenian inventory from Brillenhöhle and assemblages from the Franconian Jura in Bavaria, are needed to augment these data on raw material frequency in the region. As sites like the Felsställe indicate, no simple distribution pattern is present. The absence of microwear analyses is another limiting factor. For example, there is no indication why flakes share nearly the same mobility as blades. Generally, it is assumed that flakes were not modified into tools like blades and that they were thus less often utilized.

The time represented at sites like Buttental cave is an unknown factor. Each raw material unit represents one event or a sequence of related events. The assemblage as a whole may have accumulated over centuries in the course of many occupations, even if the lithics themselves may have been produced within one hour. The number of faunal remains with cut marks argues against one very short occupation, but a direct association with the lithics could only be provided by a better understanding of the spatial distribution at the site, which has not been documented. The main argument for viewing Buttental cave as representing one or a few fairly short occupations is provided by structural features including two chert accumulations and a hearth. The hearth was perhaps used twice, which argues for one or a few occupations. This interpretation relies on the assumption that each occupation is connected with the use of fire and that a fireplace was used only one time, as in some open-air Magdalenian sites. The restricted space in a cave or rock shelter may cause different behavior.

A hafting and retooling model (Keeley 1982) seems to account for the assemblage composition at Buttental cave. Raw material economy can be approximated by the reduction sequences inferred from refitting. Core reduction had to take into account the small nodule size of chert and radiolarite available. Preparation was limited to the platform and convex face with the aim of obtaining a large number of blades. That blades were intensively used is well known for the Magdalenian; at Buttental cave, the majority of retouched tools are on blades, and few flakes were modified.

The differing associations of blanks, cores, and tools point to a high mobility of artifacts. As expected, modified objects, mainly projectile points and blades or "knives", are the major classes transported and curated over a long period. Still, due to frequent use, these were often abandoned on sites. Debris and cores as preforms or still in the productive cycle were also transported. A good example is the exotic tabular chert from Bavaria. Within a radius of 100 to 120 kilometers from the source, cores are still present, but on the Upper Danube, blades, tools, and backed points are the last remains of this material. The discarded artifacts, especially from this easily-recognized exotic material, do not seem to be limited to one or a few tool types, however, with the exception of backed pieces, mainly points (Table 2), which points to a high mobility of these artifacts. Burins, mostly single specimens, are also made from non-local materials. Different curation and tool life expectancies have to be considered as well. Tools found on the site therefore are not necessarily related to the activities performed at the site itself.

Raw material analysis and refitting provide an estimation of the imported and locally manufactured artifacts with different qualities of manufacture. As regards imported modified artifacts, only the presence of resharpening chips and debris can testify to their use on the site. At Buttental cave the lack of water screening thus makes evaluation difficult.

Seen more closely, apparently homogeneous assemblages, even if a single occupation is probable, are constituted by different actions and cannot *a priori* be assumed to represent contemporaneous activities.

The exotic imported raw material at Buttental cave comes from distant sources — the Federsee (50–60 kilometers) and Bavaria (240 kilometers) — and indicates previous macromoves, since no intermediate sources for these raw materials have been determined. The distances involved can be reconstructed as indicating past movements, and the missing pieces as indicating anticipated mobility. The procurement of raw material was embedded in local subsistence activities. Locally available raw materials apparently not knapped on the site suggest a micromove within the same area. Missing artifacts, mainly blades and flakes, are on the same order as those brought to the site and discarded. It can be assumed that a subsequent micromove was anticipated without changing the exploitation pattern.

Distance and proportions are not sufficient to measure raw material procurement. Other factors like site function and seasonality may also affect spatial raw material distribution and its individual frequencies in the sites. The availability of raw material is assumed to be dependent on the season: the heavy vegetation cover in spring and summer affects visibility, and so does snow and ice in winter, especially in river beds and terraces. At the same time, the function of a site as hunting camp, base camp, or extraction camp determines raw material frequency.

Acknowledgments

B. Auffermann, S. Biedrowski, B. Fischer, C. Pasda, and U. Simon provided data for lithic assemblages from Bärenfelsgrotte and the Bruckersberg sites, Hohle Fels Schelklingen, Probstfels, and Burkhardtshöhle.

References Cited

Ammann, B. and A. F. Lotter
1989 Late Glacial radiocarbon- and palynostratigraphy on the Swiss Plateau. *Boreas* 18:109–126.

Eriksen, B. V.
1991 *Change and Continuity in a Prehistoric Hunter-Gatherer Society: A Study of Cultural Adaptation in Late Glacial/ Early Postglacial Southwestern Germany.* Archaeologica Venatoria 12. Archaeologica Venatoria, Tübingen.

Gamble, C.
1978 Resource exploitation and the spatial patterning of hunter-gatherers: A case study. In *Social Organisation and Settlement: Contributions from Anthropology, Archaeology, and Geography,* edited by D. Green, G. Haselgrove, and M. Spriggs, pp. 15–85. British Archaeological Reports, International Series (Suppl.) 47. BAR Publishing, Oxford.

Gorecki, P. P.
1991 Horticulturalists as hunter-gatherers: Rock shelter usage in Papua New Guinea. In *Ethnoarchaeological Approaches to Mobile Campsites: Hunter-Gatherer and Pastoralist Case Studies* edited by C. S. Gamble and W. A. Boismier, pp. 23–62. International Monographs in Prehistory, Ethnoarchaeological Studies 1. IMP, Ann Arbor.

Hahn, J.
1979 Essai sur l'écologie du Magdalénien dans le Jura souabe. In *La fin des temps glaciaires en Europe. Chronostratigraphie et écologie des cultures du paléolithique final,* edited by D. de Sonneville-Bordes, pp. 20–13. Colloques Internationaux du CNRS 271. CNRS, Paris.
1995 Die Buttentalhöhle: Eine spätjungpaläolithische Abristation im Oberen Donautal. *Fundberichte aus Baden-Württemberg* 20:1–58.

Kaulich, B.
1983 Das Paläolithikum des Kaufertsberges bei Lierheim (Gem. Appenhofen, Ldkrs. Donau-Ries). *Quartär* 33/34:2–7.

Keeley, L. H.
1982 Hafting and retooling: Effects on the archaeological record. *American Antiquity* 47:79–09.

Kind, C.-J.
1987 *Das Felsställe. Eine jungpaläolithisch-frühmesolithische Abri-Station bei Ehingen-Mühlen, Alb-Donau-Kreis.* Forschungen und Berichte zur Vor- und Frühgeschichte in Baden-Württemberg, 23. Konrad Theiss Verlag, Stuttgart.

Lauxmann, C. and A. Scheer
1986 Zusammensetzungen von Silexartefakten: Eine Methode zur Überprüfung archäologischer Einheiten. *Fundberichte aus Baden-Württemberg* 11:10–31.

Mauser, P. F.
1970 *Die jungpaläololithische Höhlenstation im Hegau (Gemarkung Bittelbrunn, Ldkrs. Konstanz).* Badische Fundberichte, Sonderheft 13. Staatliches Amt für Ur- und Frühgeschichte, Freiburg.

Pasda, C.
1990 Der Probstfels bei Beuron: Probleme bei der Auswertung eines alt gegrabenen Inventars. *Archäologisches Korrespondenzblatt* 20(1):1–9.

Peters, E.
1936 Die Buttentalhöhle an der Donau: Eine neue Magdalénienstation. *Badische Fundberichte* III:1–9.

Sturdy, D. A.
1975 Some reindeer economies in prehistoric Europe. In *Palaeoeconomy,* edited by E. S. Higgs, pp. 5–5. Cambridge University Press, Cambridge.

Weniger, G.-C.
1982 *Wildbeuter und ihre Umwelt: Überlegungen zur Mobilität jägerischer Gruppen im Jungpaläolithikum.* Archaeologica Venatoria 5. Archaeologica Venatoria, Tübingen.
1991 Überlegungen zur Mobilität jägerischer Gruppen im Jungpaläolithikum. *Saeculum* 42:8–03.

4. Fossil Mollusks and Exotic Raw Materials in Late Glacial and Early Postglacial Find Contexts: A Complement to Lithic Studies

Berit Valentin Eriksen

In this paper focus is shifted to an analysis of the use and procurement of non-local lithics and ornamental mollusks in late glacial and early postglacial southwestern Germany. The author brings together and evaluates data on ornamental objects and lithic raw material sources from 77 Late Upper Paleolithic (Magdalenian and Late Paleolithic) and Early Mesolithic (Beuronian) sites. It is argued that lithic raw materials used for tools and mollusks or other materials (such as jet and amber) used for personal adornment represent different and complementary expressions of past hunter-gatherer mobility and inter-group communication patterns. Throughout the period in question, lithic raw materials were procured actively and directly within the region. Most evidence points to a pattern of embedded procurement, probably reflecting seasonal movements throughout the region. Fossil mollusks, jet and ammonites presumably were exploited in much the same way, but the evidence is more inconclusive, and some of the deposits may have been more directly exploited. Exotic mollusks (originating from the Atlantic, the Mediterranean, or the Paris or Mainzer Basin), most likely represent different expressions of inter-group communication patterns. Some are obvious expressions of procurement through a long-distance communication, or barter, network. Others seem to represent an inter-regional communication network with direct contact through visiting between neighboring groups. The author finds a decrease in use of exotic stone raw materials in the Mesolithic assemblages, as well as changes from the late glacial to the early postglacial in the kinds of ornamental materials used, but notes that the procurement of ornamental mollusks and materials is similar to the Late Upper Paleolithic in the predominantly north-south direction of long-distance connections as well as the extremely long-distance ties to the Atlantic and the Mediterranean. *–Editors.*

Introduction

Numerous Paleolithic and Mesolithic sites from western Europe have yielded fossil or sub-recent mollusks or exotic raw materials of very 'dispersed' origins. Many of these findings obviously represent different kinds of long-distance connections. The purpose of this paper is to examine the nature, range and possible meaning of such regional and inter-regional relationships in late upper Paleolithic and Early Mesolithic central western Europe.

Keywords with respect to the interpretation of these relationships are mobility and communication. It will be argued that lithic raw materials used for tools and mollusks or other exotic materials used for personal adornment represent different expressions of mobility and inter-group communication patterns. Organic tools represent yet another expression, however, closely related to that of portable art and jewelry.

The Study Area

The case study will focus on sites associated with the Jurassic limestone formation and its immediate surroundings in northwestern Switzerland and southwestern Germany (Figure 1).

Within this area of approximately 35,000 km^2 we can differentiate several primary landscape zones (Figure 2). Most important in the present study are the mountainous region of the Jura proper (i.e. Swiss Jura, Swabian Alb and Franconian Alb) and the Black Forest foothills as opposed to the morainic lowland area of the Swiss Mittelland and Oberschwaben.

Obviously, paleoenvironmental and paleoclimatic conditions must have differed quite a bit across this area. The differences observed, however, are of minor importance with respect to the general comparative approach adopted here. The chronological table in Figure 3 does not take local deviations into consideration since—as a whole—these fertile landscapes would all have sustained a rich variety of plants and wild life throughout the period in question. From a mere subsistence-economic point of view hunter-gatherer groups of the late glacial and early postglacial could easily have subsisted all year round within even smaller parts of the study area.

On the other hand, late glacial and early post-

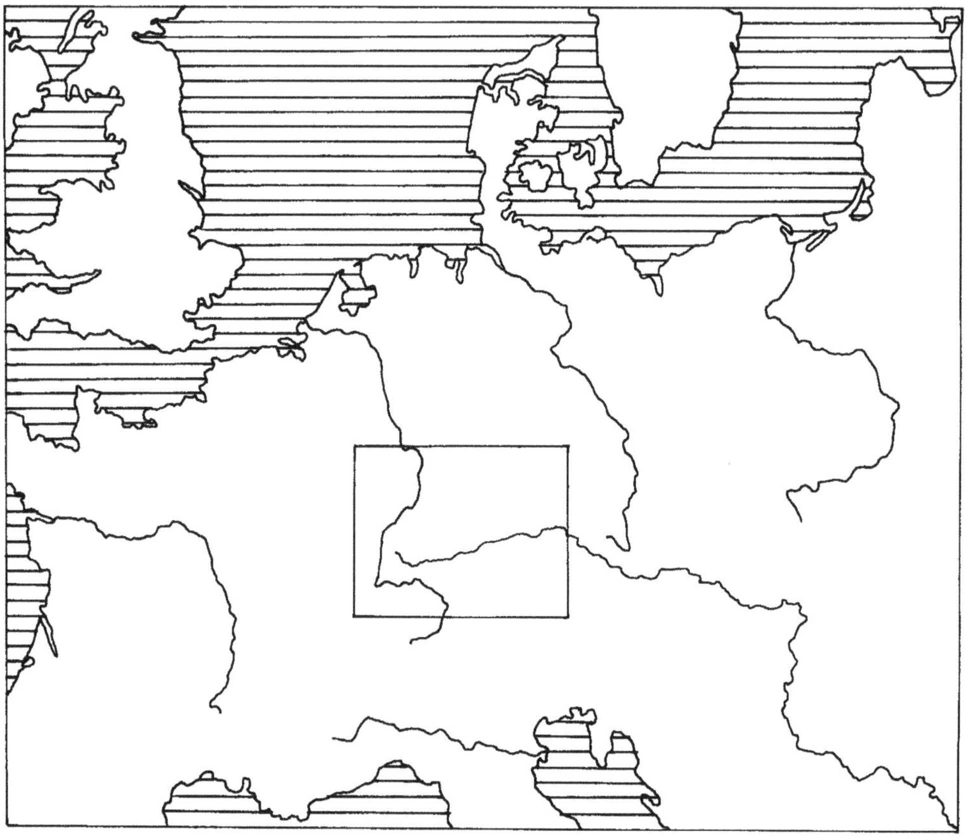

Figure 1. Location of the study area in Central Europe.

glacial settlement and mobility patterns were clearly influenced by regional differences relating to topography and hydrology (Eriksen 1996b, 1997). The Alb highland is characterized by marked karst conditions resulting in rather dry conditions on the plateau. The deep valleys cutting through the highland are generally water-bearing or would have been during the period in question. Most of the Magdalenian sites known from the study area are located in these valleys.

The Swiss Mittelland and the southern part of Oberschwaben were covered by ice during the last glaciation. During the late glacial this was a very moist region (Frenzel 1983:139) characterized by many rivers and a great number of small lakes and kettle holes (Gradmann 1956). For a long time this virgin area remained practically devoid of human occupation, but beginning in the late Allerød and throughout the early postglacial it witnessed an intensive exploitation by Late Paleolithic and Early Mesolithic hunter-gatherer groups.

The Chronological Framework

Figure 3 synthesizes our present knowledge with respect to the late glacial and early postglacial chronostratigraphy of southwestern Germany and northwestern Switzerland. Cultural development is generally conceived of as being continuous and highly endogenous in the period in question.

For methodological reasons, however, the precise dating of many sites and site levels is very difficult. This is especially problematic as regards the Late Paleolithic. Virtually all of these inventories can only be dated within a relative archaeological framework. As a result we are having severe problems in determining the absolute as well as the relative contemporaneity of especially the late glacial sites (Eriksen 1996b).

Thus, given the current state of absolute chronology, as well as the general lack of Late Paleolithic finds with fossil mollusks and exotic raw materials, it is considered justifiable to operate within a broad chronological framework. In the following analysis, the sites will therefore be divided into two groups: 1) a Magdalenian *sensu lato* comprising all sites belonging to the late glacial chronozones of Bølling, Allerød and Younger Dryas, and 2) an Early Mesolithic (Beuronian) comprising all sites belonging to the early postglacial chronozones of the Preboreal and Boreal. The Late Paleolithic findings will be specifically commented upon, whenever they differ from the 'Magdalenian norm'.

The map in Figure 4 shows the location of sites included in the study. All inventories analyzed here either contain fossil mollusks or exotic objects (e.g. jet or amber) or they have provided reliable data with respect to provenance of lithic raw materials. It must be stressed, however, that the map chiefly

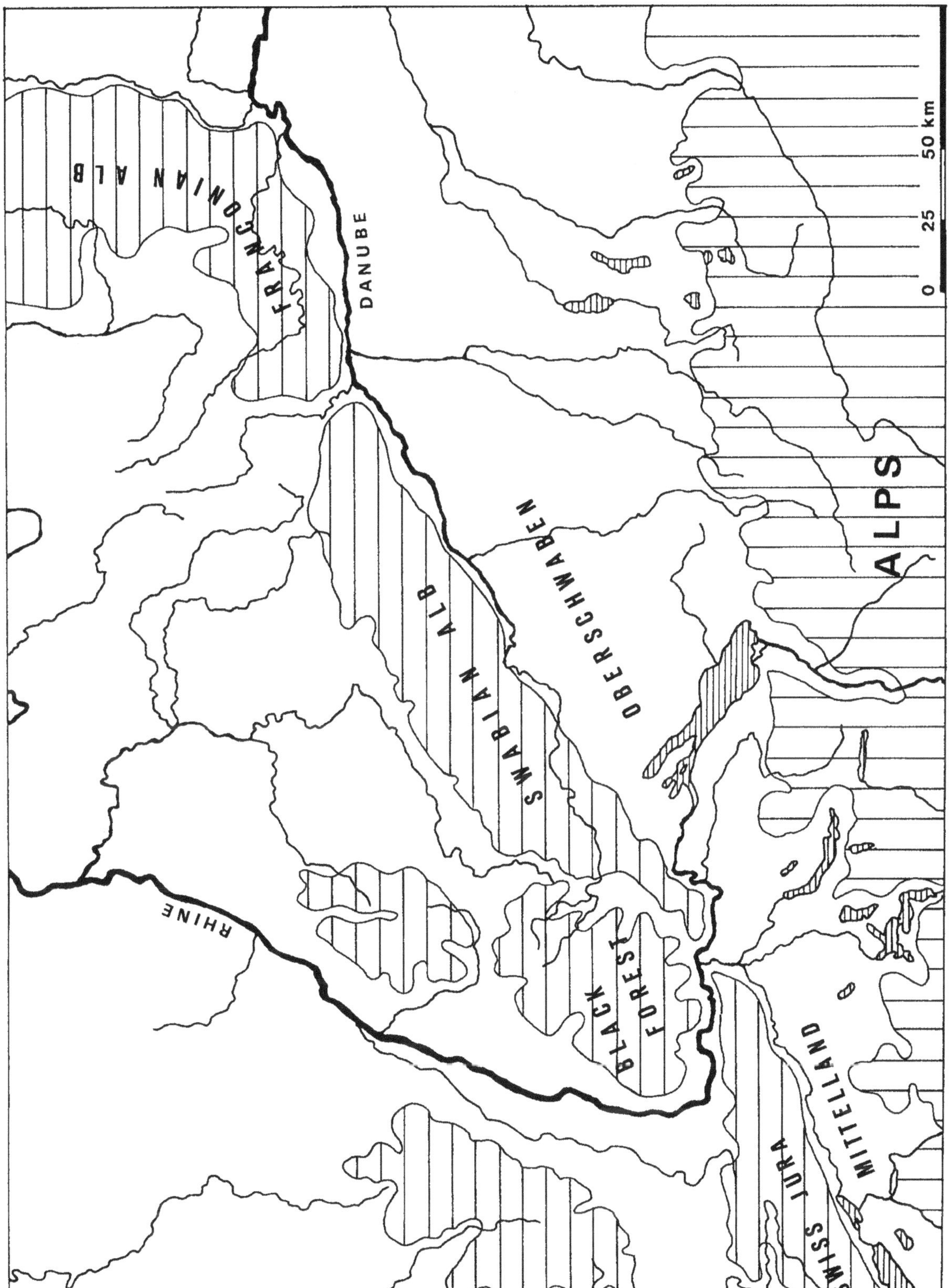

Figure 2. The study area—primary landscapes.

Years BP	Chronozones	Climate	Vegetation	Archaeological cultures
8.000				
	Boreal	warm temperate	Boreal forest (oak, elm, alder, lime, ash)	EARLY MESOLITHIC
9.000		increasing oceanic		BEURONIAN
	Preboreal	temperate	Light, open forest (pine, birch, hazel)	
10.000				
	Dryas 3	cool temperate continental	Sparse forest and park-tundra (pine - birch, willow, herbs)	
11.000				LATE PALAEOLITHIC
	Allerød	temperate	Sparse forest (pine - birch, sea buckthorn, grasses, herbs)	
12.000	(Dryas 2)	increasing oceanic		
	Bølling sensu lato	cool temperate	Park-tundra and sparse forest (birch, juniper, willow, (pine), grasses, herbs)	MAGDALENIAN sensu lato
13.000				
	Dryas 1	subarctic marked continental	Tundra and park-tundra (dwarf birch, willow, juniper, rock rose, sea-grape)	
14.000				

Figure 3. Chronostratigraphy of the Late Glacial and Early Postglacial in southwestern Germany and northwestern Switzerland.

reflects the current state of research within the region. Neither the Paleolithic nor the Mesolithic sites known constitute a random sample of the original occurrence of sites (Eriksen 1991:62f). The previously mentioned apparent migration from the Jura to the surrounding lake and river regions starting in the late Allerød thus might be at least partly due to methodological weaknesses in the data.

Lithic Raw Materials

The lithic inventories of the study area are characterized by a richly varied spectrum of raw materials. Two circumstances in particular must have influenced this variation: 1) the quality or workability of the individual materials, and 2) their accessibility and natural occurrence respectively. The map in Figure 5 synthesizes the known occurrences of local raw materials in a very generalized form. The map and the following presentation relies largely on Deecke 1933 and Hahn 1991.

An important distinction may be made between the Jurassic and the alpine raw materials. Jurassic hornstones (marked A in Figure 5) occur in a number of different varieties. They represent a generally good quality chert which is very suitable for artifact production. Most abundant is the 'Weißjura-δ' variety which may be collected in large quantities almost everywhere in the Swiss Jura, Swabian Alb and Franconian Alb. The 'Muschelkalk' hornstone (B) is also common although this is a relatively poorer quality chert. The 'Keuper' hornstone (C) is generally of a very poor quality. Only nodules from primary outcrops are suitable for artifact production. 'Keuper' hornstones, thus, are generally rare in the lithic inventories. A very fine quality chert, on the other hand, is represented by the Bavarian tabular hornstone (D) originating from the 'Weißjura-ζ' layers near Kelheim. Jasper (E) and Kimmeridgian chert (F) are also 'Weißjura' varieties of a generally high quality. However, due to their limited occurrence they are not very commonly used. Siliceous tuff (G) from Randecker Maar is finally a very distinctive raw material. It is of volcanic origin and thus has a very localized provenance. The quality, however, varies widely.

Most important among the alpine raw materials (H) is radiolarian chert. It is a highly characteristic material of a generally problematic quality. There are primary outcrops of radiolarian ores in Switzerland and Allgäu, but redeposited nodules are found almost everywhere in the alpine foreland in morainic or molasse deposits as well as in the major riverbeds. Other alpine raw materials include different varieties of quartz and quartzite.

Reservations must be made both for an insufficient knowledge of the outcrops or occurrences known in prehistoric times, as well as for the very rough, i.e. generally macroscopic, classification of individual raw material types (Weniger 1991:86).

An exact provenance determination of specific raw material sources is thus exceptional, but the evidence still allows a more general discussion. Unfortunately, the Mesolithic inventories in particular are very poorly represented in the analysis. Mesolithic artifacts often display a comprehensive patination due to heat treatment, and therefore very few Mesolithic assemblages have been analyzed with respect to raw material provenance. A total of 67 lithic inventories have been included here in a general analysis (Table 1). Of these only 10 are Mesolithic, 7 are mixed Late Paleolithic and Mesolithic, and the remaining 50 are Paleolithic (39 Magdalenian and 11 Late Paleolithic).

In the analysis, lithic raw materials occurring within certain distances from the sites were recorded with respect to their relative frequency in the inventories. A simple distinction was made between on-site, local, regional and exotic raw materials. Local raw materials (<10 kilometers) occur within the site catchment area and are thus within daily reach. Regional raw materials (10–50 kilometers) occur within the probable annual territory of the group. They either represent basic equipment brought from a previous settlement or they result from an extended trip. The so-called exotic raw materials (>50 kilometers) may theoretically represent either long-distance migrations or barter

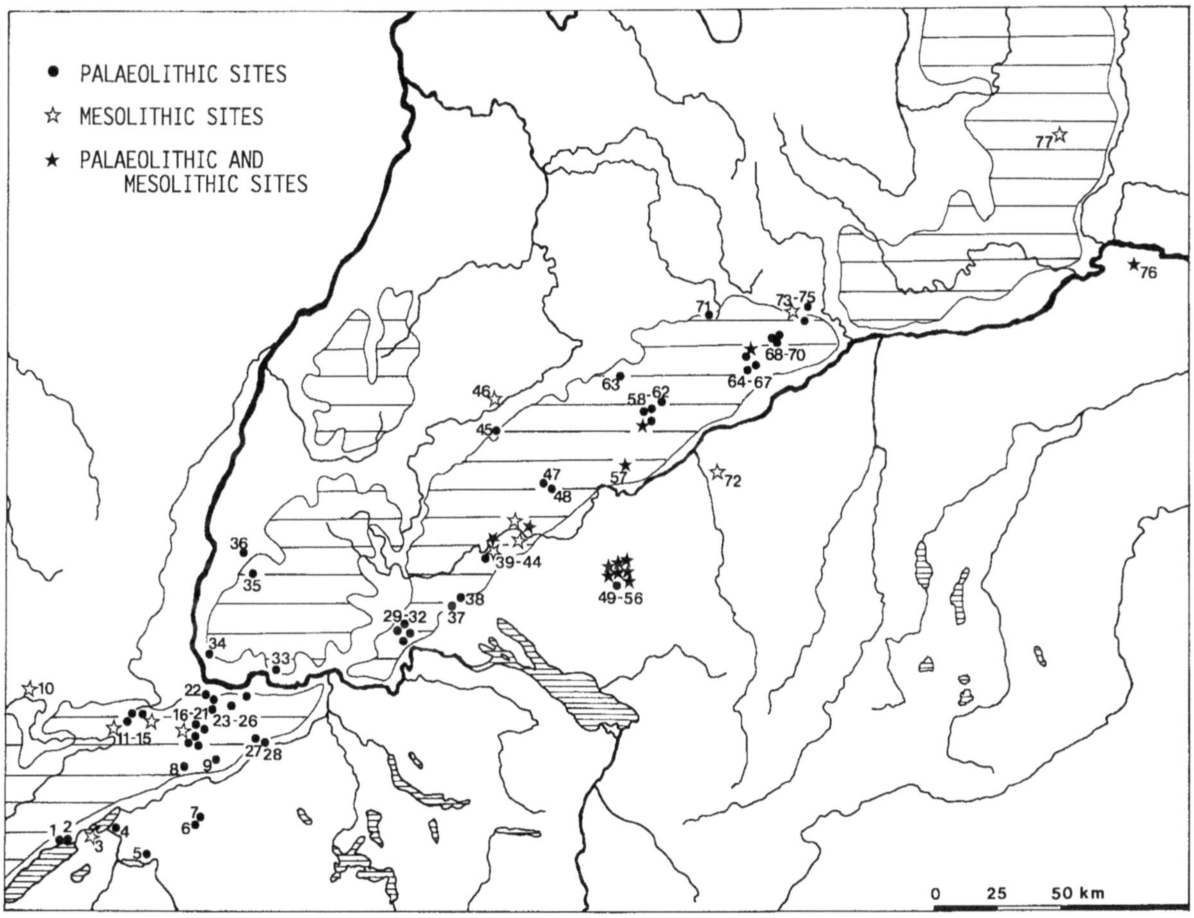

Figure 4. Location of the sites included in the study: 01 Monruz, 02 Champréveyres, 03 Gampelen-Jänet 3, 04 Lüscherz-Moos, 05 Moosbühl, 06 Hintere Burg, 07 Fürsteiner, 08 Balm bei Günsberg, 09 Rislisberghöhle, 10 Bavans, 11 Gripons, 12 Löwenburg-Ziegelacker, 13 Löwenburg-Niederfeld II, 14 Neumühle, 15 Roggenburg-Ritzigrund, 16 Chesselgraben, 17 Kohlerhöhle, 18 Kastelhöhle, 19 Birsmatten-Basisgrotte, 20 Wachtfels, 21 Büttenloch, 22 Bruederholz, 23 Birseck-Ermitage, 24 Hollenberg-Höhle 3, 25 Eremitage, 26 Bönistein, 27 Köpfli, 28 Käsloch, 29 Kesslerloch, 30 Vorder Eichen, 31 Schweizersbild, 32 Freudenthal, 33 Röthekopf, 34 Isteiner Klotz, 35 Teufelsküche, 36 Munzingen, 37 Petersfels, 38 Gnirshöhle, 39 Buttentalhöhle, 40 Jägerhaushöhle, 41 Probstfels, 42 Falkensteinhöhle, 43 Burghöhle Dietfurt, 44 Zigeunerfels, 45 Napoleonskopf, 46 Rottenburg-Siebenlinden II, 47 Annakapellenhöhle, 48 Nikolaushöhle, 49 Schussenquelle, 50 Aichbühl A, 51 Aichbühl B, 52 Aichbühl C, 53 Aichbühl D, 54 Aichbühl E, 55 Aichbühl F, 56 Henauhof NW, 57 Felsställe, 58 Hohler Fels Schelklingen, 59 Helga Abri, 60 Sirgenstein, 61 Sirgenstein Südwand, 62 Brillenhöhle, 63 Burkhardtshöhle, 64 Hohlenstein Stadel, 65 Vogelherd, 66 Spitzbubenhöhle, 67 Malerfels I, 68 Bärenfelsgrotte, 69 Spitalhöhle, 70 Klingenfelsschutzdach, 71 Kleine Scheuer Rosenstein, 72 Attenhofen, 73 Große Ofnet, 74 Hohlenstein Ederheim, 75 Kaufertsberg, 76 Sarching, 77 Bettelküche.

transactions. They generally have been transported between 50 and 200 kilometers. Obviously, this is not very exotic and it appears that all lithic raw materials may indeed have been (and very likely were) procured actively, i.e. not by trade or barter.

In the Jura inventories we observe a strong predominance of local raw materials, i.e. Jurassic hornstones. These generally account for 80 or more percent of all lithics. The inventories from the Black Forest foothills and those from the lowland area of Oberschwaben and the Swiss Mittelland complement this picture in a very interesting way. Here regional raw material predominates, while local types account for generally less than 15 percent and only appear significantly in sites situated close to the Jura. Thus, Jurassic hornstones also dominate in these regions.

Exotics are always rare, in the Jura as well as in the lowland area. Exceptions in this respect are represented by the Bruckersberg sites (Bärenfelsgrotte, Spitalhöhle and Klingenfelsschutzdach, Auffermann 1991) and the sites from Neuchâtel (Monruz and Champréveyres, Affolter *et al.* 1994 and Le Tensorer and Niffeler 1993) together with the nearby Gampelen-Jänet 3 (Nielsen 1991). Incidentally we find that the exotic raw materials primarily seem to move along a northeast-southwesterly axis following the southern limit of the Jura formation (Cattin 1990; Pousaz, ed. 1991:87, fig. 79; Hahn this volume).

The observations from this preliminary analysis emphasize the significance of material quality, i.e. workability. Jurassic hornstones are generally of a marked better quality than the alpine stones occurring in the morainic deposits of Oberschwaben and the Swiss Mittelland. Apparently the advantages of a high quality raw material seem to generally outweigh its higher costs.

It has not been possible within the present study to investigate the variation in different raw material types with respect to selective manufacture and discard patterns of specific artifact types. However, the general impression from the literature and from my own analysis of a number of inventories is that exotic raw materials (like Bavarian tabular chert) generally reflect the overall pattern of tool frequencies (see also Fisher this volume and Hahn this volume). There is no evidence of a preferential raw material selection for specific tool types. Yet, it does seem that in the Paleolithic inventories backed

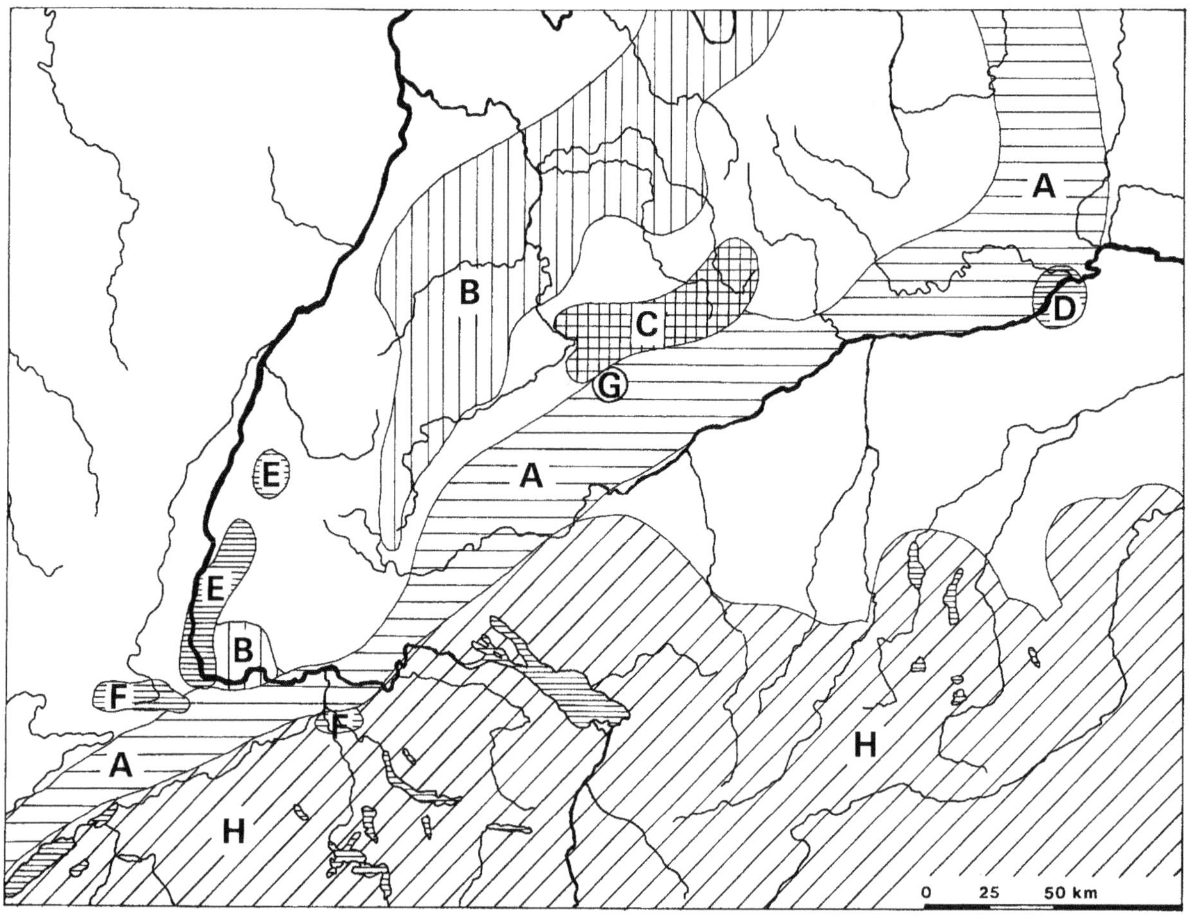

Figure 5. Important raw material provinces. A: Jurassic hornstones, B: 'Muschelkalk' hornstone, C: 'Keuper' hornstone, D: Bavarian tabular chert, E: Jasper, F: Kimmeridgian chert, G: Siliceous tuff, H: Alpine raw materials (radiolarian chert, quartz, quartzite, etc.).

Table 1. Relative Frequencies of On–Site, Local, Regional and Exotic Raw Materials in various Magdalenian (MAGD), Late Paleolithic (LP) and Early Mesolithic (EM) Inventories

Site, Level, Archaeological Dating		Provenance of lithics*				References
		on site <500 m	local <10 km	regional 10–50 km	exotic >50 km	
01. Monruz	MAGD	–	60%	–	40%	Affolter et al. 1994
02. Champréveyres	MAGD	–	59%	–	41%	Le Tensorer and Niffeler 1993
03. Gampelen–Jänet 3	EM	–	33%	51%	16%	Nielsen 1991
04. Lüscherz–Moos	LP	–	xxx	x	–	Schwab 1972
05. Moosbühl I–II	MAGD	–	xx	xx	–	Barr 1973
06. Hintere Burg	MAGD	–	x	xxx	–	Flükiger 1964
07. Fürsteiner	LP	x	95%	4%	1%	Zimmermann 1988
08. Balm bei Günsberg, 6	LP	x	67%	33%	–	Schweizer 1941
09. Rislisberghöhle, II	MAGD	–	xxx	x	–	Barr 1977
10. Bavans, 6	EM	–	x	xx	x	Pousaz, ed. 1991
11. Gripons, 4	EM	–	xxx	x	(x)	Pousaz, ed. 1991
12. Löwenburg–Ziegelacker	LP	xxx	–	–	–	Jagher and Jagher 1987
13. Löwenburg–Niederfeld II	LP	xxx	–	–	–	Jagher and Jagher 1987
14. Neumühle	LP	–	xxx	?	?	Rozoy 1978
15. Roggenburg–Ritzigrund, lower	EM	–	xxx	x	?	Jagher 1989
17. Kohlerhöhle, upper	MAGD	?	x	x	?	Lüdin 1938
20. Wachtfels, 5	LP	–	xxx	xx	–	Sedlmeier 1989
21. Büttenloch, A–B	MAGD	–	xx	xx	–	Sedlmeier 1989
22. Bruederholz	MAGD	–	xxx	x	–	Sarasin 1918
23. Birseck–Ermitage, lower	MAGD	–	xxx	x	?	Sarasin 1918
24. Hollenberg–Höhle 3	MAGD	–	34%	64%	–	Sedlmeier 1982
25. Eremitage	MAGD	–	xxx	x	–	Sedlmeier 1989
26. Bönistein	LP	–	xxx	x	–	Bandi 1947
27. Köpfli	MAGD	–	xxx	x	–	Zürcher 1969
29. Kesslerloch	MAGD	–	100%	–	–	Schmid 1977
31. Schweizersbild	MAGD	x	xx	x	–	Nüesch 1896
33. Röthekopf	MAGD	–	x	x	–	Andrée 1939
34. Isteiner Klotz	LP	–	x	x	–	Kuhn 1969
35. Teufelsküche, B–C	MAGD	–	28%	67%	?	Pasda 1994
36. Munzingen	MAGD	–	17%	82%	?	Pasda 1994
37. Petersfels (Peters)	MAGD	–	xxx	xx	x	Wiegers 1950
Petersfels (Albrecht)	MAGD	–	99%	–	1%	Albrecht 1979
38. Gnirshöhle, I	MAGD	–	xx	–	x	Albrecht et al. 1977
39. Buttentalhöhle	MAGD	x	xxx	x	x	Hahn this volume
41. Probstfels (Schmidt)	MAGD	–	83%	6%	2%	Pasda 1989
45. Napoleonskopf	MAGD	–	xx	x	(x)	Mauser 1976, Schmidt 1912
46. Rottenburg–Siebenlinden II	EM	x	60%	40%	–	Kieselbach 1993
49. Schussenquelle	MAGD	–	7%	93%	1%	Schuler 1989
50. Aichbühl A	LP,EM	–	14%	78%	8%	Eberhardt et al. 1987
51. Aichbühl B	LP,EM	–	6%	94%	–	Eberhardt et al. 1987
52. Aichbühl C	LP,EM	–	7%	92%	1%	Eberhardt et al. 1987
53. Aichbühl D	LP,EM	–	8%	88%	4%	Eberhardt et al. 1987
54. Aichbühl E	LP,EM	–	14%	81%	5%	Eberhardt et al. 1987
55. Aichbühl F	LP,EM	–	3%	97%	–	Eberhardt et al. 1987
56. Henauhof NW, 6	LP	–	17%	81%	2%	Jochim 1993
Henauhof NW, 4–5	EM	–	17%	82%	1%	Jochim 1993
57. Felsställe, IIIa–b	MAGD	99%	–	1%	–	Kind 1987
Felsställe, IIa3	EM	37%	61%	2%	(x)	Kind 1987
58. H.F. Schelklingen, Ia–c, IIa	MAGD	–	100%	(x)	–	Hahn, ed. n.d.
59. Helga Abri, IIIa–d	MAGD	–	xx	x	x	Hahn and Scheer 1983
Helga Abri, IIF1–IIF2	EM	–	100%	(x)	–	Hahn and Scheer 1983
60. Sirgenstein, I	MAGD	–	80%	20%	x	Author's registration
61. Sirgenstein Südwand, 5	MAGD	–	98%	2%	–	Author's registration, Riek 1959

Table 1. Continued

Site, Level, Archaeological Dating		Provenance of lithics*				References
		on site <500 m	local <10 km	regional 10–50 km	exotic >50 km	
62. Brillenhöhle, IV	MAGD	(x)	68%	22%	8%	Lauxmann 1988
63. Burkhardtshöhle, V	MAGD	–	92%	2%	6%	Simon 1993
64. Hohlenstein Stadel, III	MAGD	–	95%	5%	–	Hahn n.d.
65. Vogelherd, II–III	MAGD	–	95%	5%	–	Riek 1934
66. Spitzbubenhöhle, 2	MAGD	–	100%	–	–	Hahn, ed. 1984
67. Malerfels I, 3	LP	–	100%	–	–	Hahn, ed. 1984
Malerfels I, 1b	EM	–	98%	2%	–	Albrecht 1984
68. Bärenfelsgrotte, III	MAGD	–	50%	3%	32%	Auffermann 1991
69. Spitalhöhle, VIII	MAGD	–	46%	2%	12%	Auffermann 1991
70. Klingenfelsschutzdach, III–IV	MAGD	–	73%	3%	24%	Auffermann 1991
71. Kl. Scheuer Rosenstein, I–II	MAGD	–	xx	x	(x)	Maier 1936
72. Attenhofen	EM	–	50%	50%	–	Wischenbarth 1991
75. Kaufertsberg, 1 (lower)	MAGD	–	70%	30%	–	Kaulich 1983
76. Sarching '83	LP,EM	–	xx	xxx	–	Heinen 1986

*Lithic raw materials of uncertain provenance are not included in the statistics.

bladelets and backed points, in particular, appear to be made of a non-local (regional or exotic) material (Albrecht 1979:63; Kind 1987:115). Here the presence of non-local elements undoubtedly reflects re-tooling activities, and the pieces themselves represent basic equipment brought along from a previous site.

In the Early Mesolithic local hornstones also seem to have been subjected to heat treatment before the final processing into tools, especially microliths (Eriksen 1991:175f). When thermally altered, Jurassic hornstones change color from white, grey or yellow-ochreous to pink or red. At the same time their flaking abilities seem to improve significantly (Rottländer 1983:562, 1989:47f). There still remain some problems to be solved with respect to this presumable heat treatment (Eriksen 1991:180; Price *et al.* 1982:484). However, assuming that it did take place, it represents a significant technological difference between the Paleolithic and the Mesolithic flint knapping traditions. It indicates that flint knapping in the Mesolithic did require a certain degree of specialization and a thorough knowledge of raw material properties.

Heat treatment of chert or hornstone is a fairly complicated process which needs to be carefully controlled. Obviously, it adds significantly to the costs of tool manufacture and it thus may be expected to have been applied fairly selectively in prehistory (Lurie 1989:53). A preliminary analysis of some Mesolithic inventories from southwestern Germany (Eriksen 1991:178f) indicates that heat treatment probably was related primarily to the production of microliths. There is thus good reason to interpret this phenomenon as an indication of Mesolithic flintknappers trying to economize scarce high quality raw materials (Hahn 1983:370). Increased fine sedimentation of riverbeds and flourishing vegetation together with reduced erosion of outcrops must have made good quality hornstones harder to find in the early postglacial.

To summarize, we find a marked tendency to use lithic raw materials of local, or at least regional, origin during the entire period in question (of course with reservations for the relative scarcity of Mesolithic and Late Paleolithic data). The choice of raw material depends more on quality and less on distance—good quality raw material is always preferred. Exotic raw materials are usually rare and generally confined to a few, worn out blades or tools.

Most evidence from the Jura inventories points to an embedded procurement pattern (Binford 1979) probably reflecting seasonal movements throughout the region. The only significant exceptions are represented by the Magdalenian occupation of Felsställe AH IIIb (Kind 1987) and the Late Paleolithic sites of Löwenburg-Ziegelacker and Löwenburg-Niederfeld II (Jagher and Jagher 1987). At the same time, the amount of regional raw material in the inventories from the Black Forest foothills, the Swiss Mittelland and Oberschwaben indicates the complementary existence of a more direct procurement by special-purpose trips to source areas in the Alb. Admittedly, though, all distances referred to in the analysis are very short (even regarding the so-called exotic raw materials) and it is difficult to distinguish explicitly between an 'embedded' and a 'direct special-purpose' procurement pattern. In the present case study the only reasonable distinction is one of degrees.

There are only minor differences between the Paleolithic and the Mesolithic procurement patterns. The most prominent distinction is represented by the probable heat treatment in the Early Mesolithic. With the exception of Gampelen-Jänet 3 (Nielsen 1991), there appears to be slightly less exotic raw material in the Mesolithic inventories, but this observation is more or less outweighed by the occurrence of generally more regional raw material. The differences observed may be partly explained by an increased scarcity of high quality raw materials caused by environmental factors in the early postglacial. On the other hand, they may also be indicative of diachronic changes in settlement and mobility patterns (Jeske 1989). This intriguing aspect will be discussed in more detail below.

In my opinion, however, lithic raw materials are not ideal for a thorough analysis of hunter-gatherer social territories. They obviously represent a critical resource, which will be optimized with respect to investment of time and energy (Jochim 1989). Yet, at the same time they very often represent an instant 'here and now' use or production of artifacts. In the study area lithic raw materials suitable for artifact production were generally widely accessible and neither the late glacial nor the early postglacial flint knappers would have had explicit reason to 'plan ahead' except for maintaining so-called basic equipment.

In general the lithic artifact inventory of mobile hunter-gatherers will only reflect mobility within shorter periods of time (a year) and shorter distances (a few hundred kilometers) (Weniger 1991:87). A chronologically and geographically more comprehensive mobility and inter-group communication pattern would rather be reflected in the organic inventory and portable art objects and ornaments. These have a considerably longer 'lifetime' than lithic artifacts and they provide a much more reliable impression of cultural affinity or even ethnic (group) identity (Weniger 1991:99f).

Jet Ornaments and Ammonites

Jet and lignite are easy to work with flint tools and appear to have been commonly used raw materials for ornamental or art objects during the Magdalenian (Figure 6). The objects are variably referred to as being made of jet, lignite or even fossil wood, and a specific petrographic determination is rarely available (Preuschoft-Güttler 1995). The actual raw material thus may vary, but due to their morphological and typological homogeneity, the objects will be treated as a group.

Jet and lignite occur sporadically in the Posidonian slate deposits of the Lias-Epsilon (Black Jura) along the entire northwestern Alb ridge (Figure 7). Jet and lignite also occur in rich molasse deposits in Hegau, Schaffhausen and northwestern Switzerland.

The map in Figure 7 shows all Magdalenian finds of jet/lignite within the study area. The majority of the inventories contain only a few, often fragmentary pieces, but three finds in Hegau and Schaffhausen differ significantly. In Petersfels alone, more than 600 pieces have been found (Peters and Toepfer 1932). The large quantities of jet/lignite found here are very likely of local origin. As

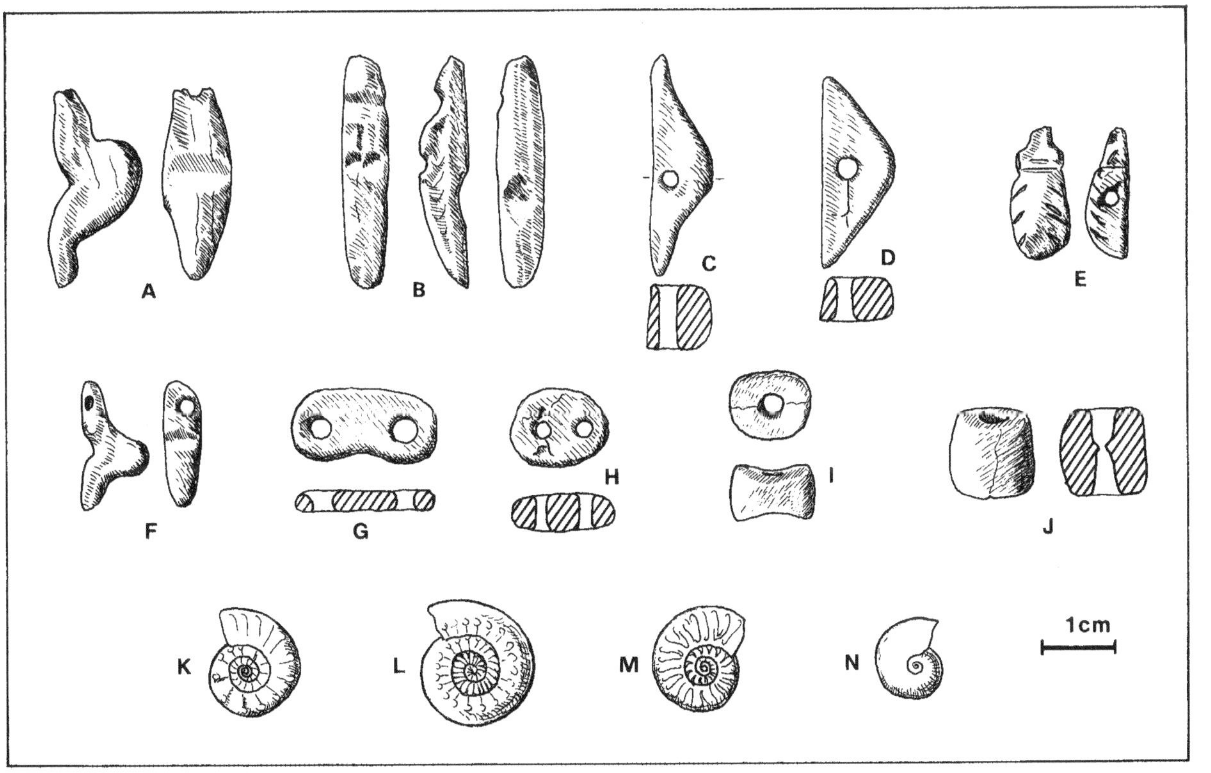

Figure 6. A-J: Magdalenian pendants in jet from Petersfels 1927-32 (after Mauser 1970). K-N: Miscellaneous ammonites from the Swabian Jura.

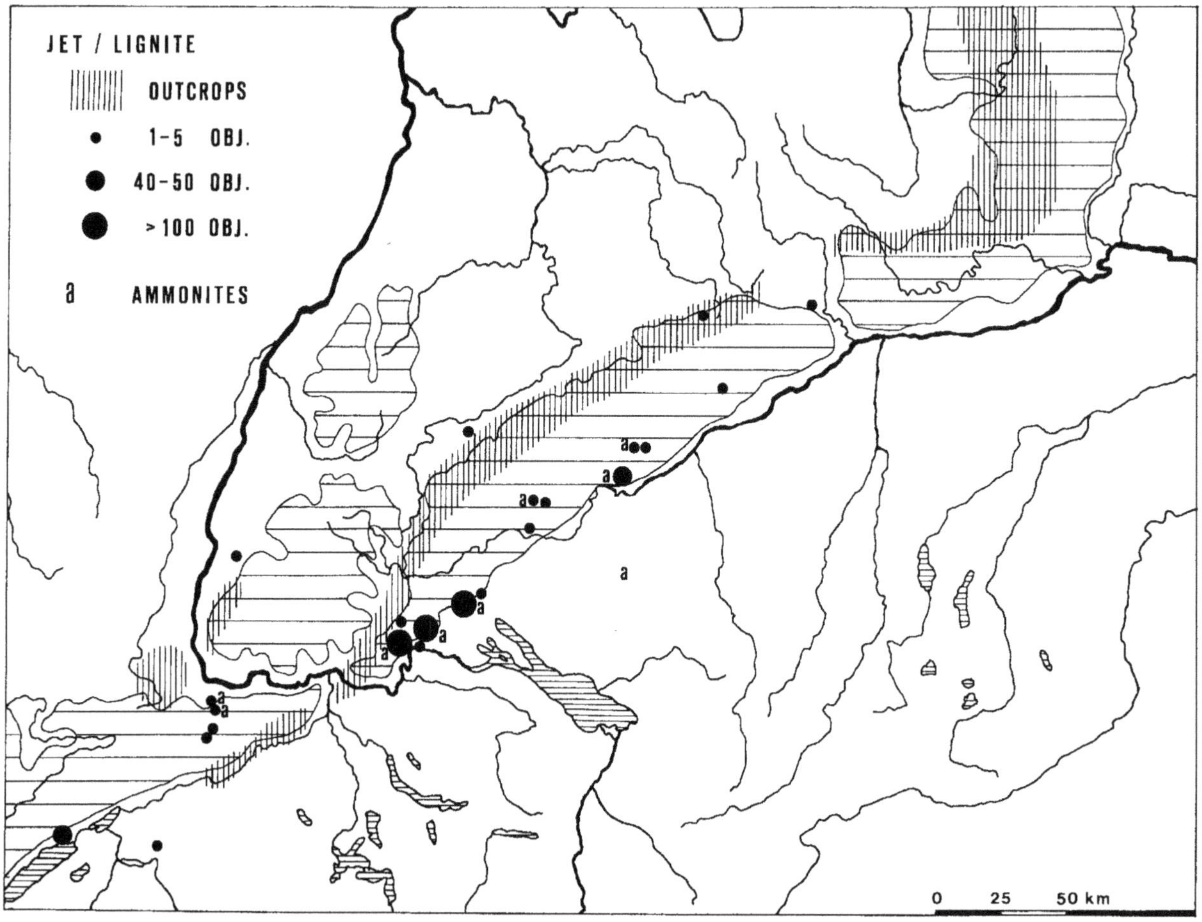

Figure 7. Outcrops of jet in the study area and Magdalenian finds with jet/ lignite objects or ammonites.

regards the remaining finds the question of provenance (and indeed sometimes the question of raw material) is more open. A precise provenance determination of individual pieces does not appear to be possible (Schmid 1977:62; Preuschoft-Güttler 1995). In any case, though, jet and lignite are local or regional raw materials that very likely were procured actively, i.e. not by trade or barter.

It is striking that they have never been found in Mesolithic or even Late Paleolithic inventories. Since we have now a fair number of recently excavated and well documented Mesolithic sites, this is very likely a valid observation, free from methodological reservations, and it represents a significant difference between the Paleolithic and Mesolithic groups.

Incidentally, the majority of pieces have been found in four sites (Kesslerloch, Schweizersbild, Petersfels and Felsställe) that all appear to have been settled in the cold season. Perhaps this is a manifestation of the existence of long-term winter base camps where people had time left for doing home crafts?

The map in Figure 7 also displays all Magdalenian finds of fossil ammonites (Figure 6). Ammonites are probably the most prominent group of Jurassic index fossils (Geyer and Gwinner 1991:132-133, fig. 69) occurring in large numbers and a variety of different species throughout the study area. The archaeological examples represent one of the more obscure find categories. They are all very small and inconspicuous and thus generally rare in the inventories. They often possess a natural central hole, but in fact it is also very easy to pierce the central part of an ammonite by pressing a small flint tip against it (Sedlmeier 1982:51). Precise species and provenance determinations of the archaeological examples are rare. Quite a few, however, are reported to be from the Lias- deposits, which makes it reasonable to treat them together with the jet ornaments (Albrecht *et al.* 1994:31f). The procurement of ammonites and jet/ lignite, in my opinion, is closely related. Incidentally, ammonites, like jet and lignite, have never been found in Mesolithic or Late Paleolithic horizons. They do, however, all occur in both Aurignacian and Gravettian inventories (Hahn 1992). The use of these raw materials, thus, appears to be strictly Paleolithic.

Ornamental Mollusks

Last but not least, ornamental mollusks occur throughout the period in question, and the Mesolithic finds, for once, appear to be fairly well represented. A total of 13 Mesolithic and 22 Magdalenian inventories contain fossil or sub-recent mollusks, and 2 Late Paleolithic inventories each contain an unspecified shell fragment. Tables 2 and 3 summarize the contents of these mollusk inventories. The map in Figure 8 displays the location and archaeological dating of the sites included.

For both the Mesolithic and the Paleolithic it is found that shells of limnic or marine Tertiary fossils were preferred to sub-recent mollusks. They have often been artificially perforated for use as pendants. Sometimes natural holes have served the same purpose. Many, however, have not been pierced and may rather be considered as blanks (Rähle 1987b:383).

The species found may be grouped by classes, including bivalves (mussels) and scaphopods (tooth shells) of a usually reasonable size (Figure 9), and gastropods (snails) which are often fairly small and inconspicuous (Figures 9-10). The retrieval of these tiny mollusks is very much dependent on the use of meticulous excavation methods like wet-screening. Thus, the majority of gastropods come from recently excavated sites. Unfortunately, most of the inventories are from caves with relatively loose gravel sediments in which the stratigraphic position of very small objects may be somewhat misleading. In a few cases the tiny gastropods very likely have 'trickled down' from overlying cultural layers. Similarly, the scant number of bivalves known from old excavations very likely represents only a small proportion of what may originally have been there.

The Magdalenian inventories display the largest variety of species (Figure 9). Bivalves play a prominent part. Fossil or even sub-recent *Glycymeris* shells are the most frequent and numerous. Gastropods are most in evidence in the recently excavated finds. Here tower shaped species like *Turritella, Tympanotonos, Potamides* and *Pirenella* occur most frequently, but the round, barrel shaped *Viviparus* are also numerous. Tooth shells (*Dentalia*) and various fossils like ammonites, belemnites and Tertiary shark teeth are also common.

In the Mesolithic inventories we find neither tooth shells nor ammonites or other fossils (Figure 10), and only one Mesolithic site within the study area has produced bivalves. The species most frequently found in Mesolithic inventories are the elegantly tower shaped *Potamides* and *Pirenella*, and the more inconspicuous *Gyraulus*.

Große Ofnet stands out quite exceptionally when compared to the other Mesolithic sites. This is partly due to the quantity of mollusks (especially *Lithoglyphus naticoides*, which is unique both in

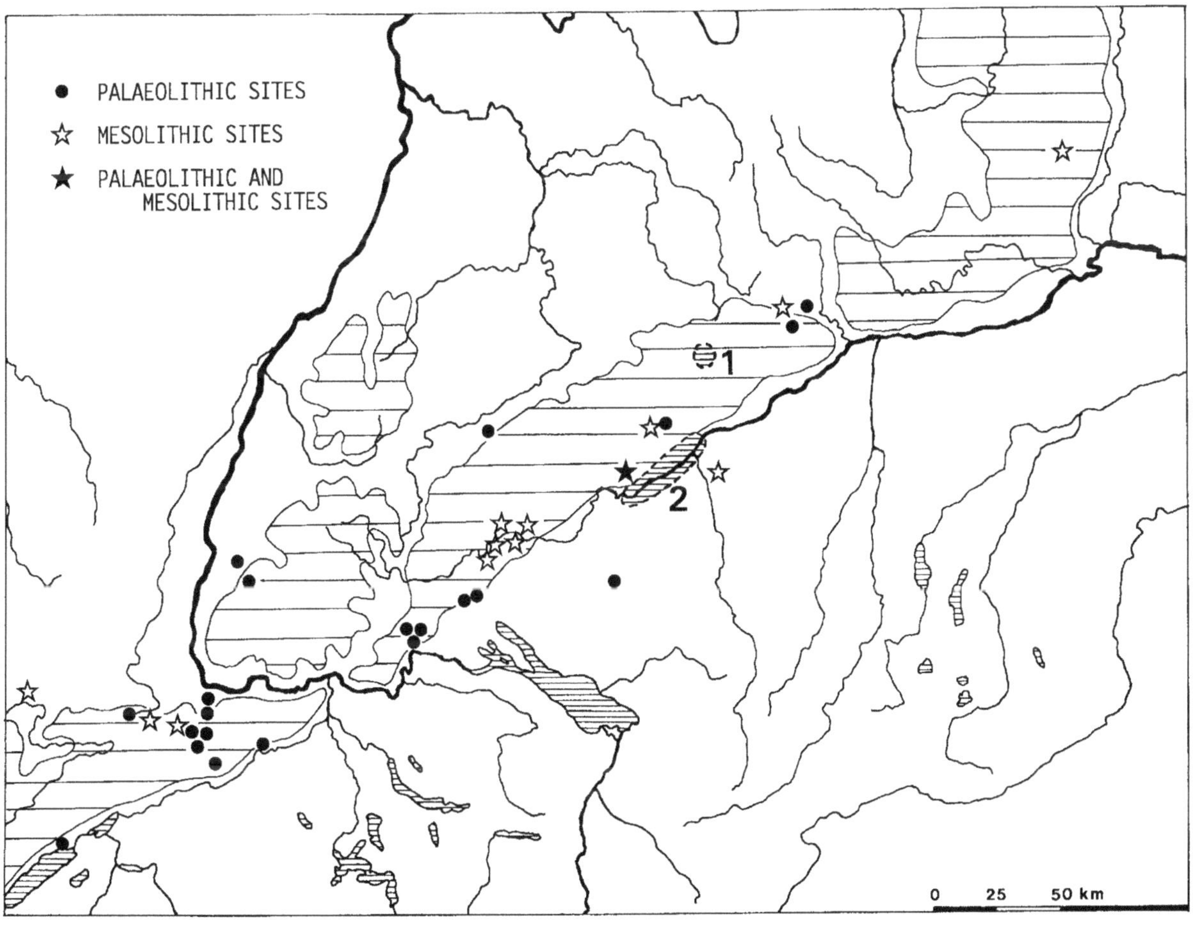

Figure 8. Location of Late Glacial and Early Postglacial sites with ornamental mollusks and domestic deposits of fossil mollusks from (1) the Steinheimer basin and (2) the Kirchberger layers on the Upper Danube.

number and provenance), and partly because of the find context. All other mollusk inventories represent stray finds associated with probable occupation floors. In Große Ofnet, however, the fossil shells were recovered from two pits containing head burials (Figure 11) where the tiny gastropods were found partly strewn in the ochreous filling, and partly attached to individual skulls, especially those of

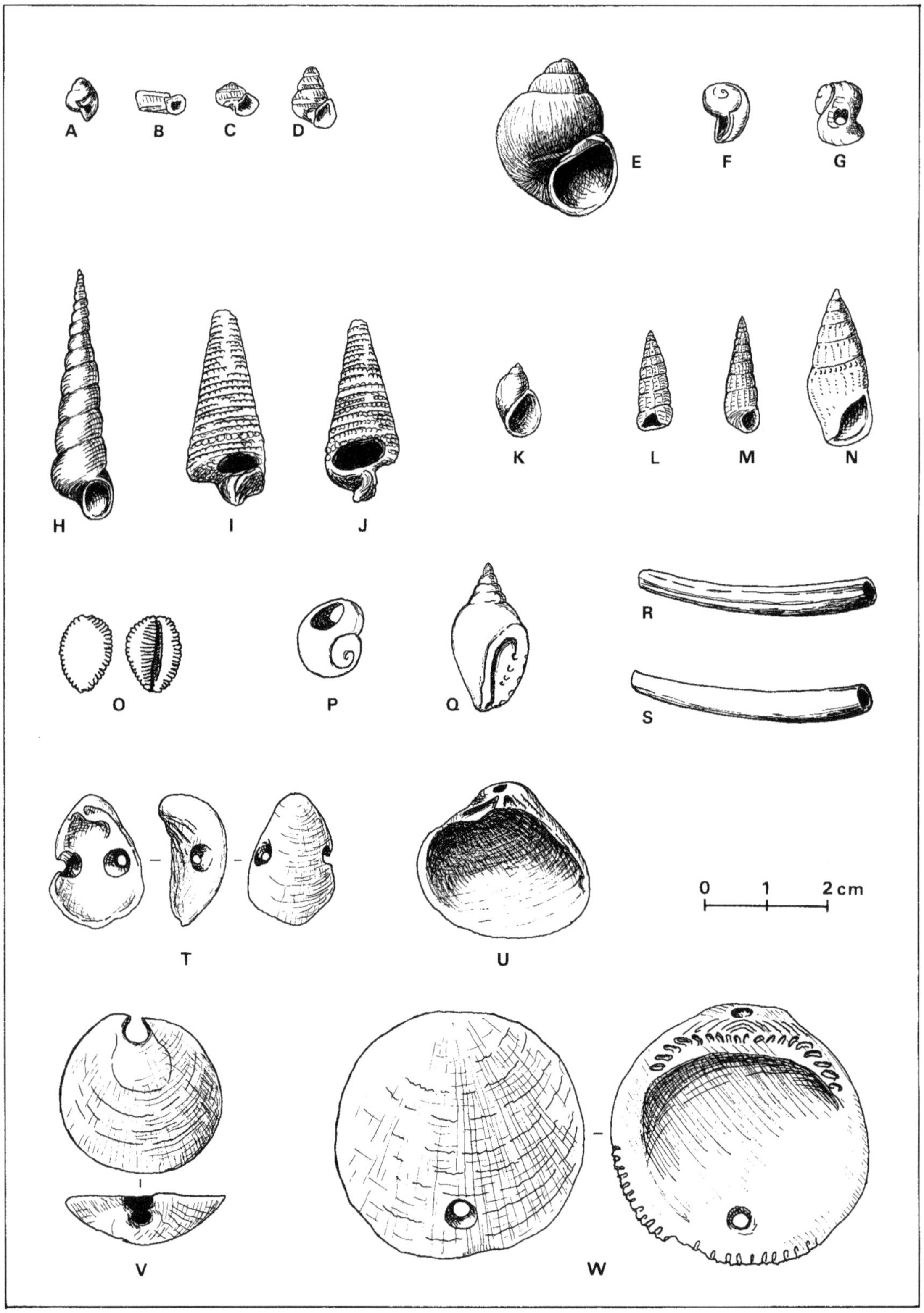

Figure 9. Ornamental mollusks from Magdalenian inventories: A-D: *Gyraulus sp.*, E-G: *Viviparus suevicus,* H: *Turritella sp.*, I-J: *Tympanotonos margaritaceus*, K: *Radix socialis*, L-N: *Pirenella sp.*, O: *Trivia sp.*, P: *Littorina obtusata*, Q: *Purpura lapillus*, R-S: *Dentalium sp.*, T: *Gryphaea arcuata*, U: *Cyrene sp.*, V-W: *Glycymeris sp.*

children and younger women (Schmidt 1912:37f). The find circumstances reveal that here the ornamental mollusks must have been part of some kind of headgear or other garment. This observation ought to be applicable to the other finds of fossil and sub-recent mollusks. In the Magdalenian as well as in the Mesolithic these objects undoubtedly were part of personal equipment.

The presumed provenance of many of the mollusks have been identified (Tables 2-3). These identifications, however, are not always quite as unambiguous as we like them to be. I have chosen to depict the most probable connections with reference to number of positive source identifications, proximity, and travel route—the latter especially with respect to pieces of Mediterranean origin.

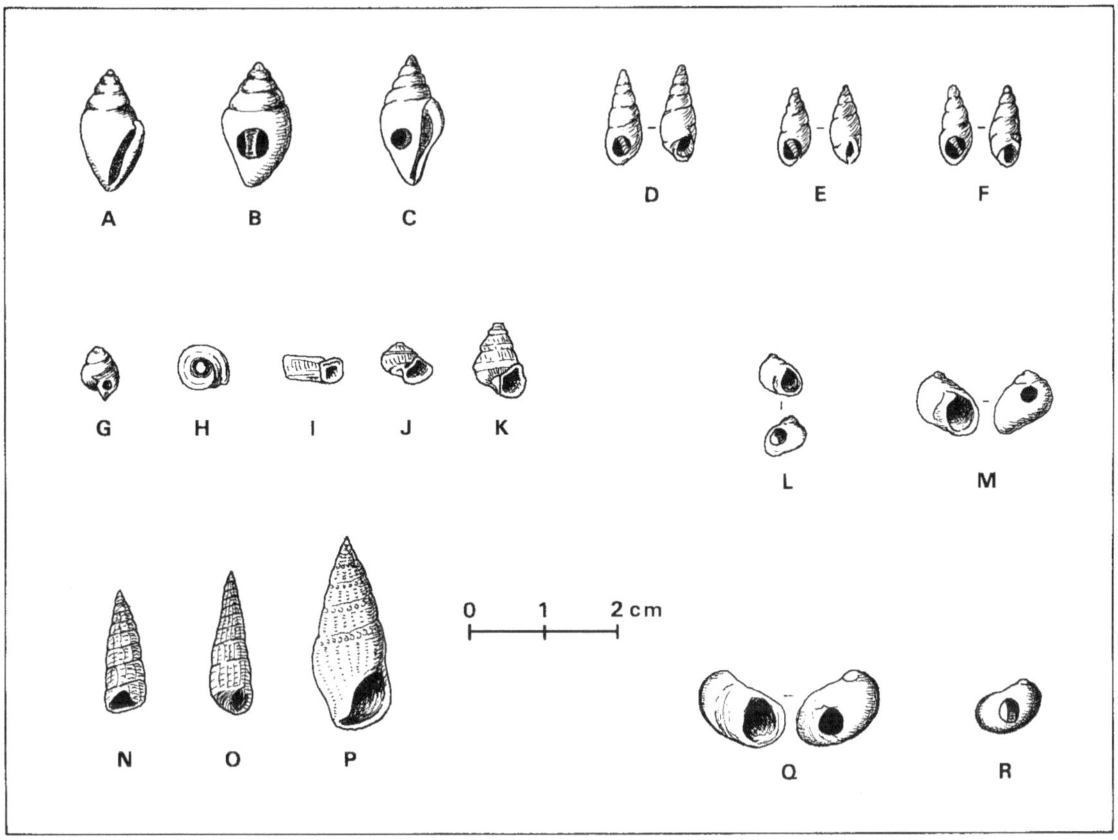

Figure 10. Ornamental mollusks from Early Mesolithic inventories: A-C: *Columbella rustica*, D-F: *Bayania sp.*, G-K: *Gyraulus sp.*, L-M: *Lithoglyphus naticoides*, N-P: *Pirenella sp.*, Q-R: *Theodoxus gregarius.*

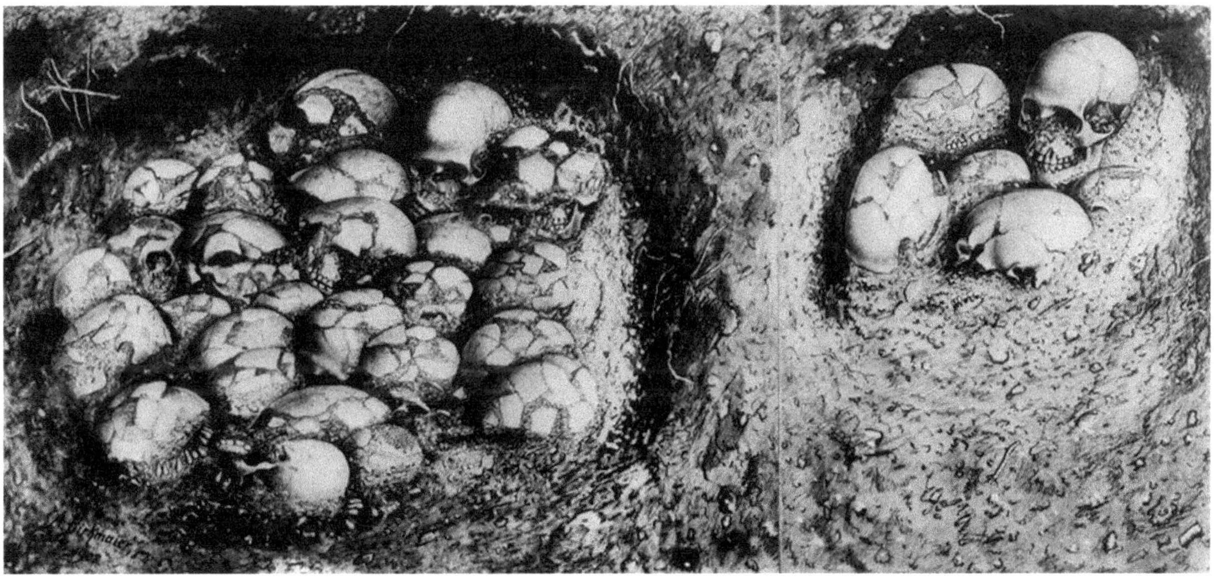

Figure 11. Große Ofnet—the head burials (after Schmidt 1912).

Table 2. Schematic Survey of Ornamental Mollusks (Gastropodae [G], Scaphopodae [S] and Bivalvae [B]) found in Magdalenian Sites* within the Study Area*

		A	B	C	D	E	F	G	H	I	J	K	L	M	N	O	P	Q	R	S	T	U	V
Mediterranean sea (sub-recent)																							
Homalopoma sanguineum	(G)	-	-	-	1	-	-	-	-	-	-	-	-	1	-	1	-	-	-	-	-	-	-
Cyclope neritea	(G)	-	-	-	-	-	-	-	-	-	-	-	-	-	-	3	-	-	-	-	2	-	-
Pirenella plicata	(G)	-	-	-	-	-	-	-	-	-	-	-	-	-	1	-	-	-	-	-	-	-	-
Pleurotoma sp.	(G)	-	-	-	-	-	-	-	-	-	-	-	-	-	1	-	-	-	-	-	-	-	-
Trochus obliquatus	(G)	-	-	-	-	-	-	-	-	-	-	-	-	-	2	-	-	-	-	-	-	-	-
Cytherea sp.	(G)	-	-	-	-	-	-	-	-	-	-	-	-	-	2	-	-	-	-	-	-	-	-
Glycymeris sp.	(B)	-	-	-	-	-	1	-	-	-	-	-	-	-	10	-	-	-	-	-	-	-	-
Atlantic sea (sub-recent)																							
Nucella / Purpura lapillus	(G)	-	-	-	-	-	-	-	-	-	-	-	-	1	-	-	2	-	-	-	-	-	-
Littorina obtusata	(G)	-	-	-	-	-	-	-	-	-	-	-	-	-	-	-	-	-	-	-	1	-	-
Astarte montagui	(B)	-	-	-	-	-	-	-	-	-	-	-	-	-	-	-	1	-	-	-	-	-	-
Paris basin (fossil)																							
Bayania sp.	(G)	-	-	-	-	-	-	-	-	-	-	-	-	-	-	1	-	-	-	-	-	-	-
Sycum sp.	(G)	-	-	-	-	-	-	-	-	-	-	-	-	-	-	-	2	-	-	-	-	-	-
Turritella sp.	(G)	-	1	-	-	3	3	-	-	-	-	-	-	-	-	-	-	-	-	-	-	-	-
Mainzer basin (fossil)																							
Proadusta meyeri	(G)	-	-	-	-	-	-	-	-	-	-	-	-	-	-	1	-	-	-	-	-	-	-
Natica sp.	(G)	-	-	-	-	-	-	-	-	-	1	-	-	-	-	-	-	-	-	-	-	-	-
Buccinum cassidera	(G)	-	-	-	-	-	-	-	-	-	1	-	-	-	-	-	-	-	-	-	-	-	-
Tympanotonos margaritaceus	(G)	-	6	-	1	1	-	14	-	-	-	-	-	-	1	-	-	-	-	-	-	-	-
Potamides lamarcki	(G)	-	-	-	-	-	-	-	-	-	1	-	2	-	-	2	-	-	-	-	-	-	-
Pirenella plicata	(G)	-	1	-	19	2	-	-	-	X	-	-	3	-	-	7	-	-	-	2	2	-	-
Polinices achatensis	(G)	-	-	-	-	-	-	-	-	-	-	-	-	-	-	-	-	-	-	-	1	-	-
Dentalium sp.	(S)	-	1	-	-	-	-	-	-	X	4	-	-	1	-	18	-	-	-	1	3	-	-
Glycymeris sp.	(B)	7	1	1	18	10	-	20	2	-	5	1	3	-	98	28	-	1	-	-	2	-	2
Cyrene sp.	(B)	-	-	-	-	-	-	-	-	-	1	-	-	2	-	-	-	-	-	-	-	2	-
Bivalvae pernostrea	(B)	-	-	-	-	-	-	-	-	-	1	-	-	-	-	-	-	-	-	-	-	-	-
Gastropodae & Bivalvae unsp.	(B)	-	-	-	-	-	-	-	-	XX	-	-	-	-	-	-	-	-	-	-	-	-	-

Table 2. Continued

		A	B	C	D	E	F	G	H	I	J	K	L	M	N	O	P	Q	R	S	T	U	V
Steinheimer basin (fossil)																							
Gyraulus trochiformis	(G)	18	-	-	-	-	-	-	-	-	-	-	-	-	-	1	-	-	1	4	7	-	-
Gyraulus sulcatus	(G)	-	-	-	-	-	-	-	-	-	-	-	-	-	-	-	-	-	-	1	1	-	-
Radix socialis	(G)	-	-	-	-	-	-	-	-	-	-	-	-	-	-	-	-	-	-	1	2	-	-
Kirchberger layers (fossil)																							
Viviparus suevicus	(G)	12	-	-	1	-	-	-	-	-	1	-	-	-	-	34	1	-	43	10	4	-	-
Brotia escheri	(G)	1	-	-	-	-	-	-	-	-	-	-	-	-	1	8	-	-	-	2	-	-	-
Melanopsis kleini	(G)	-	-	-	-	-	-	-	-	-	-	-	-	-	-	-	-	-	-	-	6	-	-
Congeria sp.	(B)	-	-	-	-	-	-	-	-	-	-	-	-	-	-	-	-	-	5	1	-	-	-
Jurassic formation (fossil)																							
Gryphaea sp.	(B)	-	-	-	-	-	-	-	-	-	-	-	-	-	4	3	-	-	-	-	-	-	-
Ostrea sp.	(B)	-	-	-	-	-	-	-	2	-	-	-	-	-	1	-	-	-	-	-	-	-	-
Various fossils	-	-	-	-	-	-	7	-	-	X	30	-	1	-	18	1	-	-	9	-	-	-	-
Unknown provenance																							
Trivia sp.	(G)	-	-	-	-	-	-	-	-	-	-	-	-	-	-	1	-	-	-	-	-	-	-
Terebratula insignis	(G)	-	-	-	-	-	-	-	-	-	-	-	-	-	1	-	-	-	-	-	-	-	-
Gastropoda indet.	(G)	-	-	-	-	-	-	-	-	1	-	-	-	-	-								

*A: Monruz (Affolter *et al.* 1994), B: Rislisberghöhle II (Sedlmeier 1988), C: Chesselgraben, lower (Sedlmeier 1988), D: Kohlerhöhle, upper (Lüdin 1938; Sedlmeier 1988), E: Kastelhöhle Nord, upper (Sedlmeier 1988), F: Birseck-Ermitage, lower (Sarasin 1918; Sedlmeier 1988), G: Hollenberg Höhle 3 (Sedlmeier 1982, 1988), H: Käsloch (Stampfli 1981; Sedlmeier 1988), I: Kesslerloch (Heierli 1907; Schmid 1977), J: Schweizersbild (Nüesch 1896; Rähle 1987b), K: Freudenthal (Karsten 1874), L: Teufelsküche (Zotz 1928; Pasda 1994), M: Munzingen (Padtberg 1925; Albrecht 1981; Pasda 1994), N: Petersfels 1927-32 (excavation) (Peters 1930; Peters and Toepfer 1932), O: Petersfels 1927-32 (screened fill) and 1974-76 (excavation) (Rähle 1983b, 1994), P: Gnirshöhle (Albrecht, Drautz and Kind 1977; Rähle 1987a), Q: Napoleonskopf (Schmidt 1912; Padtberg 1925), R: Felsställe IIIb (Rähle 1987a), S: Hohler Fels Schelklingen Ia-Ic (Rähle 1981, 1994), T: Hohler Fels Schelklingen IIa (Rähle 1981, 1994), U: Kaufertsberg 1 (Kaulich 1983), V: Hohlenstein Ederheim (Narr 1965).

Table 3. Schematic Survey of Ornamental Mollusks (Gastropodae [G] and Bivalvae [B]) found in Early Mesolithic Sites* within the Study Area

		A	B	C	D	E	F	G	H	I	J	K	L	M
Mediterranean sea (sub-recent)														
Columbella rustica	(G)	-	-	1	-	-	1	-	-	-	-	-	5	-
Atlantic sea (sub-recent)														
Cerastoderma sp.	(B)	-	-	-	-	-	-	-	-	-	-	19	-	-
Macoma balthica	(B)	-	-	-	-	-	-	-	-	-	-	2	-	-
Paris basin (fossil)														
Bayania sp.	(G)	-	3	-	-	-	-	-	-	-	-	-	-	-
Mainzer basin (fossil)														
Potamides lamarcki	(G)	-	-	-	-	-	-	2	-	-	-	-	-	-
Potamides laevissimus	(G)	4	-	-	-	1	2	-	-	-	-	-	-	-
Pirenella plicata	(G)	1	-	-	-	-	1	1	-	-	-	-	-	-
Theodoxus gregarius	(G)	-	-	-	-	-	-	-	-	-	-	-	50	-
Steinheimer basin (fossil)														
Gyraulus trochiformis	(G)	-	-	-	1	-	-	7	-	3	-	-	160	1
Gyraulus sulcatus	(G)	-	-	-	-	-	-	12	-	10	4	-	-	-
Eastern Europe/ Central Danube														
Lithoglyphus naticoides	(G)	-	-	-	-	-	-	-	1	-	-	-	4000	-
Unknown provenance														
Bivalva indet.	(B)	-	-	-	-	-	-	-	-	-	-	4	-	-

* A: Bavans 6 (Aimé and Devits 1984), B: Roggenburg-Ritzigrund, lower (Jagher 1989), C: Birsmatten-Basisgrotte, upper (Rähle 1978), D: Jägerhaushöhle 7 (Strauch 1978), E: Probstfels (Rähle 1978), F: Falkensteinhöhle (Rähle 1978), G: Burghöhle Dietfurt 250-300 (Rähle 1978, 1987b), H: Zigeunerfels A (Rähle 1978), I: Felsställe IIa3 (Rähle 1987a), J: Helga Abri IIF1-IIF2 (Rähle 1983a), K: Attenhofen (Wischenbarth 1991), L: Große Ofnet VII (Schmidt 1912; Rähle 1978; Strauch 1978), M: Bettelküche 4 (Strauch 1978).

The map in Figure 12 illustrates the probable Magdalenian connections. We find that sites in the southwestern part of the study area have considerably more long distance connections (in number of different connections, range, and number of pieces involved) than the sites in the northeast. Long distance connections are also for the most part northerly and cover distances of at least 250 kilometers. It is very probable that these connections (as illustrated) were directed primarily at the Mainzer Tertiary basin, but the possibility that they reached even further to the Paris basin or to Belgium cannot be dismissed (Sedlmeier 1988:3).

Two domestic Swabian deposits of fossil mollusks were also exploited: the Miocene lake deposits of the Steinheim basin in the Ostalb, and the so-called Kirchberger layers (Miocene molasse deposits) on the Upper Danube (Figure 8). It is notable that only Felsställe IIIb distinguishes itself by having exclusively domestic species in the mollusk inventory. As noted above, Felsställe also distinguishes itself by the fact that 99 percent of the lithics were obtained within 500 meters of the habitation area (Table 1).

Among the Mesolithic inventories, the picture remains fairly complicated (Figure 13). The domestic exploitation of fossil mollusks from the Kirchberger layers has ceased (probably because of an increased vegetation cover in the Danube river valley), but instead we observe an intensified exploitation of the Miocene deposits in the Steinheimer basin. Northerly long distance connections are still most common, but they are completely surpassed in extent (i.e., number of objects involved) by the connection between Große Ofnet and eastern Central Europe.

In conclusion, we find that there are major differences, but also clear similarities between the Magdalenian and the Mesolithic use and procurement of ornamental mollusks. The similarities include the preferred species or types of species, the chiefly northerly direction of long distance connec-

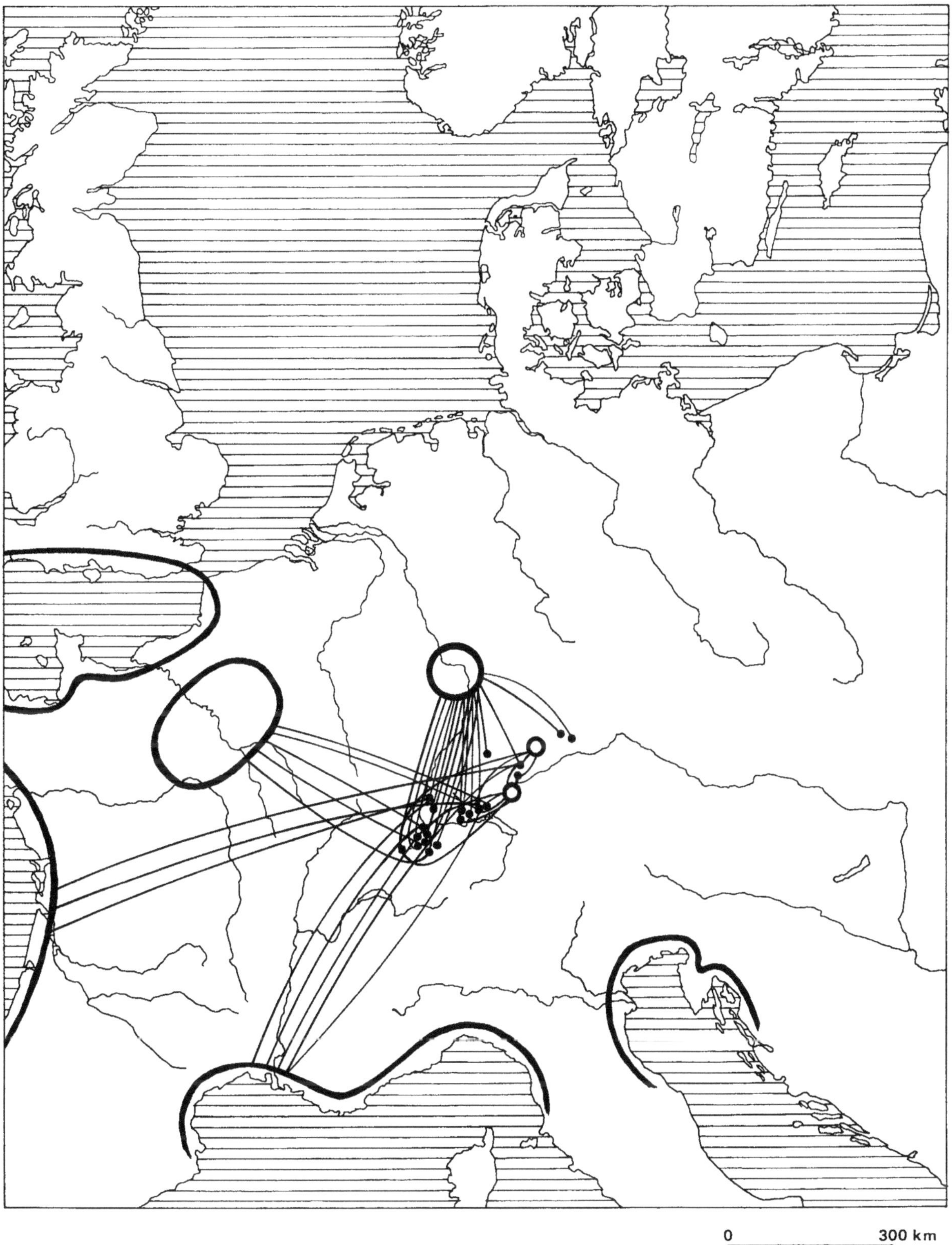

Figure 12. General map: Provenance and archaeological occurrence of ornamental mollusks in the Magdalenian. Thin lines indicate the direction of connections between archaeological sites (dots) and the most probable source areas (heavy lines) of fossil and sub-fossil shells.

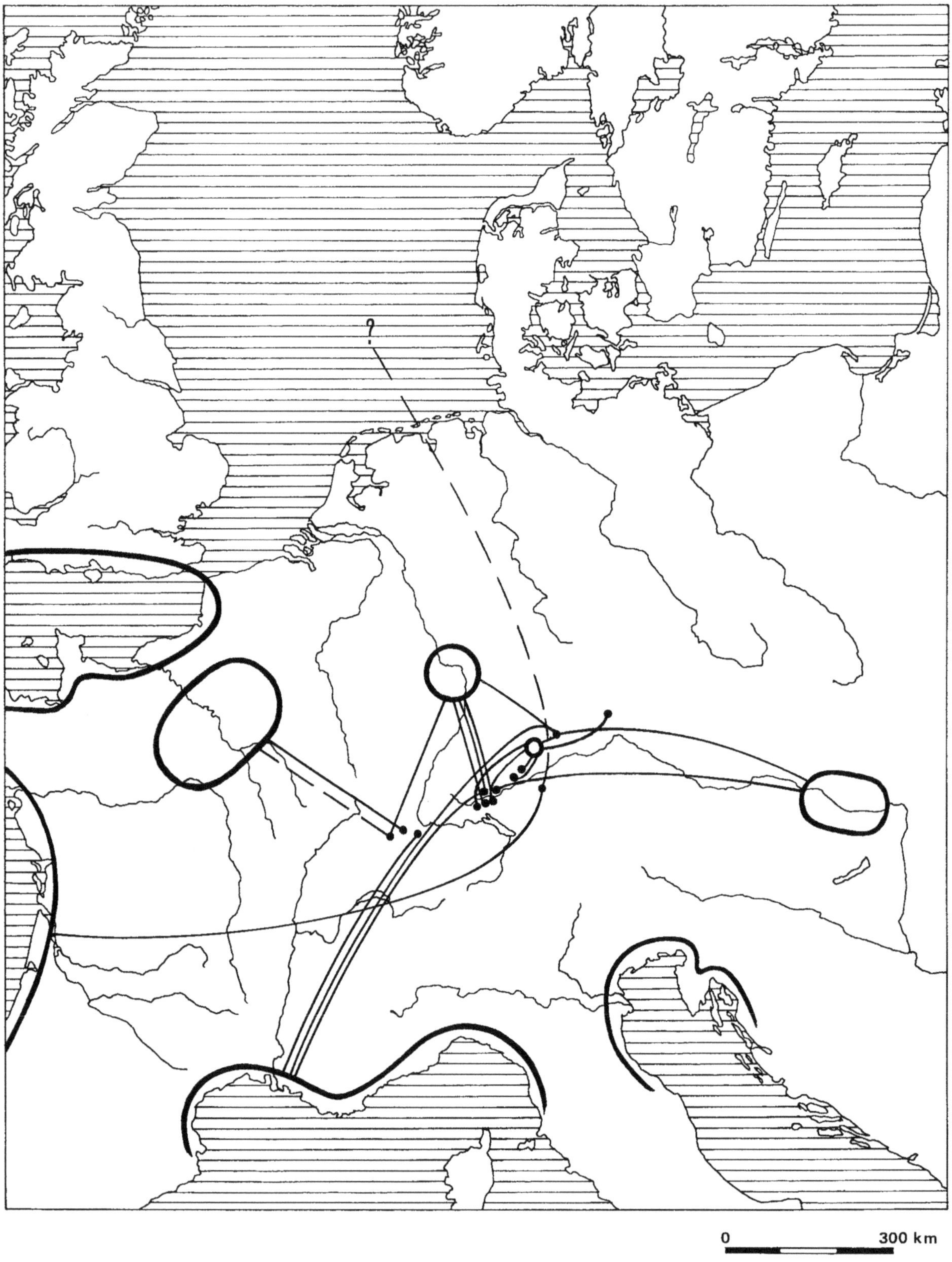

Figure 13. General map: Provenance and archaeological occurrence of ornamental mollusks in the early Mesolithic. Thin lines indicate the direction of connections between archaeological sites (dots) and the most probable source areas (heavy lines) of fossil and sub-fossil shells.

tions, the extreme long distance connections with the Atlantic and the Mediterranean, and the domestic exploitation of deposits in the Steinheimer basin.

Now the question is—do these similarities also apply to the mobility and communication patterns which are reflected in the presence of the ornamental mollusks in the inventories?

Discussion

Drawing on the work of Binford (1977, 1979), Morrow and Jefferies (1989) have presented a very useful threefold classification of general procurement patterns. They distinguish between 1) (indirect) procurement through a trade, or barter, network, 2) direct procurement by special-purpose trips to the source areas and 3) embedded procurement within seasonal movements through the region. Though initially designed for discussing lithic raw materials, this conceptual framework is also excellent for examining the procurement of 'ornamental raw materials'.

The exploitation of lithic raw materials may be attributed to the individual activities, or mobility patterns, of the past hunter-gatherer groups. This also applies to the exploitation of jet and ammonites during the Magdalenian, and to the exploitation of fossil mollusks from the Steinheimer basin and the Upper Danube. Yet, it is difficult to determine exactly how these domestic deposits of 'ornamental raw materials' were exploited.

While lithic raw material represents a critical resource, which may be optimized with respect to investment of time and energy, fossil mollusks (as well as jet and ammonites) should rather be regarded as luxury items used for personal adornment. They may be socially important, but they are in no way critical. In the mobile hunter-gatherer societies of the late glacial and early postglacial, the exploitation of local and regional deposits of such 'ornamental raw materials' most likely would have been thoroughly embedded in other (i.e. subsistence economic) activities.

It is therefore notable that within a distance of 30 kilometers from Steinheimer basin neither Paleolithic nor Mesolithic finds of ornamental mollusks have been made (Figure 8), not even in the recently excavated sites of Malerfels and Spitzbubenhöhle. If the exploitation of Miocene mollusks from Steinheimer basin was part of an embedded procurement pattern, we would expect to find those fossils in at least some of the Paleolithic and Mesolithic sites of the nearby Lone and Eselsburger valleys.

Direct procurement, on the other hand, would seem a very high investment in such tiny objects. Unfortunately, though, we lack conclusive evidence. If small special-purpose procurement camps had existed in the Steinheimer basin or along the Upper Danube, we would never be able to find them. Quarries, gravel pits and intensive cultivation would have destroyed all remains.

Jet, ammonites and the mollusks from the Upper Danube deposits are only exploited in the Magdalenian. The fossils from Steinheimer basin are exploited throughout the period in question, but the exploitation is clearly intensified in the Early Mesolithic. It is tempting to interpret these differences as another indication (cf. the discussion of lithic raw material procurement patterns through time) of diachronic changes in settlement and mobility patterns.

Previous analyses of settlement dynamics and subsistence strategies within the study area (Eriksen 1991 with references) indicate that especially the Magdalenian was characterized by a complex settlement and mobility pattern. The Magdalenians were highly adaptive and generally opportunistic hunters who had the choice of a variety of game animals. All evidence suggests that the groups stayed within the region throughout the year. In the central part of the study area (Hegaualb and Randen) they were able to practice periodic mass killings of reindeer to an extent that allowed periodic (autumn and winter) aggregation camps and probably also storage of meat.

Moreover, indications of site seasonality support a model of short seasonal movements between spring and summer residential camps in the Swabian Alb, Swiss Jura and Black Forest foothills, and autumn and winter aggregation camps in Hegaualb, Randen and the adjacent lowland (Eriksen 1996a). Once more, movements primarily seem to follow a northeast- southwesterly axis parallel to the Jura formation. On the other hand, we should also add to this model the possibility that certain groups might have been able to remain within fairly limited parts of the area, in the northeast or in the southwest, throughout the year. A fine example of such a residential winter site in the Swabian Alb is Felsställe IIIb (Kind 1987).

During the Early Mesolithic these regional differences are no longer observable. Neither is there any evidence of aggregation camps from this period. Instead we find indications of a more stable settlement structure, which apparently was accompanied by a reduction in the size of settlements (probably related both to duration of occupation and number of inhabitants). The difference in subsistence strategies between the Magdalenian and the Early Mesolithic may to some extent be reduced to a difference of degree (Eriksen 1991) caused by environmental factors. Unfortunately the evidence on seasonality offers only a rather diffuse picture, which does not support any particular model of interpretation of mobility patterns. Nevertheless, the Mesolithic groups seem to have stayed within the bounds of the study area.

The obvious changes in settlement structure, and to a certain degree subsistence strategies, are likely to be accompanied by changes in mobility strategies through time. Given the marked increase in boreal vegetation in the early postglacial, we should expect increased residential mobility in this period compared to the late glacial (Kelly 1983). It appears very reasonable to explain the differences between Magdalenian and the Early Mesolithic use and procurement of lithic raw materials and of domestic occurrences of 'ornamental raw materials' in the light of these general socioeconomic differences. Still, a few open questions remain, concerning the use of heat treatment. If Mesolithic people are characterized by high residential mobility, how do they find time to perform this allegedly complicated and time-consuming process? Heat treatment probably required some degree of advance planning.

So far, we have only discussed the local and regional procurement patterns, but as a matter of fact there are also quite a few truly exotic elements in the mollusk inventories. In the Magdalenian as well as in the Early Mesolithic the occasional sub-recent mollusks from the Mediterranean and the Atlantic are obvious expressions of procurement through a communication, or barter, network. Incidentally, these extreme long distance connections are complemented by at least two Magdalenian finds within the study area of presumably Baltic amber, at Moosbühl II (Schwab 1985) and Champréveyres (Le Tensorer and Niffeler 1993).

The exploitation of the remaining mollusk deposits represents a more complicated problem, especially with regard to the Mainzer basin which lies only some 250 kilometers from the Jura. An active exploitation of these deposits is possible, especially during the Magdalenian where connections between the two areas appear to have been quite intensive.

Two possible models of interpretation may then be considered. The first implies fixed, possibly seasonal migrations between the southwestern part of the study area and the Mainzer basin, probably following the Rhine valley. A realistic discussion of this model, though, would require evidence of clear, mutual connections, involving also lithic raw materials. But the evidence from the lithic raw material analysis rather indicates that seasonal migrations followed a southwest-northeasterly direction.

The second model (Figure 14) thus assumes that there existed Magdalenian aggregation camps in the southwestern part of the area, which more or less regularly received visitors from the Mainzer basin. As previously mentioned, possible aggregation camps are documented in this area by way of subsistence-economic and settlement dynamic analyses, and this model is thus given preference.

It is concluded that in the Magdalenian, lithic raw materials used for tools were procured actively and directly within the region. The procurement was usually embedded in the subsistence activities of the groups, thus reflecting seasonal movements throughout the region. Domestic occurrences of mollusks and other 'ornamental raw materials' generally are supposed to have been procured much the same way as the lithics, but the evidence is inconclusive, and especially the Steinheimer basin deposits may have been exploited more directly. Exotic mollusks (i.e. mollusks coming from the Atlantic, the Mediterranean, Paris or Mainzer basin), in my opinion, represent different expressions of inter-group communication patterns. Some are obvious expressions of procurement through a long-distance communication, or barter, network. Others seem to represent an inter-regional communication network with direct contact between neighboring groups.

It is very interesting to note that elements expressing an active (direct or embedded) procurement and elements expressing contact networks run transversely to one another. The seasonal migrations of the Magdalenian groups followed a southwest-northeasterly direction parallel to the Jura formation, but their contacts with other groups were primarily northerly. This observation is in no way unique among Magdalenian finds in Central and Western Europe (Floss 1994:336f).

A fine example of highly different provenance directions with respect to lithic raw material and ornamental mollusks is thus represented by Andernach find concentration II. The vast majority of lithic raw material (86.5 %) here consists of Maas flint coming from a distance of at least 100 kilometers to the northwest (Floss 1994:193f). However, the site also produced a small depot of sub-recent mollusks: 46 *Homalopoma sanguineum* and 1 *Cyclope neriteus* (Floss 1994:218) both coming from the Mediterranean, i.e. at least 800 kilometers to the south.

In the Mesolithic, lithic raw materials used for tools still were procured actively and directly within the region. As before, procurement probably was embedded in subsistence activities. The relatively few connections to the Mainzer Basin and the small number of pieces involved indicate a passive procurement (i.e., perhaps by trade or barter) of ornamental mollusks from here. The only serious exception is that of Große Ofnet.

The head burials from Große Ofnet have now been firmly dated to approximately 7,500 BP (conv. ^{14}C) (Hedges *et al.* 1989:210f). Thus, at least one of the major problems concerning this find (Naber 1974; Newell *et al.* 1990:107f) has finally been solved, but many others still remain. The find context is quite spectacular, though not entirely unique. Isolated head burials probably of Mesolithic age are known from the nearby Kaufertsberg (Kaulich 1983:93f; Schröter 1983) and Hohlenstein Stadel

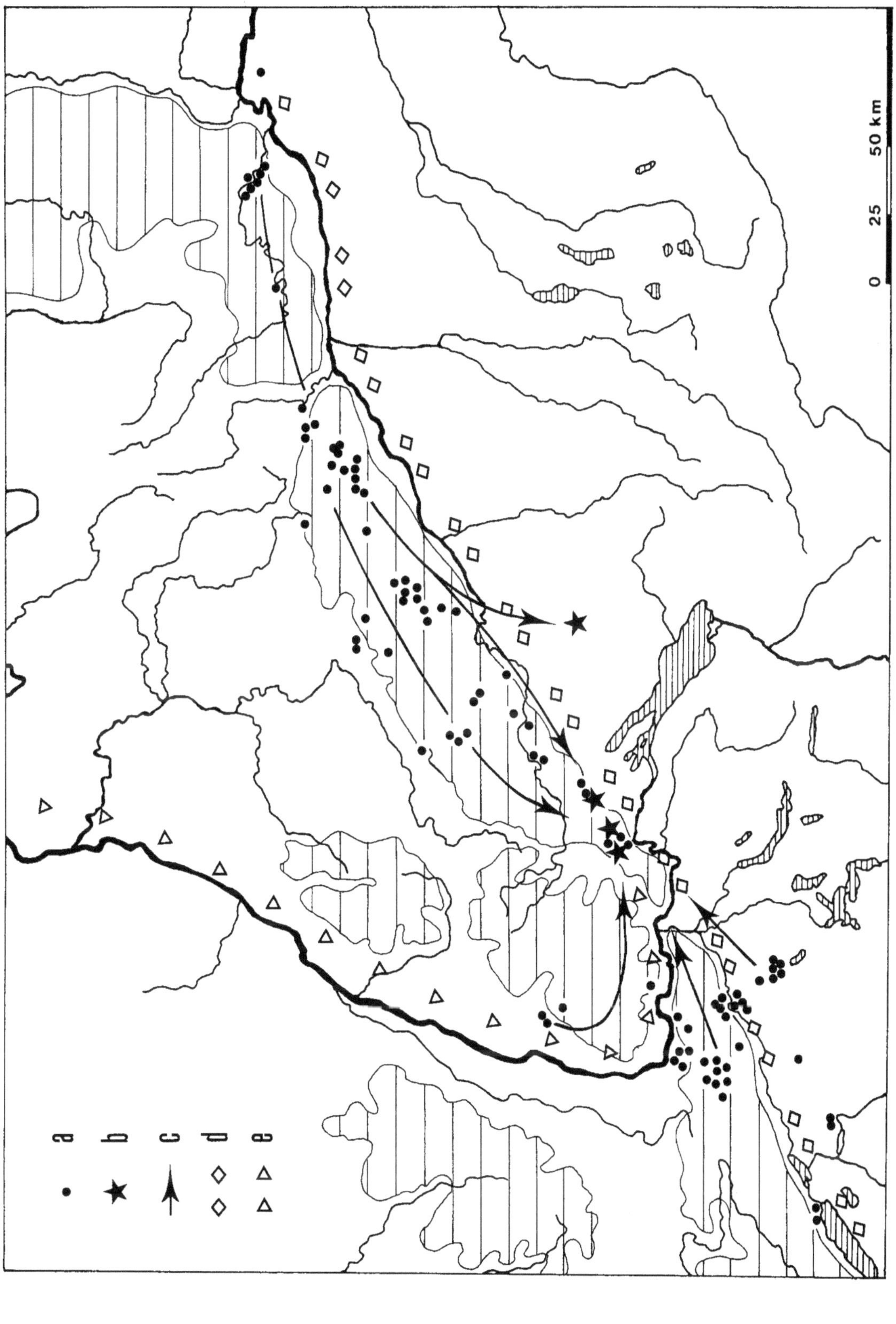

Figure 14. Model illustrating the seasonal movements of Magdalenian groups and the connection with the Mainzer Basin: A: Residential sites, B: Probable aggregation camps or other sites of major socioeconomic importance, C: Autumn-winter movements, D: Line indicating predominant direction of lithic raw material transport, E: Line indicating direction of regular inter-group communication.

(Wetzel 1938, 1961) as well as from Mannlefelsen I in Alsace (Kaulich 1983:94 with references). These sites, like Große Ofnet, are all caves or rock shelters, but this is very likely a question of conditions for preservation. Kaufertsberg and Mannlefelsen both represent solitary head burials with no grave goods. At Hohlenstein Stadel, 3 heads (adult male, adult female and little child) were found together in a pit. The woman wore a necklace of teeth from the Black Sea roach (*Rutilus frisii meidingeri*), but no ornamental mollusks. At Große Ofnet a total of 33 heads (4 adult males, 9 adult females and 20 children or juvenile females) were found in two pits (Schmidt 1912:37f). Almost all heads were lavishly adorned with ornamental mollusks or perforated red deer teeth.

The find context, the extremely large number of ornamental mollusks, and their in part truly exotic origin invites speculation. However, this is not the place for fictitious stories. In any event, the evidence from Große Ofnet is in no way inconsistent with the point made in the present discussion. The fundamental differences between the Magdalenian and the Early Mesolithic groups are socioeconomic. Ornamental mollusks and lithic raw materials each express different aspects of this basic distinction.

Acknowledgments

I dedicate this paper to my old friends and colleagues at the 'Institut für Ur- und Frühgeschichte - Abteilung ältere Urgeschichte und Quartärökologie' at the University of Tübingen. Moreover, the financial support of the Alexander von Humboldt Stiftung as well as the Danish Research Council for the Humanities is gratefully acknowledged.

References Cited

Affolter, J., M.-I. Cattin, D. Leesch, P. Morel, N. Plumettez, N. Thew and G. Wending
1994 Monruz—Une nouvelle station magdalénienne au bord du lac de Neuchâtel. *Archäologie der Schweiz* 17:94-104.

Aimé, G. and J.-A. Devits
1984 Les coquillages de parure du Mésolithique moyen des abris de Bavans (Doubs). Données récentes. *Bulletin de la Société Préhistorique Française* 81:268-269.

Albrecht, G.
1979 *Magdalénien-Inventare vom Petersfels. Siedlungsarchäologische Ergebnisse der Ausgrabungen 1974-1976.* Tübinger Monographien zur Urgeschichte 6. Archaeologica Venatoria, Tübingen.
1981 Die neuen Ausgrabungen in Munzingen 1976/1977. *Archaeologica Venatoria, Mitteilungsblatt* 2:21-23.
1984 Der spätpleistozäne und altholozäne Fundplatz Malerfels I. In *Die steinzeitliche Besiedlung des Eselburger Tales bei Heidenheim (Schwäbische Alb)*, edited by J. Hahn, pp. 90-122. Forschungen und Berichte zur Vor- und Frühgeschichte in Baden-Württemberg 17. Konrad Theiss Verlag, Stuttgart.

Albrecht, G., H. Berke, S. Haas-Campen and A. Hahn-Weishaupt
1994 Die Funde vom Petersfels in der Stätischen Sammlung Engen im Hegau. *Fundberichte aus Baden-Württemberg* 19:1-62.

Albrecht, G., D. Drautz and J. Kind
1977 Eine Station des Magdalénien in der Gnirshöhle bei Engen-Bittelbrunn im Hegau. *Archäologisches Korrespondenzblatt* 7:161-179.

Andrée, J.
1939 *Der eiszeitliche Mensch in Deutschland und seine Kulturen.* Ferdinand Enke Verlag, Stuttgart.

Auffermann, B.
1991 *Magdalénien und Mesolithikum der Bruckersbergstationen in Giengen.* Unpublished M.A. Thesis, University of Tübingen.

Bandi, H.-G.
1947 *Die Schweiz zur Rentierzeit.* Huber, Frauenfeld.

Barr, J. H.
1973 *The Late Upper Paleolithic Site of Moosbühl: An Attempt to Analyze some of its Problems.* Ph.D. Thesis, University of Bern.
1977 Die Rislisberghöhle, ein neuer Magdalénien Fundplatz im Schweizer Jura. *Archäologisches Korrespondenzblatt* 7:85-87.

Binford, L. R.
1977 Forty-seven trips: a case study in the character of archaeological formation processes. In *Stone Tools as Cultural Markers: Change, Evolution and Complexity*, edited by R. V. S. Wright, pp. 24-36. Australian Institute of Aboriginal Studies, Canberra.
1979 Organization and Formation Processes: Looking at Curated Technologies. *Journal of Anthropological Research* 35:255-273.

Cattin, M.-I.
1990 Silex-Economy on the Magdalenian Site: Hauterive-Champréveyres (Switzerland): Spatial Organization and Technology by Refitting. In *The Big Puzzle. International*

Symposium on Refitting Stone Artefacts, edited by E. Cziesla, S. Eickhoff, N. Arts and D. Winter, pp. 363-369. HOLOS-Verlag, Bonn.

Deecke, W.
1933 *Die mitteleuropäische Silices nach Vorkommen, Eigenschaften und Verwendung in der Prähistorie.* Gustav Fischer Verlag, Jena.

Eberhardt, H., E. Keefer, C.-J. Kind, H. Rensch and H. Ziegler
1987 Jungpaläolithische und mesolithische Fundstellen aus der Aichbühler Bucht. *Fundberichte aus Baden-Württemberg* 12:1-51.

Eriksen, B. V.
1991 Change and Continuity in a Prehistoric Hunter-Gatherer Society. *Archaeologica Venatoria* 12. Archaeologica Venatoria, Tübingen.
1996a Resource Exploitation, Subsistence Strategies and Adaptiveness in Late Pleistocene–Early Holocene Northwestern Europe. In *Humans at the End of the Ice Age: The Archaeology of the Pleistocene-Holocene Transition*, edited by L. G. Straus, B. V. Eriksen, J. M. Erlandson, and D. R. Yesner, pp. 101-128. Plenum Press, New York.
1996b Diskussionsbeitrag zum Spätpaläolithikum Mitteleuropas—Apologie einer urgeschichtlichen Quellenkritik. In *Spuren der Jagd–Die Jagd nach Spuren,* edited by J. Hahn, I. Campen and M. Uerpmann. Tübinger Monographien zur Urgeschichte 11. Mo Vince, Tübingen.
1997 Settlement Patterns, Cave Sites and Locational Decisions in Late Pleistocene Central Europe. In *The Human Use of Caves*, edited by C. Bonsall and C. Smith. Brit. Archaeological Reports Int. Series 667. Oxford: BAR Publishing.

Floss, H.
1994 *Rohmaterialversorgung im Paläolithikum des Mittelrheingebietes. Monographien des Römisch-Germanischen Zentralmuseums* 21. Rudolf Habelt, Bonn.

Flükiger, W.
1964 Die steinzeitliche Siedlung "Hintere Burg". *Jahrbuch für solothurnische Geschichte* 37:263-304.

Frenzel, B.
1983 Die Vegetationsgeschichte Süddeutschlands im Eiszeitalter. In *Urgeschichte in Baden-Württemberg*, edited by H. Müller-Beck, pp. 91-166. Konrad Theiss Verlag, Stuttgart.

Geyer, O. F. and M. P. Gwinner
1991 *Geologie von Baden-Württemberg.* Schweizerbart'sche Verlagsbuchhandlung, Stuttgart.

Gradmann, R.
1956 *Süddeutschland I–II.* Darmstadt.

Hahn, J.
1983 Die frühe Mittelsteinzeit. In *Urgeschichte in Baden-Württemberg*, edited by H. Müller-Beck, pp. 363-392. Konrad Theiss Verlag, Stuttgart.
1991 Erkennen und Bestimmen von Stein- und Knochenartefakten. Einführung in die Artefaktmorphologie. *Archaeologica Venatoria* 10. Archaeologica Venatoria, Tübingen.
1992 *Eiszeitschmuck auf der Schwäbischen Alb. Alb und Donau*, Kunst und Kultur 5. Süddeutsche Verlagsgesellschaft Stuttgart, Ulm.
n.d. Der Hohlenstein im Lonetal (Markung Asselfingen, Kreis Ulm). Mesolithikum und Jungpaläolithikum. Tübingen:manuscript.

Hahn, J., editor
1984 *Die steinzeitliche Besiedlung des Eselburger Tales bei Heidenheim (Schwäbische Alb).* Forschungen und Berichte zur Vor- und Frühgeschichte in Baden-Württemberg 17. Konrad Theiss Verlag, Stuttgart.
n.d. Das Achtal in der älteren und mittleren Steinzeit. Klima, Ökologie, und Archäologie einer Landschaft der Mittleren Alb im Jungpleistozän und Frühholozän. Tübingen:manuscript.

Hahn, J. and A. Scheer
1983 Das Helga Abri am Hohlen Felsen bei Schelklingen: Eine mesolithische und jungpaläolithische Schichtenfolge. *Archäologisches Korrespondenzblatt* 13:19-28.

Hedges, R. E. M., R. A. Housley, I. A. Law and C. R. Bronk
1989 Radiocarbon Dates from the Oxford AMS System: Archaeometry Datelist 9. *Archaeometry* 31(2): 207-234.

Heierli, J.
1907 *Das Kesslerloch bei Thayngen.* Neue Denkschriften der Schweizerischen Naturforschenden Gesellschaft 43. Zürcher & Furrer, Zürich.

Heinen, M.
1986 Die Bedeutung der Rohstoffanalyse an dem spätpaläolithischen, mesolithischen Freilandfundplatz Sarching '83. *Archaeologica Venatoria, Mitteilungsblatt* 10/11:6-21.

Jagher, R.
1989 Le gisement mésolithique de Roggenburg-Ritzigrund, commune de Roggenburg, canton de Berne (Suisse). In *Epipaléolithique et Mésolithique entre Ardennes et Massif Alpin*, edited by G. Aimé and A. Thévenin, pp. 105-123. Mémoires de la Société d'Agriculture, Lettres, Sciences et Arts de la Haute-Saône, Vesoul.

Jagher, E. and R. Jagher
1987 Les gisements paléolithique de la Löwenburg, commune de Pleigne. *Archäologie der Schweiz* 10:43-52.

Jeske, R.
1989 Economies in raw material use by prehistoric hunter-gatherers. In *Time, Energy and Stone Tools*, edited by R. Torrence, pp. 34-45. Cambridge University Press, Cambridge.

Jochim, M. A.
1989 Optimization and stone tool studies: problems and potentials. In *Time, Energy and Stone Tools*, edited by R. Torrence, pp. 106-111. Cambridge University Press, Cambridge.
1993 *Henauhof-Nordwest—Ein mittelsteinzeitlicher Lagerplatz am Federsee.* Materialhefte zur Archäologie in Baden-Württemberg 19. Konrad Theiss Verlag, Stuttgart.

Karsten, H.
1874 Studie der Urgeschichte des Menschen in einer Höhle des Schaffhauser Jura. *Mitteilungen der Antiquarischen Gesellschaft Zürich* 18(6):137-162.

Kaulich, B.
1983 Das Paläolithikum des Kaufertberges bei Lierheim (Gem. Appetshofen, Ldkr. Donau-Ries). *Quartär* 33/34:29-97.

Kelly, R. L.
1983 Hunter-gatherer mobility strategies. *Journal of Anthropological Research* 39:277-306.

Kieselbach, P.
1993 Die Artefakte der mesolithischen Freilandstation Rottenburg-Siebenlinden II. Unpublished M.A. Thesis, University of Tübingen.

Kind, C.-J.
1987 *Das Felsställe. Eine jungpaläolithisch-frühmesolithische Abri-Station bei Ehingen-Mühlen, Alb-Donau- Kreis.* Forschungen und Berichte zur Vor- und Frühgeschichte in Baden-Württemberg 23. Konrad Theiss Verlag, Stuttgart.

Kuhn, F.
1969 Höhlen am Isteiner Klotz I. *Archäologische Nachrichten aus Baden*, Oktober 1969:15-19.

Lauxmann, C.
1988 Die Steinartefakte des Magdalénienzeitlichen Fundhorisonts IV aus der Brillenhöhle bei Blaubeuren. Unpublished M.A. Thesis, University of Tübingen.

Le Tensorer, J.-M. and U. Niffeler
1993 Die Schweiz vom Paläolithikum bis zum frühen Mittelalter. *SPM I: Paläolithikum und Mesolithikum.* Verlag Schweizerische Gesellschaft für Ur- und Frühgeschichte, Basel.

Lurie, R.
1989 Lithic technology and mobility strategies: the Koster Site Middle Archaic. In *Time, Energy and Stone Tools*, edited by R. Torrence, pp. 46-56. Cambridge University Press, Cambridge.

Lüdin, C.
1938 Kohlerhöhle. *Jahrbuch der Schweizerischen Gesellschaft für Ur- und Frühgeschichte* 30:61-64.

Maier, H.
1936 Die altsteinzeitliche Wohnhöhle Kleine Scheuer im Rosenstein. *Mannus* 28:235-252.

Mauser, P. F.
1970 *Die jungpaläolithische Höhlenstation Petersfels im Hegau (Gemarkung Bittelbrunn, Ldkrs. Konstanz).* Badische Fundberichte Sonderheft 13. Freiburg:Staatliches Amt für Ur- und Frühgeschichte, Freiburg im Breisgau.
1976 Das eiszeitliche Jagdlager im Katzenbachtal. *Der Sülchgau* 20:35-44.

Morrow, C. A. and R. W. Jefferies
1989 Trade or embedded procurement? A test case from southern Illinois. In *Time, Energy and Stone Tools*, edited by R. Torrence, pp. 27-33. Cambridge University Press, Cambridge.

Naber, F. B.
1974 Das Ende des Ofnet-Problems? *Quartär* 25:73-84.

Narr, K. J.
1965 Die Altsteinzeitfunde aus dem Hohlenstein bei Nördlingen. *Bayerische Vorgeschichtsblätter* 30:1-9.

Newell, R. R., D. Kielman, T. S. Constandse-Westermann, A. van Gijn and W. A. B. van der Sanden
1990 *An Inquiry into the Ethnic Resolution of Mesolithic Regional Groups. The Study of Their Decorative Ornaments in Time and Space.* E.J. Brill., Leiden.

Nielsen, E. H.
1991 *Gampelen-Jänet 3. Eine mesolithische*

Siedlungsstelle im westlichen Seeland. Staatlicher Lehrmittelverlag, Bern.

Nüesch, J.

1896 *Die praehistorische Niederlassung am Schweizersbild bei Schaffhausen. Die Schichten und ihre Einschlüsse.* Zürcher & Furrer, Zürich.

Padtberg, A.

1925 *Das altsteinzeitliche Lößlager bei Munzingen.* Dr. Benno Filser Verlag, Augsburg.

Pasda, C.

1989 Das Magdalénien des Probstfels bei Beuron—Die Funde R. R. Schmidts. Unpublished M.A. Thesis, University of Tübingen.

1994 *Das Magdalénien in der Freiburger Bucht.* Materialhefte zur Archäologie in Baden-Württemberg 25. Konrad Theiss Verlag, Stuttgart.

Peters, E.

1930 *Die altsteinzeitliche Kulturstätte Petersfels.* Dr. Benno Filser Verlag, Augsburg.

Peters, E. and V. Toepfer

1932 Der Abschluss der Grabungen am Petersfels bei Engen im Badischen Hegau. *Prähistorische Zeitschrift* 23:155-198.

Pousaz, N., editor

1991 *L'abri-sous-roche Mésolithique des Gripons a Saint-Ursanne (JU/Suisse).* Cahiers d'archéologie jurassienne 2. Porrentruy:Office du patrimoine historique, Société Jurassienne d'Émulation.

Preuschoft-Güttler, A.

1995 Untersuchungen zur Präzision des Begriffs "Gagat" im Magdalénien Südwestdeutschlands und der Schweiz. Unpublished M.A. Thesis, University of Tübingen.

Price, T. D., S. Chappell and D. J. Ives

1982 Thermal Alteration in Mesolithic Assemblages. *Proceedings of the Prehistoric Society* 48:467-485.

Rähle, W.

1978 Schmuckschnecken aus mesolithischen Kulturschichten Süddeutschlands und ihre Herkunft. In *Das Mesolithikum in Süddeutschland. Teil 2: Naturwissenschaftliche Untersuchungen,* edited by W. Taute, pp. 163-168. Tübinger Monographien zur Urgeschichte 5/2. Archaeologica Venatoria, Tübingen.

1981 Schmuckschnecken aus jungpaläolithischen Fundschichten vom Hohlen Fels bei Schelklingen (Alb-Donau-Kreis). *Archäologisches Korrespondenzblatt* 11:179-180.

1983a Die Mollusken der Grabung Helga Abri bei Schelklingen mit einer Anmerkung zum Fund einiger mesolithischer Schmuckschnecken. *Archäologisches Korrespondenzblatt* 13:29-36.

1983b Schmuck aus Molluskenschalen von der Magdalénien-Station Petersfels bei Engen (Hegau). In *Naturwissenschaftlichen Untersuchungen an Magdalénien-Inventaren vom Petersfels, Grabungen 1974-1976,* edited by G. Albrecht, H. Berke and F. Poplin, pp. 154-158. Tübinger Monographien zur Urgeschichte 8. Archaeologica Venatoria, Tübingen.

1987a Die Molluskenfaunen der Grabung Felsställe bei Mühlen, Stadt Ehingen, Alb-Donau-Kreis. In *Das Felsställe. Eine jungpaläolithisch-frühmesolithische Abri-Station bei Ehingen-Mühlen, Alb-Donau-Kreis,* edited by C.-J. Kind, pp. 269-274. Forschungen und Berichte zur Vor- und Frühgeschichte in Baden-Württemberg 23. Konrad Theiss Verlag, Stuttgart.

1987b Schmuck aus Molluskenschalen von dem Abri Felsställe bei Mühlen, Stadt Ehingen, Alb-Donau-Kreis. In *Das Felsställe. Eine jungpaläolithisch-frühmesolithische Abri-Station bei Ehingen-Mühlen, Alb-Donau-Kreis,* edited by C.-J. Kind, pp. 383-385. Forschungen und Berichte zur Vor- und Frühgeschichte in Baden-Württemberg 23. Konrad Theiss Verlag, Stuttgart.

1994 Neue Funde jungpaläolithischen Schalenschmuckes vom Hohlen Felsen bei Schelklingen, Alb-Donau-Kreis, und vom Petersfels bei Engen, Kreis Konstanz. *Fundberichte aus Baden-Württemberg* 19:95-98.

Riek, G.

1934 *Die Eiszeitjägerstation am Vogelherd.* Franz F. Heine, Tübingen.

1959 Ein magdalénienzeitlicher Rastplatz unter der Südwand des Sirgensteinfelsens. *Fundberichte aus Schwaben* N.F. 15:30-42.

Rottländer, R. C. A.

1983 *Einführung in die naturwissenschaftlichen Methoden in der Archäologie.* Archaeologica Venatoria 6. Archaeologica Venatoria, Tübingen.

1989 *Verwitterungserscheinungen an Silices und Knochen.* Archaeologica Venatoria 8/2. Archaeologica Venatoria, Tübingen.

Rozoy, J.-G.

1978 *Les derniers chasseurs.* Bulletin de la société archéologique champenoise. Société archéologique champenoise, Reims.

Sarasin, F.
1918 *Die steinzeitlichen Stationen im Birstal zwischen Basel und Delsberg.* Neue Denkschriften des Schweizerischen Naturforschenden Gesellschaft 54/2, Zürich.

Schmid, E.
1977 Die Umwelt der Jäger vom Kesslerloch. In *Die Kultur der Eiszeitjäger aus dem Kesslerloch,* edited by H.-G. Bandi, J. Bürgi, K. Gerhardt, H. Müller-Beck and E. Schmid, pp. 56-62. Seekreis, Konstanz.

Schmidt, R. R.
1912 *Die diluviale Vorzeit Deutschlands.* Schweizerbart'sche Verlagsbuchhandlung, Stuttgart.

Schröter, P.
1983 Zum Schädel vom Kaufertsberg bei Lierheim (Gem. Appetshofen, Ldkr. Donau-Ries). *Quartär* 33/34:99-105.

Schuler, A.
1989 Das Magdalénien der Schussenquelle. Die Steinartefakte der Grabung von Oskar Fraas (1866). *Archäologisches Korrespondenzblatt* 19:11-22.

Schwab, H.
1972 Lüscherz-Moos. Eine epipaläolithische Fundstelle im Bernischen Seeland. *Jahrbuch des Bernischen Historischen Museums* 49/50:207-230.
1985 Gagat und Bernstein auf dem Rentierjägerhalt Moosbühl bei Moosseedorf (Kanton Bern). *Jahrbuch des Bernischen Historischen Museums* 63/64:259-266.

Schweizer, T.
1941 Die Azilien-Station "Unter der Fluh". *Jahrbuch für solothurnische Geschichte* 14:259-269.

Sedlmeier, J.
1982 *Die Hollenberg-Höhle 3. Eine Magdalénien-Fundstelle bei Arlesheim, Kanton Basel-Landschaft.* Basler Beiträge zur Ur- und Frühgeschichte 8. Habegger Verlag, Derendingen-Solothurn.
1988 Jungpaläolithischer Molluskenschalen-Schmuck aus nordwestschweizerischen Fundstellen als nachweis für Fernverbindungen. *Archäologisches Korrespondenzblatt* 18:1-6.
1989 Jungpaläolithikum und Spätpaläolithikum in der Nordwestschweiz. Ph.D. Thesis, University of Bern.

Simon, U.
1993 Die Burkhardtshöhle—eine Magdalénienstation am Nordrand der Schwäbischen Alb. Unpublished M.A. Thesis, University of Tübingen.

Stampfli, H. R.
1981 Die Fauna der Magdalénien-Station Winznau-Käsloch. Ergebnisse einer musealen Nachforschung. *Archäologie der Schweiz* 4(2):36-41.

Strauch, F.
1978 *Gyraulus trochiformis* als Schmuckschnecke aus mesolithischen Kulturschichten Süddeutschlands (Jägerhaus-Höhle, Große Ofnet, Bettelküche). In *Das Mesolithikum in Süddeutschland. Teil 2: Naturwissenschaftliche Untersuchungen,* edited by W. Taute, pp. 161-162. Tübinger Monographien zur Urgeschichte 5/2. Archaeologica Venatoria, Tübingen.

Weniger, G.-C.
1991 Überlegungen zur Mobilität jägerischer Gruppen im Jungpaläolithikum. *Saeculum* 42(1):82-103.

Wetzel, R.
1938 Die Kopfbestattungen und die Knochentrümmerstätte des Hohlensteins im Rahmen der Urgeschichte des Lonetals. *Verhandlungen des deutschen Gesellschafts für Rassenforschung* 9:193-212.
1961 Der Hohlestein im Lonetal. *Mitteilungen des Vereins für Naturwissenschaft und Mathemathik in Ulm (Donau)* 26:21-75.

Wiegers, F.
1950 Rohstoffversorgung im Paläolithikum. *Prähistorische Zeitschrift* 34(1):225-230.

Wischenbarth, P.
1991 Eine mesolithische Freilandfundstelle mit Konchylienfunden im Lkr. Neu-Ulm. *Archäologisches Korrespondenzblatt* 21:203-210.

Zimmermann, T.
1988 Die spätpaläolithische Fundstelle Seeberg-Fürsteiner. Die Silexaufsammlung von Walter Flükiger. Unpublished M.A. Thesis, University of Tübingen.

Zotz, L. F.
1928 Die paläolithische Besiedlung der Teufelsküchen am Ölberg beim Kuckucksbad. *Prähistorische Zeitschrift* 19:1-53.

Zürcher, A.
1969 Die spätjungpaläolithische Freilandstation Winznau-Köpfli. *Jahrbuch für solothurnische Geschichte* 42:138-201.

5. Retooling and Raw Material Economies: Technological Change in Late Glacial and Early Postglacial Southern Germany

Lynn E. Fisher

This chapter is based in the same geographic area as the previous one, but narrows the focus to a selected group of Magdalenian, Late Paleolithic and Early Mesolithic assemblages to examine the effects of broad patterns of technological change on raw material economy. It is argued that changes in late glacial and early postglacial stone tool production complicate comparisons of raw material assemblages. Specifically, increase in frequency of microlithic tools and the corresponding decline in importance of larger retouched implements made on flakes and blades may indicate a trend towards reduced potential use-life of stone implements. This can be expected to alter the rate at which stone raw material is discarded by mobile foragers. This chapter explores this possibility and its effects on raw material assemblages through analysis of sixteen lithic assemblages in southern Germany. A comparison of use of local and transported raw materials among unretouched and retouched pieces in Magdalenian, Late Paleolithic, and Early Mesolithic assemblages suggests a strong association between production of small microlithic elements and use of locally available raw materials. This association appears in Magdalenian assemblages with abundant small backed bladelets and Early Mesolithic assemblages characterized by small geometric microliths. There is some indication that imported materials, more common in Magdalenian and Late Paleolithic assemblages in this study, are associated with larger flake and blade tools with a longer potential use life. The author suggests that low frequencies of imported raw materials on Early Mesolithic sites may be attributable to a focus on producing small microliths with a short potential use-life rather than to changes in the scale of forager mobility. *–Editors.*

Introduction

> Raw materials and manufacturing strategies are the result of careful choices made within the wider context of the tool-using behavior, which in turn is a solution to particular problems. (Torrence 1989:64)

Changing patterns of stone raw material transport can offer important insights into the scale and organization of hunter-gatherer mobility in prehistory. A reduction in frequencies of non-local raw materials in some early postglacial as compared to late glacial contexts in Europe suggests the intriguing possibility that inferences can be made about changing size or organization of social and economic territories. However, in many cases it may be difficult to draw conclusions about mobility in a direct way from simple measures such as the distances over which stone raw materials are transported. This is because raw material transport is also affected by factors other than mobility. In this paper I discuss acquisition and use of stone raw materials in the context of broader patterns of technological change in late glacial and early postglacial Germany. I argue that stone raw material transport is affected by tool use life, re-tooling strategies, and more broadly, the varied roles stone tools can play in a wider technological system.

The archaeological record of Late Upper Paleolithic and Mesolithic Europe shows a complex series of changes in tool design and in the role of stone tools in the larger technology. A variety of small abruptly retouched ("backed") chipped stone elements (microliths in a broad sense) increased in importance during this period, until, in Early Mesolithic times, they made up a majority of retouched tools. At the same time, the frequency of larger blade or flake tools declined. Reasons for these changes are incompletely understood and probably diverse. Cultural processes that might help to explain the shift toward microlithic tools include 1) a long history of innovation in lightweight, long-distance hunting weapons (such as atlatl-launched spears and the bow and arrow) in the context of varying conditions for hunting large animals (Myers 1989; Straus 1991) and 2) innovation in special-purpose tools in the context of long-term subsistence diversification (Vierra 1995). Cultural processes underlying a relative decline in the larger blade and flake tools are less commonly investigated, but may suggest toolmakers began relying less on stone tools for some kinds of tasks.

Whatever the causes, the increased use of composite implements introduces new problems for analysis. Microliths in the Late Upper Paleolithic and Mesolithic played a different role in stone technologies than did larger tools on flakes and blades. For example, microliths cannot be resharpened or re-used, but must be replaced when broken or damaged. As a result, increased use of composite implements points to a shift in the nature of stone tool maintenance, such that raw material economies involve less resharpening, recycling and reuse of stone tools and more frequent replacement of small stone components. This must affect stone raw material acquisition and transport. Curation of tools that can be recycled and re-used is an important mechanism of raw material transport. Accordingly, a decline in such tools might reduce the frequency of non-local raw materials even in the absence of any other change.

This paper examines late glacial and early postglacial changes in raw material use for tool making and re-tooling, based on a broad contrast between microlithic elements (including backed bladelets, backed points, geometrics, and micropoints) and larger tools made on blades or flakes. Microlithic and non-microlithic tools can be understood as two contrasting domains within the larger chipped stone technology, with different implications for production, use, and maintenance of tools. As the relative importance of these two aspects of chipped stone technology vary, the characteristics of lithic assemblages and raw material acquisition can be expected to change.

Mobile Hunter-Gatherers, Microliths, and Raw Material

Explaining Late Paleolithic and Mesolithic shifts in the kinds and proportions of microliths produced, with particular emphasis on the geometric microliths characteristic of Mesolithic technologies throughout Europe, has been a major focus of recent research in European prehistory. Approaches taken in explaining microlithic technology have long focused on projectile technology, particularly the development and spread of the bow and arrow. Recently, Vierra (1995) has proposed that geometric microliths might be interpreted as part of a general trend towards increasingly complex composite food getting tools in an increasingly crowded hunter-gatherer socioeconomic landscape. Both approaches generally regard microliths as a novel feature of Mesolithic technologies, and the primary focus of explanations of technological change at the Paleolithic/Mesolithic transition. In the southern German Magdalenian, Late Paleolithic, and Early Mesolithic, it is impossible to be sure exactly what were the uses of microlithic elements. However, it is clear that microlithic tools are present in all three technocomplexes, and that technological change in the late glacial and early postglacial in the region involved change in these as well as non-microlithic components of the industry. What follows represents an attempt to place change in raw material economies in the context of these broader trends in chipped stone technology.

Archaeologists have made great strides in pursuing questions about prehistoric cultures through analysis of chipped stone assemblages (e.g., Torrence, ed. 1989). However, the search for broadly applicable analytical concepts linking variation in lithic assemblages with dimensions of variation in prehistoric cultural systems is hampered by the sheer complexity of relationships between lithic assemblages and activities. For example, although mobility should be one factor affecting the character of lithic assemblages, effects are not likely to be uniform or easily recognized (Bamforth 1986, 1990). Intervening variables that are relatively independent of mobility strategies, such as hafting (Keeley 1982), tool use life (Ammermann and Feldman 1979; Shott 1989), abundance and quality of raw material sources, and location of raw material relative to settlement locations (Bamforth 1990), may shape assemblage characteristics in more readily detectable and predictable ways. Most importantly, stone tools can be used in a very wide range of different activities, each with its own schedule of use and repair. These considerations require us to consider raw material acquisition within a technological system that relates tasks to artifact forms, reduction strategies, and tool maintenance tactics in the context of a particular local resource distribution.

Two general perspectives on the role of tools in cultural systems have emerged in recent archaeological literature. Both regard the form, frequency, and distribution of stone tools and their manufacture debris as explicable in terms of problem solving by toolmakers, but they differ in a focus on different kinds of problems. One might be termed the raw material package view, the other the tool design perspective.

A number of important studies have focused on the tactics used by toolmakers to provide stone for tool needs in mobile settlement systems. Tools and cores may often serve as portable raw material packages that make stone for tool making available when needed even under conditions of limited time, scarce raw materials, and/or unpredictable tool needs (Kelly 1988; Kuhn 1994; Morrow 1995). The role of chipped stone in this model can be generalized to production of tool edges for a variety of tasks. This approach, focused on what Kuhn has termed the "technological delivery system" (1995), is attractive to archaeologists because it avoids the thorny issue of specific tool functions and allows comparison of whole chipped stone assemblages by means of easily measured variables such as tool edge per unit volume. There is considerable support from ethnographic and archaeological studies for the view of transported tools, flakes, and cores as stone raw material packages in mobile toolkits (Gould et al. 1971; Kuhn 1994; Perlès 1992:437; Shott 1986).

The raw material package view provides one clear model of what the stone toolkit of mobile foragers might look like. Tools designed with transport constraints in mind should be flexible implements with a wide range of potential functions and/or the capacity to be reshaped to meet unpredictable needs (Bleed 1986; Nelson 1991; Shott 1986). Tools in a mobile toolkit should thus have a long potential use-life and should be designed so as to be recyclable even if broken or damaged. As a general rule, these requirements are more easily met by relatively large stone tools, because a few large tools provide a larger number of potential tool edges per gram of stone tool carried than would many small tools (Morrow 1995; Nelson 1991:71). In southern German Upper Paleolithic and Mesolithic toolkits, blade tools, robust flake tools allowing of resharpening, and cores fit the expectations of this model of the mobile toolkit well.

Within this perspective, microlithic tools represent a problem to be explained: if large stone tools represent a useful balance between raw material procurement and tool needs, why would mobile hunter-gatherers adopt smaller tools?

A second trend in recent literature on stone technology focuses on the design of food getting implements (Torrence 1983, 1989; Bleed 1986; Myers 1989; Vierra 1995). These approaches have tended to take a much broader view of the utility of tools by exploring their role as problem-solving responses within subsistence systems. A tool-design analysis has the advantage of providing a more direct conceptual link to models of prehistoric foraging behavior, but requires the inference that the lithic assemblage in question consists of or at least contains food-getting tools. The design of tools closely linked to foraging activities is likely to be determined by a variety of effectiveness criteria (e.g., lethalness and accuracy of hunting weapons, reliability, and maintainability). Decisions concerning raw materials and how to make stone available when needed should be made with these function-specific goals in mind.

Both of these perspectives have moved away from earlier assumptions of a simple, one-to-one correspondence between prehistoric tasks and the tool forms recognized by archaeologists to place tool function in a broader ecological context. However, they offer concepts of differing usefulness for specific archaeological contexts. Predictions about mobile toolkits based on the "raw material package" view must be applied with caution to chipped stone assemblages in which food-getting tools predominate, because such general design goals as flexibility and long potential use life will be only part of a complex set of specifications. Conversely, the tool design perspective is less clearly applicable to assemblages that consist largely or entirely of maintenance tools (cf. Kuhn 1995). Taken together, though, these two approaches to technology offer a useful perspective on likely organizational contrasts between maintenance and food-getting tools. Though tool function is impossible to establish in a detailed way for the samples that comprise this study, a comparison of microlithic and non-microlithic tools should provide a means to explore the effects of varying tool production goals on raw material economies.

Nelson argued that the requirements of transporting tools may often be more important in determining tool form than the specific tasks for which a tool is used (1991:76). It may be that this argument is best applied to subsets of assemblages rather than whole assemblages. Production and maintenance of stone tools is a goal-oriented activity embedded in other activities. Comparisons between assemblages require careful attention to the activity context of lithic production.

Late Glacial and Early Postglacial in Southern Germany: Climate Change and Hunter-Gatherer Prehistory

With a rich database created by decades of archaeological research and work in Quaternary paleoecology, study of the European late glacial and early postglacial offers important opportunities to contribute to anthropological theory concerning the dynamics of settlement and technological change in foraging societies. The archaeological record of late glacial and early postglacial settlement in Europe suggests a broad range of variability in settlement and subsistence strategies (Jochim 1980), and a complex pattern of response to regional patterns of climate and demographic change. In southern Germany, unstable climatic conditions, low initial population densities, and migration from other regions created tremendous variation in natural and social conditions for hunter-gatherer adaptation. Some assessment of this context is necessary in order to investigate patterns of stone raw material use.

The study area in southern Germany (Figure 1) centers on the Swabian and Franconian Jura, a rolling limestone plateau that runs across southwest Germany and continues north into the western part of Bavaria. Many of the known late glacial and early postglacial sites in this region are found in caves and rockshelters along the Danube and in the river valleys draining the Jura limestone plateau. Open-air sites are also known in the rolling moraine landscape of the Alpine Foreland to the south, and the hilly country of the Main and Neckar drainages to the north.

Late glacial archaeological complexes in the area begin with the Magdalenian, which spans the en-

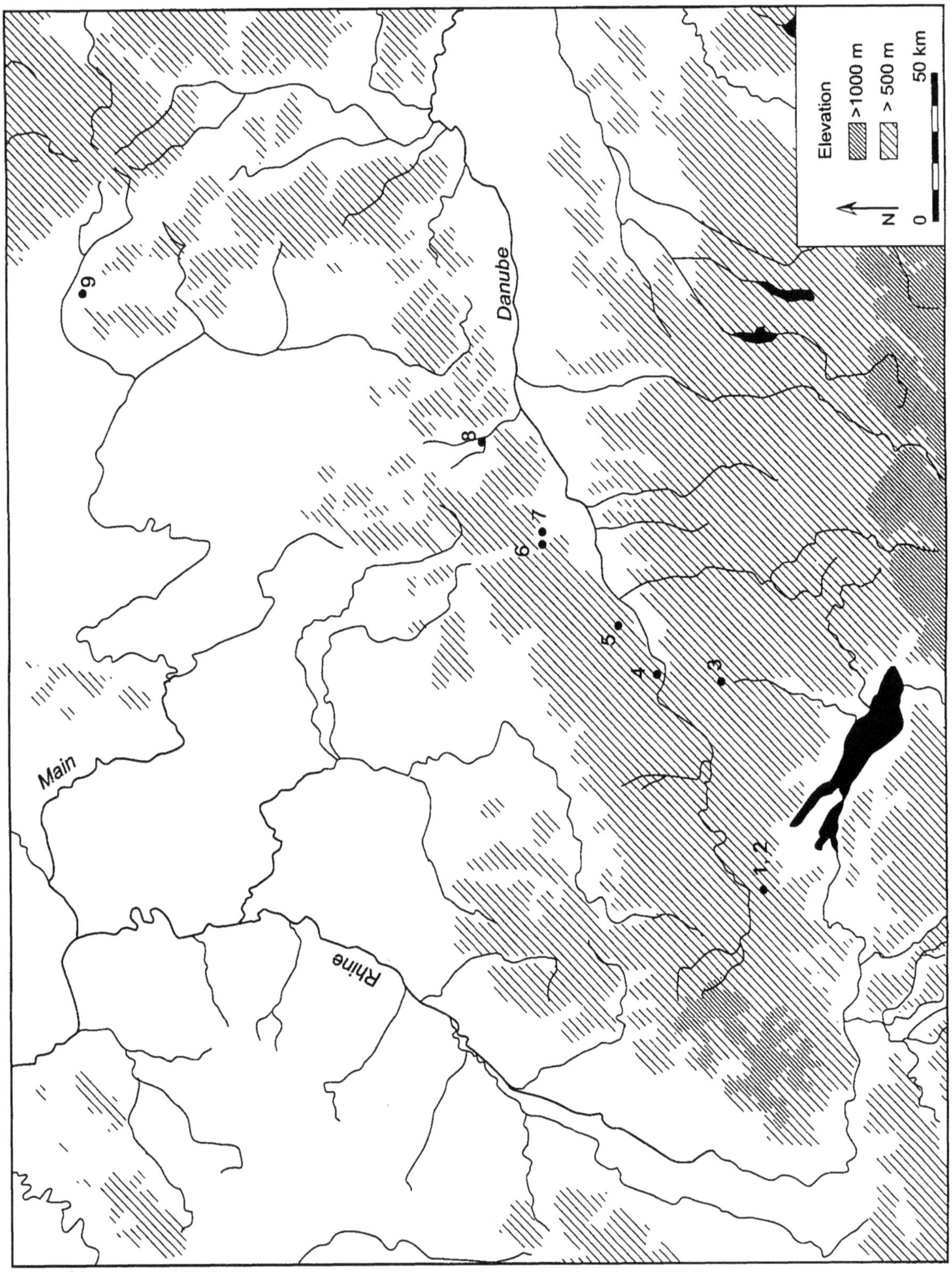

Figure 1. Location of sampled sites in southern Germany. Numbers correspond to Table1.

tire late glacial period. Radiocarbon dates on Magdalenian sites in southern Germany range from about 15,500–11,500 radiocarbon years B.P., but are concentrated in the Bølling chronozone, between about 13,000 and 12,000 B.P. Many southern German Magdalenian sites have been investigated, most frequently cave and rockshelter sites in the narrow, deep valleys draining the southern edge of the Swabian/Franconian plateau (Eriksen 1991; Hahn 1981; Weniger 1982). The Magdalenian is followed by a little understood "Late Paleolithic" industry characterized by backed points, backed bladelets, and short scrapers, known in the southern German area through many surface collections (Eriksen 1991). Only a handful of Late Paleolithic occupation layers in caves (Taute 1972) and excavated open-air sites (Jochim 1995) can be attributed to the Late Paleolithic, and radiometric dates for these sites are so far unavailable. Their stratigraphic position and faunal assemblages indicates that the Late Paleolithic can be attributed to the Allerød and the Younger Dryas. Early Mesolithic occupations with geometric microliths appear in southern Germany during the Preboreal, but available radiocarbon dates suggest sites are rare until the early Boreal (9,000 radiocarbon years B.P.).

Climate and landscape change north of the Alps took place in several episodes during 8,000 years of unstable conditions and rapid change. Alpine glaciers retreated quickly about 16,000 B.P., and southern Germany was free of ice and largely free of permafrost from then on (Lang 1962). Treeless herbaceous plant communities were invaded by shrub juniper and then tree birch following a rapid increase in warmth sometime between 13,000 and 12,300 ^{14}C years B.P. (Guiot 1987:116; Kaiser and Eicher 1987:298; Lister 1988:133; Ammann and Lotter 1989; Broecker 1994). This advance of shrub and woodland plants marks the beginning of a period of vegetation development referred to as the Late Glacial Interstadial, comprising the Bølling and Allerød, here not separated by any very marked Dryas II cooling. During the Late Glacial Interstadial (ca. 13,000–11,000 B.P.), a complex and relatively unstable mosaic of open ground, grass/herb communities, and woodland developed. Temperatures remained fairly stable until a period of cooling at about 10,800 ^{14}C years B.P. (the Younger Dryas episode—Guiot 1987:116; Broecker 1994). Pine, expanding northward from Glacial refugia, appears in southern Germany during this period in birch-pine forests, and about 11,800 B.P. began to exceed birch in pollen assemblages. Pine-birch forests then dominated the area until the deciduous woodlands rich in hazel shrub were established about 9,000 B.P. (Frenzel 1983:134).

Diversity and abundance of animal species declined during the late glacial. Some species, such as mammoth and woolly rhino, became locally extinct early in the late glacial (Eriksen 1991:37), and a number of other species, including musk ox, bison, ibex, wild horse, and wild ass, disappeared by the end of the period. Reindeer persist in the area into the Preboreal, after 10,000 B.P. (Hahn and Steppan 1993), but are not abundant in archaeological faunal assemblages after the Magdalenian. Very low densities of animal game probably characterized the latter part of the late glacial and the earliest Holocene, between about 11,800 and 9,000 B.P. By this period, gregarious species including reindeer and horse were rare or absent, and the available fauna in the pine forests of the region must have been dispersed and mobile. It is not until the later Preboreal, shortly before 9,000 B.P., that two species very important to Mesolithic adaptations appear: hazel, which proliferates at the end of the Preboreal along with the species of the mixed oak forest, and wild pig.

Human population density was probably very low at the beginning of the period. A small number of radiocarbon dates from Magdalenian sites in southern Germany indicate movement of people into the area soon after the retreat of glaciers (c. 16,000 B.P.), but the majority of dates fall during the period 13,000–12,000 radiocarbon years B.P. This suggests that late glacial populations became established during the Late Glacial Interstadial (Housley et al. 1997). It is probably misleading to compare site densities for the Magdalenian, Late Paleolithic, and Early Mesolithic, since the archaeological record of the Magdalenian derives almost entirely from cave and rockshelter excavations, while the bulk of evidence for the Late Paleolithic and Early Mesolithic comes from surface finds, unfortunately not often integrated into regional syntheses (Gumpert 1927). However, the low density of Late Paleolithic compared to Early Mesolithic finds in plowzone collections in the regions north and south of the limestone plateau may suggest population increase after about 9,000 B.P.

Hunting of both large mammals and small game were important in late glacial and early postglacial southern Germany. Faunal assemblages from Magdalenian, Late Paleolithic, and Early Mesolithic contexts suggest that a wide variety of food resources were exploited (Hahn 1981; Eriksen 1991). Important large game species included reindeer and horse in the late glacial, and roe deer, red deer and wild pig in the Early Holocene. Late glacial faunal assemblages also contain abundant bones of arctic hare and grouse while Early Holocene sites show evidence of increased importance of fishing and fowling in rich postglacial riverine and lakeshore habitats. Available evidence indicates that human groups were highly mobile within this region during the late glacial and early postglacial.

Settlement and subsistence systems for the Magdalenian period in southern Germany have been a focus of study using many different approaches over the years, including the seasonality and catchment studies of Sturdy and Gamble and seasonal round models developed by Hahn (1979, 1981), Weniger (1982), and Eriksen (1991, this volume). The Magdalenian in this region is characterized by variable assemblage size, from small lithic scatters to dense accumulations of chipped stone, bone, and other materials. Larger assemblages tend to be associated with reindeer and with relatively open locations near large valleys or in the region's moraine lowlands. Smaller sites with diverse faunal assemblages predominate in cave or rockshelter sites in the narrow upland valleys. This pattern has been argued to reflect seasonal aggregation and dispersal (Weniger 1982; Eriksen 1991), with repeated and seasonally patterned occupation of both upland and lowland locations (Jochim 1991).

Late Paleolithic and Mesolithic settlement is more difficult to characterize due to a smaller number of excavated sites with good faunal preservation. However, one interesting pattern suggested by existing evidence is a reduction in the degree of redundancy characterizing hunter-gatherer use of locations. Very large accumulations of cultural material such as are seen in some Magdalenian sites are not known from these later periods, and patterning in seasonal use of locations is less clear (Eriksen 1991; Jochim 1991).

This case study is based on analysis of excavated assemblages stored in museums in southern Germany. Assemblages were chosen for analysis with the goals of 1) representing a variety of Magdalenian assemblages, including larger and smaller assemblages in contrasting environmental settings, and 2) investigating several Late Paleolithic and Early Mesolithic assemblages for comparison. Where possible, attempts were made to include assemblages that have already been the subject of descriptive publications, and for which radiocarbon dates exist. However, samples from two important sites each with two or more late glacial levels (Kaufertsberg, Helga Abri) were included even lacking radiocarbon assays. The assemblages and available radiocarbon determinations are listed in Tables 1 and 2.

Samples and methods

In order to compare a relatively large number of assemblages, samples of chipped stone artifacts were chosen from larger assemblages (Magdalenian levels from Felsställe, Petersfels, Helga Abri), while smaller assemblages were analyzed in their entirety. Sampling methods varied for each site depending on the specific cataloguing and curation arrangements for each collection. Three spatially and vertically separated levels/areas from the complex open-air site at Petersfels (Albrecht 1979) were included; for each, all cores and retouched tools were analyzed, and a random sample of 25% of debitage was selected from each square meter. Samples from the very large Magdalenian assemblage from Felsställe (>600,000 chipped stone artifacts, Kind 1987) were chosen by counting and sorting debitage within 4 selected square meter units by size and raw material, and then selecting a random sample of at least 400 flakes and blades from each unit for detailed analysis. All retouched pieces from the sampled squares were included. All materials recovered from

Table 1. Definition of Late Glacial and Early Postglacial assemblages from southern Germany[a]

Site[b]	Assemblage	Definition
1. Petersfels	P2	Petersfels 3, AH 2 (sample)
(Albrecht exc.)	P3	Petersfels 1, AH 3 (sample)
	P4	Petersfels 1, AH 4
2. Gnirshöhle	G1	Gnirshöhle I
	G2	Gnirshöhle II
3. Henauhof-West	HW	Henauhof-West, sandy peat
4. Felsställe	F2	Level 2a3 zones 1 and 3
	F3a	Level 3a (sample, 4 sq-m. units)
	F3b	Level 3b (sample, 4 sq-m. units)
5. Helga Abri	H2	Helga Abri IIf7/8 (sample, 4 sq-m. units)
	H3	Helga Abri IIIa/b (sample, 4 sq-m. units)
6. Spitzbubenhöhle	SP	Spitzbubenhöhle level 2
7. Malerfels	MA	Malerfels level 1b
8. Kaufertsberg	KO	Kaufertsberg, upper level
	KU	Kaufertsberg, lower level
9. Schräge Wand	SW	Schräge Wand level 7

[a] Sources describing excavations and collections included in this study include: Petersfels, Albrecht 1979; Gnirshöhle I, II, Albrecht et al. 1977; Henauhof-West, Jochim 1995; Felsställe, Kind, ed. 1987; Helga Abri, Hahn and Scheer 1983, 1984; Spitzbubenhöhle, Hahn, ed. 1984; Malerfels, Albrecht 1984; Kaufertsberg, Kaulich 1983; Schräge Wand, Naber 1968.
[b] Numbers correspond to site identification numbers in Figure 1.

Table 2. Radiocarbon Dates for the Late Glacial and Early Postglacial Assemblages[a]

Assemblage	Material	Age[b, c]
Felsställe level 2a3 (F2)	charcoal	8,190 ± 65
Malerfels I, level 1b (MA)	charcoal	8,605 ± 210
	charcoal	9,075 ± 100
	charcoal	9,560 ± 250
Petersfels 3, level 2 (P2)	bone	11,700 ± 90
Felsställe level IIIa (F3a)	bone	12,680 ± 120
Felsställe level IIIb (F3b)	bone	12,134 ± 365 (avr)
Petersfels 1, level 3 (P3)	bone	12,513 ± 285 (avr)
Gnirshöhle II (G2)	bone	12,670 ± 320 (avr)
Kaufertsberg, lower (KU)	reindeer antler	12,610 ± 90 (AMS)
Gnirshöhle I (G1)	bone	12,830 ± 100
Petersfels 1, level 4 (P4)	bone	12,898 ± 290 (avr)
Spitzbubenhöhle 2 (SP)	bone	13,840 ± 120
	bone	15,230 ± 100

[a] Sources for radiocarbon dates: Felsställe, Jaguttis-Emden 1987:377, 380; Gnirshöhle I, II, Albrecht, pers. comm.; Kaufertsberg, Housley *et al.* 1997:30; Malerfels, Albrecht, 1984:99; Petersfels, Albrecht 1979:49 and Jaguttis-Emden 1983:50; Spitzbubenhöhle, Hahn 1984:51.
[b] (avr) = averaged value based on more than one sample.
[c] (AMS) = AMS ^{14}C determination. All others are conventional radiometric ^{14}C determinations.

the two main concentrations (zones 1 and 3) within the smaller Early Mesolithic assemblage from Felsställe (Kind 1987) were included in the analysis with no sampling. The Magdalenian and Late Paleolithic levels from Helga Abri have not yet been described in detail (but see Hahn and Scheer 1983). To include these important Magdalenian and Late Paleolithic levels, all material recovered from the late glacial levels[2] was counted, and four square meters were selected for detailed analysis. All lithic material from the four selected units was included in the analysis.

Several different approaches to chronological ordering of late glacial and early postglacial assemblages are possible. Assemblages can be dated typologically by attribution to one of the archaeological complexes recognized for the region: Magdalenian, Late Paleolithic, and Beuronian A, B, or C. Alternatively, assemblages can be biostratigraphically ordered by their position in a sequence of regional pollen zones used as a measure of climate change. Both schemes have disadvantages for regional comparison. The archaeological complexes extend over significant blocks of time. This is particularly unsatisfying for an understanding of late glacial archaeology, since the Magdalenian, without any agreed-upon typological subdivision, spans several thousand years of rapid environmental change. A biostratigraphic ordering of assemblages presents difficulties because varying local conditions and uneven preservation of organic remains such as charcoal, microfauna, and pollen make correlation between assemblages uncertain. Most importantly, these coarse measures of time tend to lock archaeologists into an implicit assumption of episodic change, while many patterns of cultural change are more likely to have been continuous in nature, cross-cutting the boundaries of typological units.

In order to explore long-term trends in use of stone raw materials, the assemblages will be presented in the tentative chronological ordering shown in Table 3. Relationships between technological variables and time will be explored using a coarse and a fine chronology. The coarse chronology relies on typological and biostratigraphic evidence to order assemblages into four groupings: Magdalenian, later Magdalenian, Late Paleolithic, and Early Mesolithic. The finer chronology adds radiocarbon dates where available to produce a tentative ordering of most of the assemblages from earlier to later.

Several kinds of problems must be acknowledged in applying this scheme. Some assemblages lack dates, and many are represented only by a single radiocarbon determination. Special problems are

Table 3. Chronological Ordering of Assemblages

Arch. Complex	Chronozone	Assemblage[a]
Early Mesolithic	Boreal	F2
	Preboreal/Early Boreal	SW
		MA
Late Paleolithic	Allerød/	HW
	Younger Dryas	KO
		H2
Later Magd.	Later Bølling /Allerød	H3
		P2
		F3a
		F3b
Magdalenian	Oldest Dryas/Bølling	P3
		KU
		G1
		P4
		G2
		SP

[a]See Table 1 for identification of assemblages.

associated with chronology in the Magdalenian and Late Paleolithic. Most Magdalenian radiocarbon dates fall during the early part of the Late Glacial Interstadial (Bølling), which co-occurs with a plateau in the radiocarbon calibration curve. This makes interpretation of radiocarbon determinations in this period uncertain. The coarse and fine chronologies adopted here for the Magdalenian are intended to serve as a starting point for evaluating patterns during this long period of marked climate and environmental change. Magdalenian assemblages probably dating to the later part of the Late Glacial Interstadial (e.g., Petersfels 3 AH 2, Felsställe levels 3a and 3b, Helga Abri 3a/b) will be regarded as "Late Magdalenian" sites. The intent of this division is to assess trends in chipped stone technology within the Magdalenian, not to suggest any broadly applicable subdivision of the Magdalenian.

Dates for the Late Paleolithic assemblages are lacking, and it is not known if these all fall into the Allerød chronozone or Younger Dryas. A program of radiocarbon dating for Late Paleolithic sites in this region is urgently needed. Archaeological chronologies for the Early Magdalenian allow a somewhat more fine-grained division. No radiocarbon determinations are available for Schräge Wand, but its Beuronian A/B assemblage places it with high probability in the Preboreal or early Boreal. Radiocarbon dates for Malerfels I and from the Beuronian C level at Felsställe (2a3) accord well with regional chronologies.

Though later work may alter the relative order of some of these assemblages, these orderings do accord well with stratigraphy and established regional chronologies. Clearly, further information may change the relative orderings of these assemblages. Conclusions reached here will require further examination and testing on other assemblages, and in light of better chronological information.

Activity Contexts and Tool Production in Late Glacial and Early Postglacial Southern Germany

Trends in the form, frequency, and production techniques of microlithic tools are an important element of late glacial/early postglacial archaeological chronologies in southern Germany. Simple backed bladelets are characteristic of Magdalenian sites, but also appear in Late Paleolithic and Early Mesolithic assemblages (Hahn 1991; Taute 1971). Backed points with curved or angled back occur in many Magdalenian assemblages but are especially common in the Late Paleolithic (Hahn 1991). Geometric microliths and triangular micropoints with retouched bases are characteristic of the Early Mesolithic Beuronian complex, where they are associated with the microburin technique (Taute 1971). The Beuronian (Early Mesolithic) is divided into earlier periods (Beuronian A and B) characterized by isosceles triangles, triangular micropoints, and narrow trapezes, and a later period (Beuronian C) characterized by narrower, elongated scalene triangles, simple backed bladelets, and small micropoints (Taute 1971). Microburins, geometrics, and triangular micropoints also appear occasionally in Late Paleolithic assemblages (Hahn 1991).

These detailed descriptive typologies are useful for comparing sites within archaeological complexes, but make it difficult to assess longer-term trends, since any comparison across time periods inevitably compares frequencies of different types. In order to focus on long-term trends in composite technology, we will apply a simple crosscutting classification of retouched items. First, "microliths" in the broadest sense—small lithic elements very likely to have been components of composite tools—are contrasted to larger formal tools made on flakes or blades. Second, it proves useful to divide microliths into two categories—backed bladelets vs. pointed microliths (including simple backed points, geometric microliths, and triangular micropoints). Both divisions are broad organizational categories intended to characterize assemblages in terms of tool making and tool maintenance activities.

In southern Germany, microlithic elements are produced by an array of simple techniques. Throughout the late glacial and early postglacial, backing (abrupt retouch along the lateral margin of a piece) and/or oblique truncation is used to produce straight (or more rarely, curved) blunted edges that shape blanks into a variety of cutting edges, points, and barbs. Bladelets may be snapped to produce straight segments lacking the irregularities of proximal and distal ends. In the Early Mesolithic of southern Germany, the "microburin" technique of notching and striking bladelets to segment them is common. The abruptly retouched or backed edge or edges may serve several functions, such as 1) stabilizing the fit of a flake or bladelet in a haft, 2) creating a point that combines the sharpness of an unretouched edge with the stability of a blunted edge (Fischer 1989; Rozoy 1989), and 3) regularizing variable flakes or bladelets into desired standardized shapes (Perlès 1992).

Microlithic elements in the samples considered here are as follows.

1. Backed bladelets (usually blade segments) with a single relatively straight backed edge, and sometimes end or other lateral retouch.

2. Pointed microlithic elements, including

a. simple backed points created by a single backed edge forming a point (≤45 degrees) with an unretouched edge (G.E.E.M.1972:364). The backed edge is often curved or angled, and the piece may be symmetrical or asymmetrical (Bohmers and Wouters 1956:29; Hahn 1991:209).

b. geometric microliths, including triangles, occasional lunates, and narrow trapezes.
c. triangular micropoints, with a point (≤45 degrees) formed by the intersection of a single backed edge with an unretouched edge, and a straight, convex or concave retouched base. Following Taute, in extremely asymmetrical points one of the basal angles may reach 90 degrees, but if a basal angle exceeds 90 degrees the piece is a scalene triangle (1971:45).

Backed bladelets in the Magdalenian and Late Paleolithic samples are very small, with mean widths ranging from 5 to 7 millimeters, while backed points are typically somewhat larger, but still distinctly smaller than non-backed tools.

Tool Functions and Activity Scheduling

Though the function of stone tools is notoriously difficult to make generalizations about, it is probably safe to assume that microlithic (composite) and non-microlithic (hafted or hand-held) tools form groups with broadly contrasting functions. Non-microlithic blade tools including endscrapers, dihedral burins, burins on end retouch or on a broken end, borers, and laterally retouched blades are characteristic of Magdalenian assemblages in southern Germany. The subsequent Late Paleolithic and Early Mesolithic complexes contain somewhat fewer non-microlithic blade tools and characteristic short scrapers. In contrast with microlithic elements, this group of tools can be characterized as relatively large, robust pieces with stable modified edges that can stand up to relatively long wear. As a group, these long-use-life tools probably have a range of functions such as processing (scraping, planing, whittling, boring) hide, wood, bone or other materials. Many were probably not directly involved in acquiring food.

Microliths are better candidates for food-getting tools, including weapons for hunting, fishing, and fowling. It has long been thought that backed armatures of various types were designed for use as projectile points, lateral cutting edges, and/or barbs. Though the number of possible ways to haft and use these tiny stone elements makes conclusions about their function uncertain (Clarke 1976; Zvelebil 1985; Finlayson 1990; Vierra 1995), it is clear from rare finds of organic remains around Europe that at least some backed bladelets, backed points, and geometrics were in fact used as projectile tips and barbs or, alternatively, as cutting implements (Aquilas-Wauters 1956; Malmer 1969; Moss 1983; Moss and Newcomer 1982; Myers 1989; Owen 1988; Ströbel 1959; Symens 1986; Vaughan 1985, 1987). Very broadly, it is reasonable to conclude that microliths, known to have been projectile weapons for large game hunting in at least some instances, could also have provided cutting edges and barbs for a variety of food-getting tools, including fishing and fowling weapons as well as, possibly, implements more geared toward processing plants or other resources.

Tool Production, Re-tooling and Raw Material Economy

These broad contrasts in tool design also have useful implications for tool production, re-tooling and raw material economy. Blade tools with terminal working ends, or more broadly any robust flake or blade tools allowing of resharpening, are relatively flexible implements with long potential use lives. The retouched working edge can be quickly repaired or remade into another tool without the necessity of producing new blanks (Perlès 1992:437). This forms a useful contrast with composite tools incorporating microlithic elements, for which re-tooling is not a simple matter of retouching an existing tool, but requires production of new blanks to replace those lost or damaged.

Non-microlithic tools can also be very versatile implements, either through reshaping or through use of a single blank for a tool with two working ends. Multiple tools (blades with two modified working ends) are characteristic of the Magdalenian in southern Germany, but occur in lower frequencies in the Late Paleolithic and Early Mesolithic. Composite tools may have considerable built-in versatility (a good example is the arrow, which might be used in hunting a wide range of prey sizes [Churchill 1993] and applied expediently to a variety of other tasks such as cutting, scraping, drilling). However, microliths lack the property of being extensively recyclable into different kinds of tool edges, as from a borer to a burin.

These two broad classes of stone tools should differ in the costs of tool maintenance or repair. To repair composite tools, the toolmaker must have raw material on hand, either in the form of spare components or a nodule/core. Though production of small microlithic stone tools may be quick compared to the manufacture of a larger stone tool or particularly of an organic implement, small stone tools are likely to have a shorter use life than either larger, re-workable stone tools or more resilient organic implements, and therefore may require considerable time and materials (stone, hafting materials) for frequent replacement of stone insets. To achieve the sort of maintainable "plug-in-and-pull-out" technology often envisioned in the case of Mesolithic microliths (Clarke 1976), it would presumably be necessary to produce stone elements of a standardized size and shape, such that they would fit into an existing haft (Owen 1988). Production of standardized stone elements might be assumed to require considerable investment in time, skill, and high-quality raw materials.

If flexibility and use life of stone components

decline, we might predict an increased demand for stone. This is because any given volume of stone raw material should be exhausted relatively rapidly by a technological system requiring many, small "throwaway" stone tools and fewer large, reusable, multi-purpose stone tools (Morrow 1995). Wood, antler or bone hafts may be curated, while the lithic components must be replaced, perhaps after a single use. Though composite tools incorporating microliths might be easier to repair than barbed antler projectile points (Clarke 1976; Myers 1989), these considerations suggest that their increasing use probably introduced new costs into lithic raw material acquisition systems.

For microlithic tools, raw material acquisition should be determined by the scheduling of food-getting activities requiring composite tools. In the case of larger blade and flake tools, costs of raw material acquisition should be lower, and raw material acquisition may focus primarily on ensuring the availability of a flexible supply of raw material to fulfill varied needs as they arise. These considerations suggest that an examination of relationships between these broad tool classes, blank size, and stone raw material types should shed light on the technological context of raw material transport.

Tool Production Contexts: Variation in Microlith Frequencies

The relative frequency of microlithic elements as opposed to larger, long-use-life tools in late glacial and early postglacial assemblages should thus provide some clues about the kinds of lithic production and maintenance activities undertaken at different locations. Here we will examine the relationship between microliths and other tool forms as a factor shaping raw material economies.

The relative frequency of backed elements in tool assemblages varies considerably in the samples studied (Table 4). Only one small Magdalenian assemblage among those studied (SP, Spitzbubenhöhle 2) lacks backed pieces. As a group, the Magdalenian assemblages show the widest range of relative frequencies of backed elements. Among the nine Magdalenian assemblages with backed pieces, the relative frequency of backed elements ranges from 15.3% to 46.6%. The highest percentages as well as the largest numbers of backed elements among Magdalenian assemblages appear at Petersfels (P2, Petersfels 3 AH 2; and P3, Petersfels 1 AH 3), where backed bladelets are common and backed points rare. This is presumably related to the specific activities carried out at the Petersfels site, which appears to have been repeatedly visited for hunting and butchering reindeer, horse, and hare (Albrecht 1979). Backed points appear more commonly, though never outnumbering backed bladelets, among the Magdalenian assemblages Kaufertsberg-Unten (KU), Helga Abri 3a (H3a) and Helga Abri 3b (H3b). Backed points in the southern German Magdalenian and Late Paleolithic are variable in form, including examples with curved or angled retouched edges.

The Late Paleolithic assemblages (see Table 4) also contain variable percentages of backed pieces, but none of the three in this sample match the very

Table 4. Form and Frequency of Backed Elements in Late Glacial and Early Postglacial Assemblages

Archaeological Complex	Assemblage	Backed pieces	% of tools	Total tools	BB	PT	MI	GE	VA	BR
Early Mesolithic	F2	38	55.9	68	4		5	12	1	16
	MA	28	82.4	34			10	15	3	
	SW	47	73.4	64			21	15	6	5
Late Paleolithic	HW	8	29.6	27	4	3	1			
	KO	51	32.1	159	39	11		1		
	H2	4	21.0	19	1	3				
Later Magdalenian	H3	31	18.4	168	24	7				
	P2	102	46.6	219	102					
	F3a*	15	22.0	67	15					
	F3b*	17	15.3	111	16	1				
Magdalenian	P3	43	43.0	100	43					
	KU	35	16.7	210	31	4				
	G2	4	33.0	12	4					
	P4	2	22.2	9	1	1				
	G1	7	24.1	29	7					
	SP	0	—	16						

Headings in this table refer to the following tool types: BB, backed bladelet; PT, backed point; MI, triangular micropoint, end-retouched micropoint; GE, geometric microlith (triangle, crescent, trapezoid); VA (other microlith, atypical); BR (broken microlith, unidentified).

*The comparison between Kaufertsberg and other sites is problematic because the early excavation of this rockshelter took place without screening, and the large size and high frequency of blades suggests small items and flakes were not collected and and curated. It is very likely that the typically small Magdalenian backed bladelets are under-represented. Nevertheless it is interesting that this collection does show markedly higher proportions of backed implements in the upper, Late Paleolithic level than in the lower, Magdalenian level.

high frequencies found in the Magdalenian levels at Petersfels. Among Late Paleolithic assemblages, backed points are more numerous, but still do not outnumber backed bladelets except in the case of the very small sample from the Late Paleolithic level at Helga Abri (H7).

Microlithic elements make up the majority of retouched pieces in all of the Early Mesolithic assemblages considered. These include triangular micropoints and geometrics in the two Beuronian A/B assemblages, and a wider variety of forms in the single Beuronian C assemblage.

Overall, microlithic elements make up a large proportion of retouched tools 1) on some Magdalenian sites with large numbers of backed bladelets, and 2) in all the Early Mesolithic assemblages, in which backed elements fall into two broad categories—micropoints and geometric microliths. late glacial (Magdalenian and Late Paleolithic) assemblages with comparatively low or moderate percentages of backed bladelets tend to have more variable forms of backed elements, including backed points, which range widely in size and shape.

The size of microlithic elements shows a complex trend over time (Figure 2). Backed bladelets, common on Magdalenian assemblages but rare in later times, are made on narrow, thin bladelets throughout the time period considered here, and show no trend towards smaller width or thickness over time. Other microlithic elements considered as a group (including simple backed points, triangular micropoints, and geometrics) do show a trend of decreasing width, length, and thickness over the late glacial and early postglacial. Assuming that these pointed forms included projectile points, this suggests that toolmakers throughout the late glacial and early postglacial consistently chose to develop smaller projectile armatures. This trend begins in the later Magdalenian and continues through the Late Paleolithic and Early Mesolithic.

The gradual increase in reliance on composite tools, the development of new forms and new manufacture techniques for shaping microliths, and the consistent reduction in size of pointed microliths clearly indicates a reorganization of food-getting technologies (insofar as they contain stone elements) during the late glacial and early postglacial. This is accompanied by some changes in other aspects of the stone technology.

Borers, burins, end-retouched or laterally retouched pieces, scrapers, splintered pieces, and varia (other retouched pieces or fragments) made on blades or flakes are common to Magdalenian, Late Paleolithic, and Early Mesolithic assemblages in southern Germany. Over the course of the late glacial, blade tools become somewhat less common, and retouched pieces in general show some reduction in size. Blade tools are most characteristic of the Magdalenian assemblages (Figure 3). The Late Magdalenian assemblages show a reduction in the frequency of blade tools and a somewhat higher frequency of tools made on flakes, which make up from 10 to 30% of the non-backed tools The Late Paleolithic and Early Mesolithic assemblages as a group, on the other hand, show an increase in the proportion of fragmentary pieces whose blank type could not be clearly identified.

Raw Material Economies in the Context of Trends in Tool Design

Hafted implements are likely to be discarded at locations where re-tooling is going on. Thus, if microliths make up a relatively large proportion of retouched tools on a site, we can infer that re-tooling implements or weapons was an important activity on the site, and would be likely to influence decisions about raw material acquisition. Examining raw material economy on these sites compared to those where fewer microlithic elements are discarded should provide clues to the kinds of problems in raw material economies introduced or influenced by a tool-production context emphasizing manufacture or re-tooling of composite tools.

On the basis of the above considerations, we can formulate some useful questions about raw material economies on late glacial and early postglacial sites. First, are long-use-life tools discarded far from their geological source area more often than microlithic tools? If the use life of transported tools is an important factor leading to movement of raw materials, we would expect to find that much of the evidence for raw material transport is specifically due to transport of relatively large blade and flake tools with high potential for resharpening and recycling. This proposition can be explored by comparing the types of objects made on non-locally obtainable raw materials on late glacial and early postglacial sites.

Second, how did toolmakers at late glacial and early postglacial sites address the problem of obtaining adequate raw material for re-tooling implements incorporating microlithic implements? If re-tooling of microlithic armatures imposed high demands on raw material economies, we expect to find either that re-tooling takes place in locations where raw material is abundant, or that raw material is transported in the form of cores or other raw material packages to supply the need. This can be approached by examining raw material use and production of microlithic elements in assemblages with varying frequencies of microlithic elements. High frequencies of microlithic elements are likely to indicate locations where re-tooling was an important activity and may have affected raw material economies.

Availability of Stone Raw Materials in Southern Germany

Raw materials for chipped stone manufacture in southern Germany derive largely from the Triassic, upper Jurassic, and Cretaceous sedimentary rocks that make up the southern and central German and Swiss hills and plateaus, and to a lesser extent from the calcareous alps, localized volcanic deposits, and a few other minor sources. Although

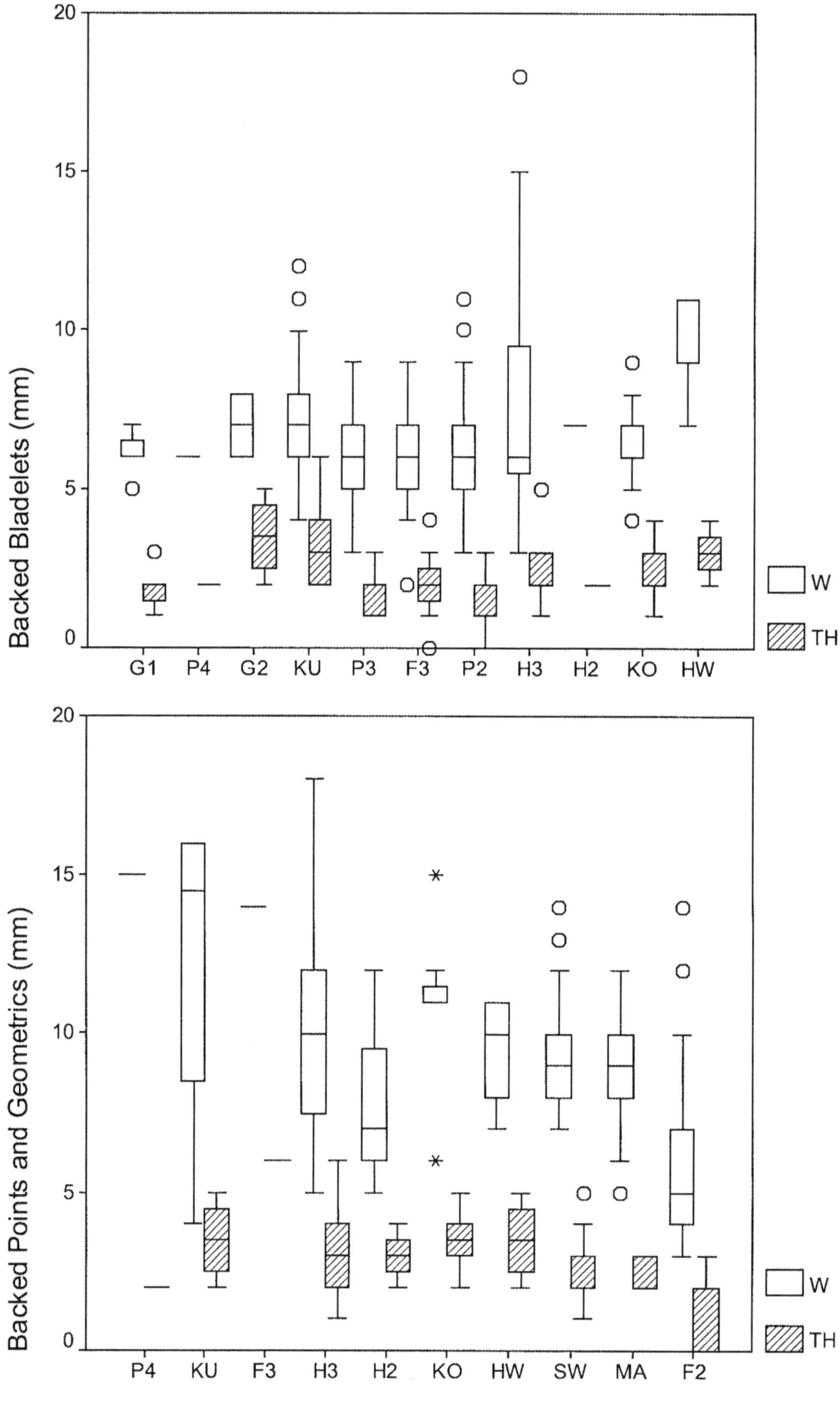

Figure 2. Dimensions of microlithic elements in Late Glacial and Early Postglacial assembalges.

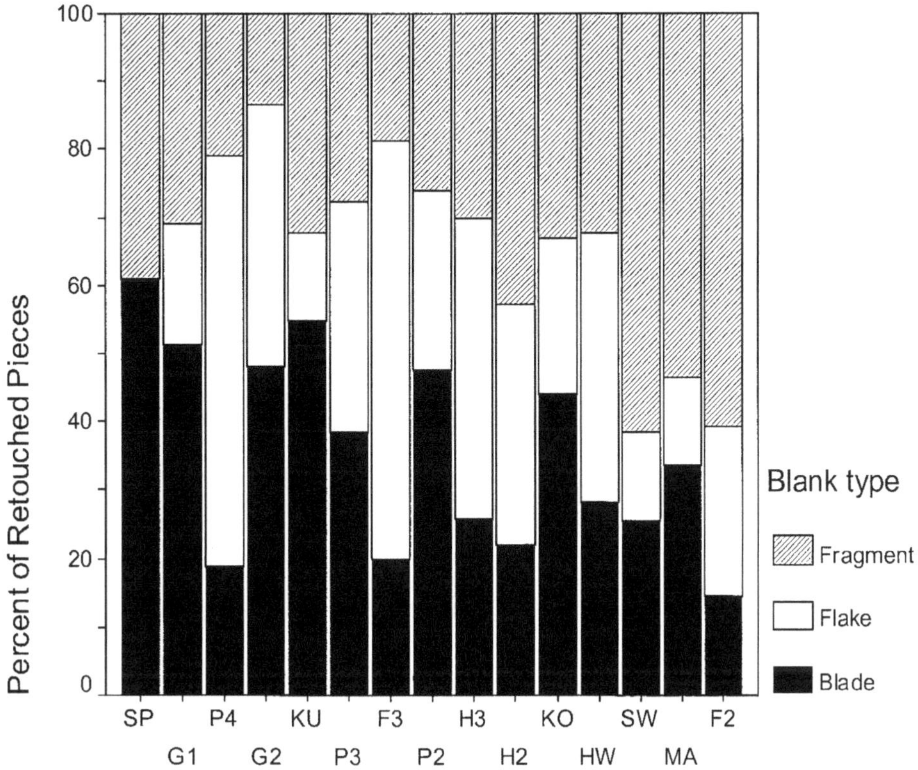

Figure 3. Relative frequency of blades and flakes among retouched items in Late Glacial and Early Postglacial assemblages.

outcrops of chert-bearing rocks in primary geological situation can be found, the most common chert sources are in secondary deposits including moraines, river gravels, and weathered clays. Residual gravels from Tertiary and earlier Pleistocene river transport make up widespread surface deposits throughout the hill and plateau landscape of southern Germany. These old terraces bear cherts and radiolarites derived from surrounding uplands. The distribution of sources in southern Germany is extremely complex because the Tertiary and Quaternary geologic history of the area involves movement of great masses of sediments from the Alps, the Black Forest, the Bavarian crystalline rock massif, and the central hill and plateau area into the deep sedimentary basins of the Rhine and the Alpine Foreland. The generalized geological map in Figure 4 shows the location of raw material-bearing deposits.

A variety of materials are visually identified by regional prehistorians based on color, banding, texture, characteristics of cortex, and fossil inclusions. Information on regional variations within widespread chert-bearing deposits is available only for a few areas, particularly some Bavarian cherts that were intensively exploited in Neolithic times. Source locations of materials can therefore be identified in most cases only to the entire area covered by chert-bearing rocks or sediments. A good deal of work has been done to build up comparative collections of materials from known sources and to map outcrops and secondary sources (Burkert 1996; Fritsch and Neubauer 1987; Lipinski and Löhr 1987; Löhr 1987; Löhr and Schönweiß 1987). As this work continues, it is to be hoped that a much more detailed picture of raw material use and transport will appear.

Chert is common in the upper Jurassic limestones of the Swabian and Franconian plateaus, and Jurassic chert forms the backbone of prehistoric stone raw material economies in the region. Most common is a white or gray nodular chert varying from poor-quality, very coarse chert to a fine-grained chert with excellent flaking characteristics (Deecke 1933:34). Nodules are mostly small (from walnut- to head-sized), found in varying concentrations in outcrops of the parent rock as well as in slope deposits or loamy cover sediments of the plateau surface near the parent material (Binsteiner 1992; Deecke 1933:36; Geyer and Gwinner 1991:153-4; Gwinner 1976:62-63; Müller 1958; Rutte 1965:76). These white/gray Jurassic cherts are available in many places on the limestone plateaus. An investigation of raw material sources in the central Swabian plateau near the Upper Paleolithic site Geißenklösterle by Burkert located several productive concentrations of Jurassic cherts within a few kilometers of the site (1996:279). Burkert described collection of usable Jurassic cherts as a quick and easy process (1996:281).

A distinctive weathered form of the same nodular cherts was also widely used in prehistoric times. Weathered cherts have a yellow-gray to beige color

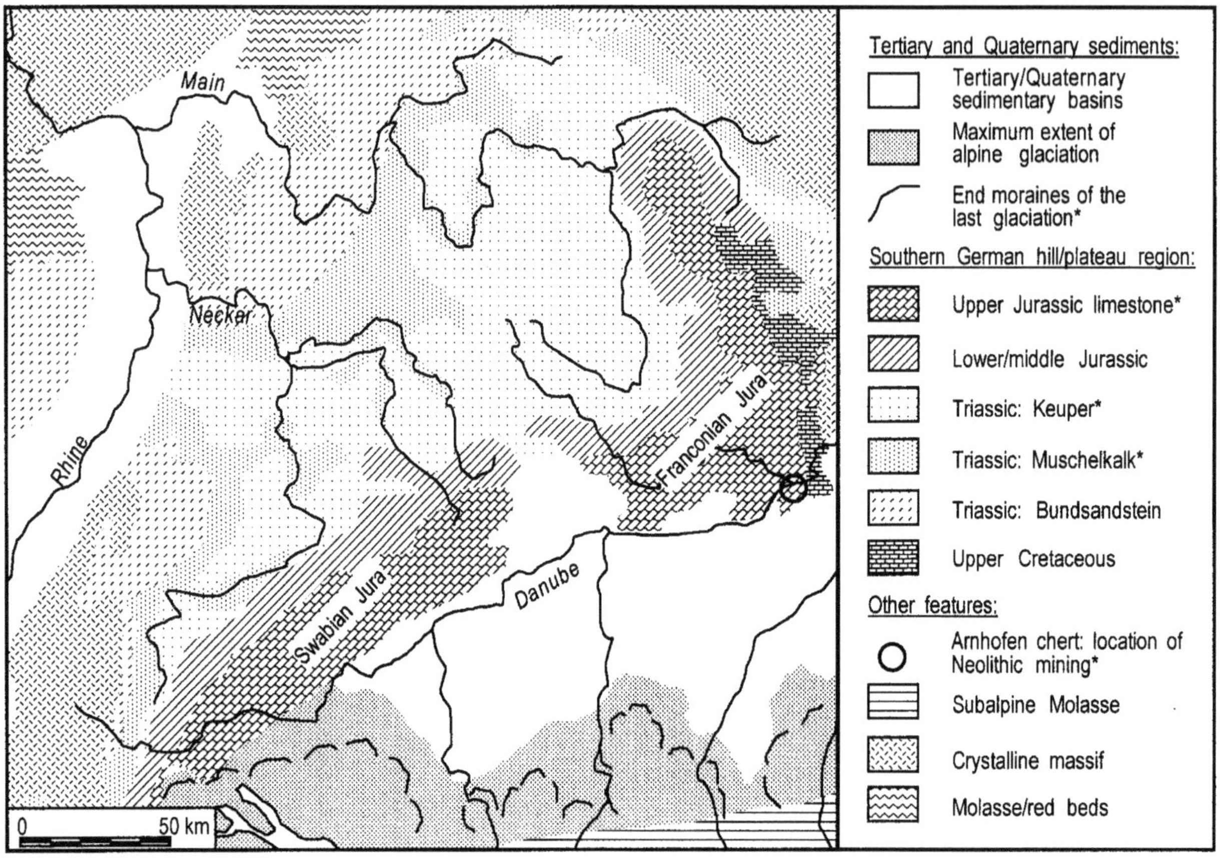

Figure 4. Surface geological deposits of southern Germany.

due to longer exposure to weathering (Burkert 1996). These include coarse- and fine-grained varieties, and may often be flawed by incipient fractures. Weathered chert nodules are found, sometimes in very dense concentrations, in residual clays, Tertiary molasse deposits, and river gravels in many locations (karst depressions, valley fill) on or adjacent to the limestone plateaus (Beurer 1971; Geyer and Gwinner 1991:323). Compared to the unweathered white nodular cherts, the distribution of the weathered varieties can generally be regarded as more localized. Concentrated accumulations of chert from the upper Jurassic limestones in residual clay/loams can be found on plateau tops near valley margins (Burkert 1996), and in remnants of Tertiary and Quaternary river gravels in the eastern Swabian Jura (Auffermann 1991:19, 1991; Mall 1968).

Another concentrated source of Jurassic cherts appears in a variety of gray tabular cherts in bedded limestones in the southern part of the Franconian plateau between Regensburg and Eichstätt (Binsteiner 1990:4; Herrmann 1975:155, 1983). Primary deposits of these chert-bearing limestones are covered by a thick complex of Quaternary gravels and Tertiary sands. Secondary deposits in residual clays along the southern edge of the Franconian plateau were intensively exploited in prehistory (Binsteiner 1990). These gray banded cherts formed in a series of quiet bays along an ancient shoreline where varying conditions produced distinctive chert formations (Binsteiner 1989, 1990). Cherts near Arnhofen (Binsteiner 1990; Blank 1994; Engelhardt and Binsteiner 1988; Moser 1978, 1981), for example, have a distinctive blue-gray to gray-white color, fine banding, and often excellent fracture characteristics, with tabular pieces several centimeters thick. The relatively thick tabular form or elongated/flattened nodule shapes characteristic of these deposits is well suited to blade production. This material was used and widely transported during Late Upper Paleolithic and Mesolithic times, and was mined in the later Neolithic for specialized blade production (Auffermann 1996; Binsteiner 1992; Freund 1974/5; Naber 1981; Weinig 1989).

Overall, varieties of Jurassic chert are abundant. These are the dominant raw materials found in most Paleolithic and Mesolithic assemblages both on the Swabian/Franconian plateau and in surrounding regions. Though less abundant, more variable in quality, and less commonly transported beyond the local area in prehistory, chert can also be found north of the Jurassic limestone area, in the hill/plateau region of the Neckar/Main drainage. White to beige cherts from the Keuper formations occur in small nodules and often are flawed by numerous fracture planes (Löhr and Schönweiß 1987:128). Where such flaws are absent, the material has good to excellent flaking characteristics. Within the area where Keuper cherts occur naturally, they are among the most common material on Mesolithic sites (Hahn 1983; Krumbeck 1927:190; Löhr and Schönweiß 1987; Naber 1970). These cherts are found both in primary deposits (primarily in sandstones of the upper Keuper north of the eastern Swabian Alb, and to the north and east into Bavaria) and in residual deposits and gravels on the terraces of the Neckar and Main rivers, where they can be found mingled with Jurassic cherts, cherts and petrified wood from the neighboring Muschelkalk and Bundsandstein formations, and black radiolarian chert ("Lydite") (Löhr and Schönweiß 1987:131). Lydite, a very hard cryptocrystalline material marked by veins of light-colored crystalline quartz (Krumbeck 1927:188; Rutte 1965; Weinig 1989; Zotz and Freund 1973:46), was the most widely used of these other materials. It appears in the study area in the form of small river cobbles rarely reaching a maximum size between 10-15 centimeters (Krumbeck 1927:189).

In the alpine foreland to the south, raw material for chipped stone tools is limited to small, dispersed nodules of radiolarite, quartzite, and other materials of alpine origin found in moraines and river gravels. The most important here are radiolarites ranging in color from dark brownish red to green or blue-green, derived from upper Jurassic limestones of the Swiss and Allgäu Alps (Deecke 1933:26-29). Although the flaking characteristics of this very fine-grained material can be excellent, most pieces are flawed by quartz veins or fractures (Deecke 1933), and appear in the form of small, worn pebbles and cobbles only rarely ranging in size up to a maximum dimension of about 10 centimeters (Krumbeck 1927:278). Radiolarite can be found in molasse, moraine, outwash, and river gravels in the lowlands north of the Alps, and possibly on a high Danube terrace in the southern central part of the Swabian Jura (though a search by Burkert [1996:280] produced no radiolarite on the plateau). The most abundant sources are gravel terraces in the alpine foreland.

Observers who have tried to assess the abundance of alpine radiolarites find that pieces of a usable quality and size are extremely difficult to locate (Burkert 1996:281; Deecke 1933:26-29). Radiolarite does, however, appear in Upper Paleolithic and Mesolithic assemblages. Burkert characterized the search for usable radiolarite as either largely determined by chance or requiring a great expenditure of time (1996:281). Compared to other raw materials discussed here, radiolarite must be regarded as rare and dispersed.

Somewhat more abundant in the alpine-derived fluvioglacial deposits are small (rarely up to 10 centimeters maximum dimension) rounded cobbles of a black, grainy chert that Burkert has identified as a metamorphosed alpine radiolarite (1996). Though more abundant than radiolarite, the black chert also is flawed by a common blocky fracture.

Other raw materials of local importance in Paleolithic assemblages include a fine-grained, honey-colored silicified tuff from the Randeck crater on the northern Swabian Alb (Riek 1957, 1959), and a beige to yellow-brown fine-grained Cretaceous quartzite ("Mörnsheimer Bryozoanhornstein") appearing in large relict blocks (up to 1 cubic meter) on the Alb plateau between the Danube and the Altmühl rivers southeast of the Nördlinger Ries (Lemcke 1988:12) and less commonly in the crater itself (Beurlen et al. 1978:175). Descriptions by Beurlen et al. (1978) and Gall et al. (1977) suggest that this material is locally abundant and is also available in large pieces, as most sources in the area are not. Although finer-grained pieces of this material might be confused with yellow-brown Jurassic chert, it seems from available evidence that this material was not widely transported in prehistory, but appears mostly in assemblages in and around the Ries crater (e.g., Magdalenian and Late Paleolithic levels at Kaufertsberg, Kaulich 1983 and this study).

Quartz ("Bergkristall") of alpine origin is occasionally found in small amounts on Stone Age sites in southern Germany. Deecke noted the presence of this material in gravels of the upper Rhine south of Strasbourg (1933:3).

A generalized description of raw material sources and their accessibility and abundance and in southern Germany can be summarized as follows. On the Swabian and Franconian limestone plateaus, raw material (Jurassic chert, silicified tuff) for chipped stone tool production is abundant and could generally have been located with little time or effort. Though generally found in cover sediments and other weathered deposits, localized chert concentrations where quantities of usable chert can be found exist in a number of areas. The Ries crater offers abundant local raw material in the form of the Cretaceous quartzite, and is also close to sources of Jurassic chert. Finally, in the broad, rolling moraine area south of the Danube, stone raw material is scarce. Radiolarites, quartzites and alpine cherts found in river gravels would have been time-consuming to seek out and are often unsuitable for flaking.

Raw Material Transport

Transported raw materials are recognized only in cases where distinctive raw materials are found outside the areas in which they occur naturally. This must give an underestimate of stone raw material transport, particularly within the large areas where varied Jurassic cherts and radiolarites are found. Keeping these problems is mind, some observations can be made about raw material transport in the study area.

With the exception of Henauhof-West (HW), all of the sites considered here are located in environments where local raw materials are relatively abundant, though varying in quality. As would be expected, HW is the only assemblage consisting largely of material not available within a few kilometers of the site. This single example illustrates a general pattern for the region: the higher quality Jurassic cherts were transported into surrounding regions. The other assemblages considered here consist largely, in some cases overwhelmingly, of locally available raw materials. However, in Magdalenian, late Magdalenian, and Late Paleolithic occupations at two sites, transported materials make up moderate proportions of total assemblages, and suggest that carrying raw material, in some cases over long distances, played a role in regional raw material economies.

Tables 5 and 6 compare the frequency of locally available (within 10 kilometers of the site), regionally available (10-50 kilometers), and long-distance transported (>50 kilometers) raw materials among microlithic and non-microlithic retouched tools with proportions for unretouched flakes and blades. In each assemblage, some material was identified only as "other" raw material, not attributable to a known source.

The six earlier Magdalenian assemblages resemble each other in their almost complete reliance on material types potentially obtainable within the local area of each site (<10 kilometers). Raw materials that can be positively identified as deriving from deposits more than 10 kilometers away were found only in Gnirshöhle I (G1) and Kaufertsberg, lower level (KU). In the case of Gnirshöhle this is limited to three unretouched flakes and blades of a Cretaceous flint of unknown origin. The very large Magdalenian chipped stone assemblages from Felsställe (F3a, F3b) consist almost entirely (>99%; Kind 1987) of brown, weathered Jurassic chert available within a few hundred meters of the site. The Petersfels assemblages (P4, P3, P2) are more diverse in raw material composition, but consist largely of white/gray and brown Jurassic cherts, with some radiolarite, all of which could be found within 10 kilometers of the site location. The preserved assemblage from Kaufertsberg, lower level (KU), by contrast, contains a modest amount of transported materials, which make up about 10% of unretouched flakes and blades, with a slightly higher percentage (12.6%) among non-microlithic tools. Here the non-local raw materials include a variety of types of stone available within 50 kilometers of Kaufertsberg, including radiolarite, quartzite, lydite, and Keuper chert. All of the material from sources more than 50 kilometers distant from Kaufertsberg is high-quality tabular chert from southern Franconian sources.

Among the four later Magdalenian assemblages considered, raw materials used were also largely

Table 5. Raw Material Transport in Late Paleolithic and Early Mesolithic Assemblages*

Assemblage, by raw mat'l group		Unretouched flakes & blades		Backed tools		Non-microlithic tools	
		n	%	n	%	n	%
Early Mesolithic							
F2	Local	1658*	95.1	47*	85.4	48*	97.9
	Long-distance	1*	0.1	0*	—	0*	—
	Other	84*	4.8	8*	14.5	1*	2.0
SW	Local	455	87.5	40	85.1	17	100.0
	Long-distance	11	2.1	0	—	0	—
	Other	54	10.4	7	14.9	0	—
MA	Local	201	97.6	27	100.0	5	100.0
	Regional	5	2.4	0	—	0	—
Late Paleolithic							
HW	Local	19	17.9	1	12.5	4	21.0
	Regional	81	76.4	7	87.5	13	68.4
	Long-distance	3	2.8	0	—	1	5.3
	Other	3	2.8	0	—	1	5.3
KO	Local	201	85.2	28	54.9	91	84.2
	Regional	11	4.7	10	19.6	5	4.6
	Long-distance	13	5.5	6	11.8	6	5.6
	Other	11	4.7	7	13.7	6	5.6
H2	Local	32*	59.2	3	100.0	11*	73.3
	Regional	2*	3.7	0*	—	2*	13.3
	Long-distance	14*	25.9	0*	—	1*	6.7
	Other	6*	11.1	0*	—	1*	6.7

*Asterisks in table mark values that are sample counts, not assemblage totals.

Table 6*. Raw Material Transport in Magdalenian Assemblages

Assemblage, by raw mat'l group		Unretouched flakes & blades		Backed tools		Non-microlithic tools	
		n	%	n	%	n	%
Later Magdalenian							
H3	Local	660*	73.3	17*	54.8	76*	55.5
	Regional	63*	7.0	8*	25.8	25*	18.2
	Long-distance	139*	15.4	2*	6.5	22*	16.1
	Other	38*	4.2	4*	12.9	14*	10.2
P2	Local	567*	98.4	90	88.2	114	97.4
	Long-distance	1*	0.2	0	—	0	—
	Other	8*	1.4	12	9.8	3	2.6
F3a	Local	1760*	99.8	12*	85.7	52*	100.0
	Other	4*	0.2	2*	14.3	0*	—
F3b	Local	4091*	99.8	14*	87.5	94*	100.0
	Other	8*	0.2	2*	12.5	0*	—
Magdalenian							
P3	Local	368*	97.6	43	100.0	56	100.0
	Other	9*	2.4	0	—	0	—
KU	Local	258	88.6	28	80.0	149	85.1
	Regional	13	4.5	3	8.6	18	10.3
	Long-distance	17	5.8	0	—	4	2.3
	Other	3	1.0	4	11.4	4	2.3
G1	Local	77	90.6	7	100.0	21	95.4
	Long-distance	3	3.5	0	—	0	—
	Other	5	5.9	0	—	1	4.5
P4	Local	50	98.0	2	100.0	7	100.0
	Other	1	2.0	0	—	0	—
G2	Local	34	89.5	3	75.0	8	100.0
	Other	4	10.5	1	25.0	0	—
SP	Local	110	100.0	0	—	13	81.2
	Other	0	—	0	—	3	18.8

* Asterisks in table mark values that are sample counts, not assemblage totals.

from local sources. Non-local stone was identified only in the Magdalenian levels from the Helga Abri rockshelter (H3), located in the central Swabian Alb. The Helga Abri Magdalenian assemblage is, like Kaufertsberg, very diverse in its raw material composition. Materials deriving from sources 10-50 kilometers away include radiolarite and other moraine-derived materials and a silicified tuff from the Randeck crater. Longer distance transport is again limited to Bavarian tabular chert. Overall, transported materials make up about a quarter by count of the total raw material in this sample. Regionally transported materials make up only 7% of unretouched flakes and blades, but 25.8% of the 31 backed pieces, and 18.2% of the 137 non-microlithic tools. In a sample of 767 unretouched flakes and blades from the H3 assemblage, 139 (15.4%) were derived from the same tabular chert sources, approximately 170 kilometers from Helga Abri. This material was identified less frequently (only 2 examples, or 6.5%) among backed points and bladelets from H3, but was more common among non-microlithic tools (22, or 16.1%).

The Late Paleolithic assemblages considered in this study provide somewhat more evidence of raw material transport. All three of the samples (HW, Henauhof-West; H7, Helga Abri IIF7; and KU, Kaufertsberg, upper level) produced some non-local materials. Henauhof-West stands out as an assemblage consisting largely of regionally transported materials. This must be understood as related to the location of this small Late Paleolithic lakeshore site in the raw-material poor moraine area of Oberschwaben. The majority of unretouched and retouched pieces in the assemblage are made on Jurassic cherts not available in the immediate vicinity of the site. The nearest sources of Jurassic chert, which make up the majority of materials in the assemblage, are on the Swabian Alb approximately 15 kilometers away. A small number of unretouched flakes and blades and a single non-microlithic tool provide evidence of longer-distance transport at Henauhof-West (HW). These include 2 pieces of Bavarian tabular chert (ca. 200 kilometers to nearest natural occurrence), a single unretouched flake of Randeck tuff (60 kilometers), and one non-microlithic tool made on a Cretaceous quartzite found in the southern Franconian area, about 110 kilometers from Henauhof.

Helga Abri IIF7 (H7) and Kaufertsberg, upper level (KO), on the other hand, are located in areas with relatively abundant local raw materials, and show a pattern of primary use of locally available materials. Smaller amounts of regionally transported materials and longer-distance imports are present, in both cases among both unretouched and retouched pieces.

The assemblage from Kaufertsberg, upper level (KO) contains a similar array of raw materials as that found in the underlying Magdalenian level at this site, but showed somewhat more evidence of transported raw materials. Regionally available materials identified in this assemblage included Keuper chert, radiolarite and quartzite from moraine sources, and lydite, available in Main river gravels. These materials made up a small percentage of unretouched flakes and blades (11, or 4.7%) and non-microlithic tools (5, or 4.6%), but 10 (19.6%) of the 51 backed bladelets and points. Longer-distance transports again include only the Bavarian tabular chert, available about 90 kilometers from Kaufertsberg. A small percentage of unretouched flakes and blades (5.5%) and non-microlithic tools (5.6%) were made on Bavarian chert, but 11.8% of backed pieces.

The small sample analyzed from the Late Paleolithic levels at Helga Abri (H2) contained a relatively high frequency of material transported more than 50 kilometers, again only the high-quality Bavarian chert, identified as the material of 14 (25.9%) of 54 unretouched flakes and blades, but only one retouched piece. This sample also produced a small number of unretouched and retouched pieces made on regionally available materials similar to those represented in the Magdalenian levels at Helga Abri. Though the sample is small, these data suggest use of diverse local and transported raw materials.

The three Early Mesolithic assemblages, on the other hand, showed very little evidence of raw material transport. Only Schräge Wand (SW) produced 11 unretouched flakes and blades (2.1%) made on Bavarian tabular chert, available about 150 kilometers from this location. A single flake of the same material, approximately 190 kilometers from the source area, was identified in the sample of 1743 unretouched flakes and blades analyzed from F2, the Early Mesolithic assemblage from Felsställe. No other materials could be positively identified as deriving from sources more than 10 kilometers distant from these three Early Mesolithic sites.

Data on raw material composition from these 16 archaeological assemblages presents several different kinds of patterns. First, although stone raw material types available within about 10 kilometers of each location were the primary source of material in most cases, the abundant and relatively high-quality Jurassic cherts were transported beyond their source areas into surrounding regions. In the raw material poor environment at Henauhof (HW) these made up the majority of tools, flakes, and blades. In areas with adequate raw material sources, on the other hand, local materials make up the majority of assemblages.

Magdalenian, late Magdalenian, and Late Paleolithic assemblages from two of the sites studied (Kaufertsberg and Helga Abri), however, indicate a good deal of movement of raw material over considerable distances. One specific raw material type, a high quality Bavarian chert, was brought to locations in the Ries basin and the eastern and central Swabian Alb, at distances up to 200 kilometers, and used primarily for the production of relatively regular, unretouched blades and non-microlithic blade tools. The total number of objects made on Bavarian chert in these samples is small, but makes up a fairly large proportion of unretouched flakes and blades in the assemblages from Helga Abri (H3, H2), and moderate proportions of retouched pieces at Helga Abri (H3) and Kaufertsberg (KO).

The high-quality Bavarian chert, though rare in other assemblages, was identified in two out of three Early Mesolithic assemblages, all three Late Paleolithic assemblages, and two of ten Magdalenian assemblages. This indicates that this raw material was regionally important in all three .

Finally, all assemblages from Helga Abri and Kaufertsberg also show a pattern of shorter-distance transport of materials obtainable between 10 and 50 kilometers of the site. For these two sites, this includes radiolarites, quartzite, and lydite from gravel sources such as moraines and streambeds, and secondary local sources including Keuper chert and Randeck tuff. These highly variable sources make up larger proportions of retouched tools than unretouched flakes and blades.

Drawing conclusions about raw material transport from these diverse assemblages is challenging for several reasons. First, as acknowledged above, considerable regional transport of stone materials might go unrecognized because raw material sources cannot be identified with accuracy. Second, the diversity of raw material assemblages within technocomplexes makes it a challenge to identify trends through time. These samples suggest that the maximum distances over which raw materials are transported remain similar throughout the late glacial and early postglacial, while the diversity and abundance of non-local materials appears highest in the late Magdalenian and Late Paleolithic, and lowest in the Early Mesolithic. The following sections make an attempt to understand the diversity of patterns in raw material transport in late glacial and early postglacial southern Germany by placing raw material evidence in the context of inter-site variations in frequency of microlithic and non-microlithic tools.

Raw Material Transport in Context of Tool Production

Most of the Magdalenian and Late Paleolithic assemblages with non-local raw materials show somewhat higher percentages of transported material among retouched than unretouched pieces. This suggests tool curation is one of the mechanisms leading to variation in raw material composition between assemblages. The three Early Mesolithic assemblages included in this study, by contrast, contained only a very small number of chipped stone pieces made on raw materials not locally available, and all were unretouched flakes or blades. This suggests that tool curation played a less important role in Early Mesolithic contexts. This possibility can be explored by examining in greater detail what seems to be a consistent association in the Magdalenian and Late Paleolithic assemblages studied here between a group of relatively robust, non-microlithic tool classes and non-local raw materials.

Relative frequencies of backed implements within tool assemblages (Table 4) were presented above as a measure of the relative importance of production of microlithic and non-microlithic tools. The Magdalenian and late Magdalenian assemblages with evidence for transported raw materials (H3 and KU) have low relative frequencies of backed implements. They also are the only two Magdalenian assemblages among those studied that produced more than one backed point. The overwhelming use of locally available raw materials at the Magdalenian assemblages with the highest frequencies of backed bladelets (P3, P2), on the other hand, may suggest an association between the production of abundant microlithic implements and a focus on local raw materials. This general inter-site pattern may help to place the scarcity of transported stone raw materials in the Early Mesolithic assemblages in a useful light. That is, although non-local stone is rare on these Early Mesolithic sites, it appears to be no more common on Magdalenian sites with high frequencies of small microlithic elements (backed bladelets). In the case of a toolkit consisting largely of very small implements with little or no potential for extending use-life through resharpening, it is perhaps not surprising that very few stone implements are discarded far from their source.

Within-site patterns of association between non-microlithic tool classes and non-local raw materials could be explored only for the small number of Magdalenian and Late Paleolithic assemblages showing moderate quantities of transported stone. Cross-tabulations and the chi-squared statistic were used to test the null hypothesis that non-local raw materials were equally well represented among unretouched flakes and blades, microlithic tools, and non-microlithic implements. Results of this analysis are summarized in Table 7. Among the four assemblages compared, only one (HW) shows equal representation of non-local material among unretouched flakes and blades and microlithic and non-microlithic tools. The other three all show significant associations between retouched tools and non-local raw material. Much of the chi-square values reported derive from much higher than expected values for local material among unretouched flakes and blades and for regional stone types among non-microlithic tools at both KU and H3. KO, on the other hand, shows a strong association between backed points and regional stone raw materials (mostly radiolarite).

Table 7. Test of Association between Artifact Classes and Raw Material Transport

Assemblage	X^2	df	Inference, probability
HW	1.07	4	Accept null hypothesis, $p = .9$
KO	24.11	4	Reject null hypothesis, $p << .001$
H3	41.81	4	Reject null hypothesis, $p << .001$
KU	10.6[1]	4	Reject null hypothesis, $p < .05$

[1]Two of nine expected values fall below 5.

These results generally indicate that retouched tools were more commonly discarded far from their original raw material than unretouched pieces. This provides some support for the notion that curation of retouched tools is an important factor shaping raw material assemblages in late glacial contexts in southern Germany. Do these data provide any support for the hypothesis that this pattern has more to do with larger, robust tools with long potential use-lives than microlithic elements?

A closer look at these three assemblages suggests that patterns may be more complex. First, although somewhat more non-local stone was identified among non-microlithic tools than backed pieces at KU (Kaufertsberg, lower level) and H3 (Helga Abri Magdalenian), the reverse is true in the KO assemblage (Kaufertsberg, upper level) (see Tables 5 and 6). Second, non-microlithic tools made on non-local materials are not consistently larger than those made on local materials. Comparison of the types of tools made on non-local stone reveals no single, consistent pattern among the three assemblages for which in-depth examination is possible.

Figure 5 shows a comparison of the width of retouched and unretouched blades produced on both local and non-local materials at three of the late glacial assemblages (KU, H3, and KU). Blades made on "imported" (from sources > 50 kilometers distant) and regional materials (sources 10-50 kilometers distant) at H3, Helga Abri 3a/3b (late Magdalenian), are more or less consistently broader than those made on local stone. The picture is more

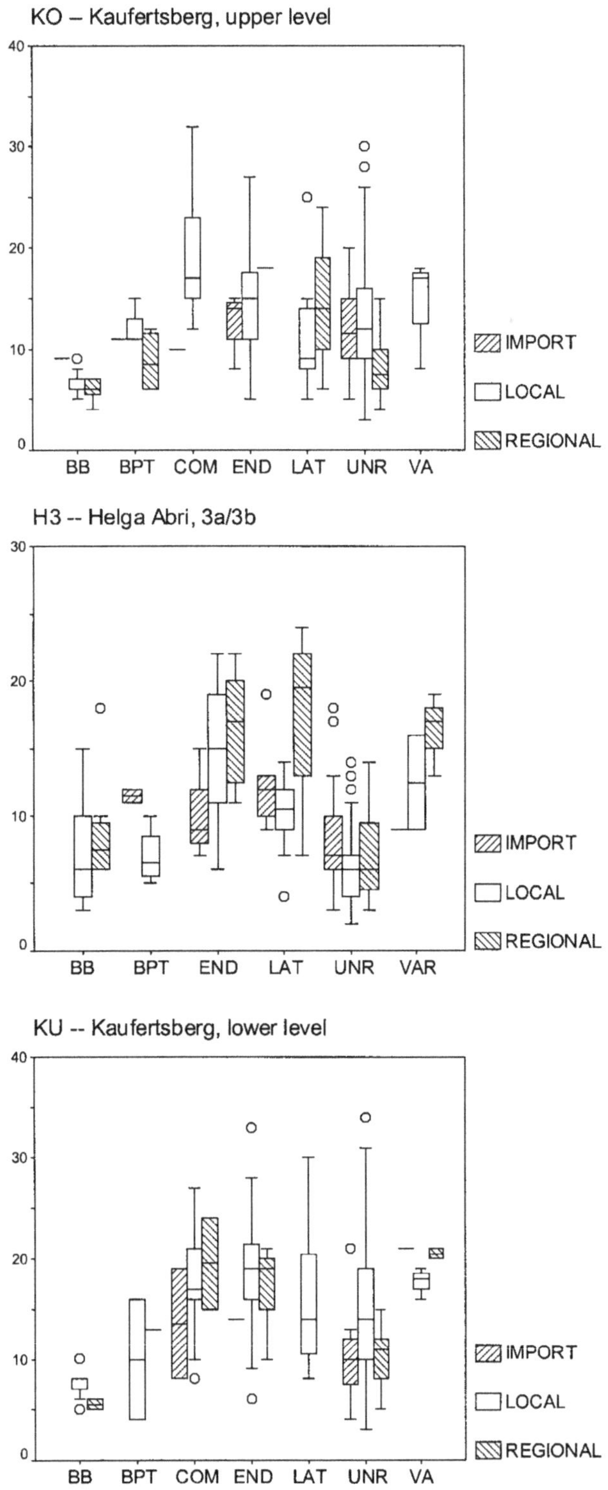

Figure 5. Comparison of width (mm) of retouched and unretouched blades produced on local and non-local stone from three Late Glacial assemblage: KO, H3, KU. Artifact classes depicted are: BB, backed bladelet; BPT, backed point; END, blade tool with terminal working end; COM, combination tool; LAT, laterally retouched blade; UNR, unretouched blade; and VA, other retouched blades.

complex in both Kaufertsberg assemblages, however. Both are characterized by very broad, robust blades made on a distinctive local Cretaceous quartzite. This raw material, which occurs in relatively large blocks in the area to the west of Kaufertsberg, may be particularly suited to the production of broad, strong blades. Stone brought from more distant sources includes Keuper chert, radiolarite, and lydite, all found in small nodules, as well as the more distant Bavarian tabular chert, which was used to produce slender blades.

Non-local raw materials were identified most frequently among a group of tool classes made on broad, thick blades at KU (Kaufertsberg, lower level). In this example, the small number of non-microlithic tools made on non-local materials includes 6 burins, 2 combination tools, and 3 endscrapers, as well as two other retouched blades and a borer. Although the assemblage contains 22 laterally retouched blades made on narrower blades, all of these were made using locally available stone. Burins, combination tools, and endscrapers belong to a group of tools made on relatively broad, thick blades in the KU assemblage (mean widths 18.6, 17.7, and 20.6 mm respectively). At H3 (Helga Abri, 3a/3b), though, this pattern is not repeated. Instead, the most common tools made on non-local stone are laterally retouched blades (13 on Bavarian tabular chert, 5 on regional sources including radiolarite), which are relatively narrow tool forms (mean width 13 mm). At KO, narrow backed bladelets and backed points are made on an array of local and non-local raw materials, as are blade tools with terminal working ends, laterally retouched blades, and unretouched blades.

In summary, although intensive use of local raw material is associated with high frequencies of backed bladelets in these late glacial assemblages, a pattern of consistent use of non-local raw materials for larger implements cannot be shown in all three of the diverse raw material assemblages considered. The complexity of associations between tool classes, raw material transport, and blank size is probably due to a number of factors, including the specific qualities of raw materials used.

Conclusion

The traditionally recognized periods of late glacial and early postglacial southern Germany are defined largely on the basis of armatures and other retouched tools. Some of the other trends in chipped stone technologies documented here, however, suggest considerable continuity between these periods, and an early appearance of a trend towards smaller tools. This provides an interesting perspective for comparing production of microlithic tools in the context of different technologies over the course of the late glacial and early postglacial.

The relationship between trends in raw material use, tool production, and mobility during the late glacial and Early Holocene needs to be understood in relation to at least two separate aspects of a changing relationship between tools and tasks. First, we note an increase in use of composite tools. This trend may be driven by factors such as selection for efficient projectiles under conditions of a late glacial decrease in reliability of returns from hunting large mammals. This process appears to be a gradual one beginning early in the late glacial. Inter-site variability in the frequency of backed bladelets in Magdalenian sites suggests that re-tooling took place in relatively concentrated episodes, during which local raw materials were relied on for bladelet production. Late Paleolithic assemblages vary less in the frequency of backed implements, and also show greater diversity of backed point forms. In these three assemblages, backed bladelets, backed points, and non-backed blade and flake implements were produced using a diverse array of local and non-local raw materials. Early Mesolithic assemblages resemble some Magdalenian assemblages in that they have a high relative frequency of microlithic tools associated with a scarcity of non-local raw materials.

Second, there is a decrease in emphasis on large stone tools with high potential use life. From the patterns examined here, I suggest that the much lower relative frequency of non-microlithic flake and blade tools in Early Mesolithic assemblages is an important factor shaping raw material economies during the Early Holocene. Archaeological assemblages cannot be assumed to reflect a narrow or unvarying set of production goals. Instead, it may be necessary to partition assemblages according to some index of the uses to which tools are put. Chipped stone may play a role in many different kinds of activities. In cases where the functional roles played by chipped stone change over time, broad comparisons of assemblage characteristics such as raw material composition may not yield clear results.

Evaluating hypotheses about culture change during the late glacial and early postglacial will require consideration of multiple long-term trends. Assessing the timing and direction of trends in chipped stone technology is made difficult by coarse chronological control and by changing relationships between tools and tasks. Much work remains to assess these relationships in the southern German late glacial and early postglacial.

Acknowledgements

Dr. Gerd Albrecht, Prof. Joachim Hahn, Dr. Claus-Joachim Kind, Prof. Hans-Peter Uenze, and Prof. Christian Züchner kindly allowed me access to museum collections and provided me with documents and ample assistance. This study was supported by a German Academic Exchange Direct Stipend, by a Sigma Xi Grant-in-Aid of research, and by Rackham School of Graduate Studies, University of Michigan. Dr. Berit V. Eriksen provided constructive comments on an earlier draft of this paper.

References Cited

Albrecht, G.

1979 *Magdalénien-Inventare vom Petersfels. Siedlungsarchäologische Ergebnisse der Ausgrabungen 1974-1976.* Tübinger Monographien zur Urgeschichte 6. Archaeologica Venatoria, Tübingen.

1984 Der spätpleistozäne und altholozäne Fundplatz Malerfels I. In *Die steinzeitliche Besiedlung des Eselburger Tales bei Heidenheim (Schwäbische Alb)*, edited by J. Hahn, pp. 90-122. Forschungen und Berichte zur Vor- und Frühgeschichte in Baden-Württemberg 17. Konrad Theiss Verlag, Stuttgart.

Albrecht, G., D. Drautz, and C.-J. Kind

1977 Eine station des Magdalénien in der Gnirshöhle bei Engen-Bittelbrunn im Hegau. *Archäologisches Korrespondenzblatt* 7:161-179.

Ammann, B., and A. F. Lotter

1989 Late-Glacial radiocarbon- and palynostratigraphy on the Swiss Plateau. *Boreas* 18:109-126.

Ammermann, A. and M. Feldman

1979 On the 'making' of an assemblage of stone tools. *American Antiquity* 39:610-616.

Aquilas-Wauters, R. F.

1956 Une pointe de La Gravette fichée dans un fragment de machoire de *Cervus giganteus. Bulletin de la Société Royale Belge d'Anthropologie et de Préhistoire* 67:31-36.

Auffermann, B.

1990 *Magdalénien und Mesolithikum der Bruckersbergstationen in Giengen.* Unpublished Masters Thesis, Tübingen.

1991 Magdalénien und Mesolithikum der Bruckersbergstationen in Giengen. *Archäologische Informationen* 14(2):291-293.

1996 Zur Frage der Tauschbeziehungen im süddeutschen Magdalénien: Das Beispiel Plattenhornstein. In *Spuren der Jagd—Die Jagd nach Spuren,* edited by I. Campen, J. Hahn, and M. Uerpmann, pp. 273-278. Tübinger Monographien zur Urgeschichte 11. Archaeologica Venatoria, Tübingen.

Bamforth, D. B.
1986 Technological efficiency and tool curation. *American Antiquity* 51:38-50.
1990 Settlement, raw material, and lithic procurement in the Central Mojave Desert. *Journal of Anthropological Archaeology* 9:70-104.

Beurer, M.
1971 *Kieselsäureanreicherungen in den oberjurassischen Sedimenten der Schwäbischen Alb.* Beihefte des geologischen Jahrbuches 109. Hannover.

Beurlen, K., H. Gall and G. Schairer
1978 *Die Alb und ihre Fossilien: Geologie und Paläontologie der Schwaben- und Frankenalb.* Kosmos, Stuttgart.

Binsteiner, A.
1989 Die neolithische Abbau auf Jurahornstein von Baiersdorf in der südlichen Frankenalb. *Archäologisches Korrespondenzblatt* 19:331-337.
1990 Das neolithische Feuersteinbergwerk von Arnhofen, Ldkr. Kelheim (with contributions by J. Riederer and B. Engelhardt). *Bayerische Vorgeschichtsblätter* 55:1-56.
1992 Die Rolle der Knollenhornstein im Neolithikum Bayerns. *Archäologisches Korrespondenzblatt* 22:355-357.

Blank, R.
1994 Plattenhornstein-Artefakte im südwestfälischen Raum. Ein Beitrag zur Distribution süddeutscher Hornsteine im Neolithikum. *Archäologisches Korrespondenzblatt* 24:29-39.

Bleed, P.
1986 The optimal design of hunting weapons: Maintainability or reliability. *American Antiquity* 51(4):737-747.

Bohmers, A. and A. Wouters
1956 Statistics and graphs in the study of flint assemblages. III. A preliminary report on the statistical analysis of the Mesolithic in Northwestern. *Palaeohistoria* 5:27-38.

Broecker, W.
1994 Massive iceberg discharges as triggers for global climate change. *Nature* 372:421-424.

Burkert, W.
1996 Die Herkunft des Jurahornstein-Rohmaterials im Gravettien der Geißenklösterle-Höhle bei Blaubeuren unter Berücksichtigung der Rohstoffvorkommen des nördlichen Obershwaben. In *Spuren der Jagd—Die Jagd nach Spuren,* edited by I. Campen, J. Hahn, and M. Uerpmann, pp. 279-284. Tübinger Monographien zur Urgeschichte 11. Archaeologica Venatoria, Tübingen.

Churchill, S. E.
1993 Weapon technology, prey size selection, and hunting methods in modern hunter-gatherers: Implications for hunting in the Palaeolithic and Mesolithic. In *Hunting and Animal Exploitation in the Later Palaeolithic and Mesolithic of Eurasia,* edited by G. L. Peterkin, H. M. Bricker, and P. Mellars, pp. 1-10. Archeological papers of the American Anthropological Association 4.

Clarke, D. L.
1976 Mesolithic Europe: The economic basis. In *Problems in Economic and Social Archaeology,* edited by G. de G. Sieveking, I. H. Longworth and K. Wilson, pp. 449-481. Duckworth, London.

Deecke, W.
1933 *Die mitteleuropäischen Silices nach Vorkommen, Eigenschaften und Verwendung in der Prähistorie.* Gustav Fischer Verlag, Jena.

Engelhardt, F. and A. Binsteiner
1988 Vorbericht über die Ausgrabungen 1984-1986 im neolithischen Feuersteinabbaurevier von Arnhofen, Ldkr. Kelheim. *Germania* 66:1-29.

Eriksen, B. V.
1991 *Change and Continuity in a Prehistoric Hunter-Gatherer Society: A Study of Cultural Adaptation in Late Glacial-Early Postglacial Southwestern Germany.* Archaeologica Venatoria 12. Archaeologica Venatoria, Tübingen.

Finlayson, B.
1990 The function of microliths: Evidence from Smittons and Starr, SW Scotland. *Mesolithic Miscellany* 11(1):2-6.

Fischer, A.
1989 Hunting with flint-tipped arrows: Results and experiences from practical experiments. In *The Mesolithic in Europe* edited by C. Bonsall, pp. 29-39. John Donald Publishers, Edinburgh.

Freund, G.
1974/5 Ein jungpaläolithischer Depotfund aus der Sesselfelsgrotte im unteren Altmühltal. *Jahrbuch für fränkischen Landesforschung* 34/35:17-36.

Frenzel, B.
1983 Die Vegetationsgeschichte Süddeutschlands im Eiszeitalter. In *Urgeschichte in Baden-Württemberg,* edited by H.-J. Muller-Beck, pp. 91-165. Konrad Theiss Verlag, Stuttgart.

Fritsch, B. and D. Neubauer
1987 Die Silexvorkommen in Beziehungen zu den früh- und mittelneolithischen

Fundstellen des Hegaus. *Archäologische Informationen* 10:161-166.
Gall, H., R. Hüttner, and D. Müller
1977 *Erläuterungen zur Geologischen Karte des Rieses 1:50,000.* Geologica Bavarica 76. Geologisches Landesamt Bayern.
G.E.E.M.
1972 Epipaléolithique-Mesolithique: Les armatures non géometriques. *Bulletin de la Société Préhistorique Française* 69:364-375.
Geyer, O. F. and M. P. Gwinner
1991 *Geologie von Baden-Württemberg, 4. Auflage.* Schweizerbart'sche Verlagsbuchhandlung, Stuttgart.
Gould, R. A., D. Koster, and H. Sontz
1971 The lithic assemblage of the Western Desert Aborigines. *American Antiquity* 36:149-169.
Guiot, J.
1987 Late Quaternary climate change in France estimated from multivariate pollen time series. *Quaternary Research* 28:100-118.
Gumpert, C.
1927 *Fränkisches Mesolithikum: Die Steinzeitlichen Besiedlung der Fränkischen Rezat und oberen Altmühl im Tardenoisien.* Mannus-Bibliothek Nr. 40. Verlag von Curt Kabizsch, Leipzig.
Gwinner, M. P.
1976 *Origin of the Upper Jurassic Limestones of the Swabian Alb (Southwest Germany).* Contributions in Sedimentology 5. Stuttgart.
Hahn, J.
1979 Essai sur l'écologie du Magdalénien dans le Jura souabe. In *La fin des temps glaciaires en Europe*, edited by D. de Sonneville-Bordes, pp. 203-211. Colloques internationaux du C.N.R.S. 271. Centre National de la Recherche Scientifique., Paris.
1981 Abfolge und Umwelt der jüngeren Altsteinzeit in Südwestdeutschland. *Fundberichte aus Baden-Württemberg* 6:1-27.
1983 Die frühe Mittelsteinzeit. In *Urgeschichte in Baden-Württemberg*, edited by H. Müller-Beck, pp. 363-392. Konrad Theiss Verlag, Stuttgart.
1984 Die spätpleistozäne Jagdstation Spitzbubenhöhle: Stratigraphie und archäologische Funde. In *Der steinzeitliche Besiedlung des Eselsburger Tales bei Heidenheim (Schwäbische Alb)*, edited by J. Hahn, pp. 21-54. Forschungen und Berichte zur Vor- und Frühgeschichte in Baden-Württemberg 17. Landesdenkmalamt Baden-Württemberg, Stuttgart.
1991 *Erkennen und Bestimmen von Stein- und Knochenartefakten: Einführung in die Artefaktmorphologie.* Verlag Archaeologica Venatoria, Tübingen.
Hahn, J., editor
1984 *Der steinzeitliche Besiedlung des Eselsburger Tales bei Heidenheim (Schwäbische Alb).* Forschungen und Berichte zur Vor- und Frühgeschichte in Baden-Württemberg 17. Landesdenkmalamt Baden-Württemberg, Stuttgart.
Hahn, J. and A. Scheer
1983 Das Helga-Abri am Hohlenfelsen bei Schelklingen: Eine mesolithische und jungpaläolithische Schichtenfolge. *Archäologische Korrespondenzblatt* 13:19-28.
1984 Eine magdalénienzeitliche Siedlungsstruktur im Helga-Abri am Hohlefels, Gemeinde Schelklingen, Alb-Donau-Kreis. *Archäologische Ausgrabungen in Baden-Württemberg 1983*, pp. 35-37.
Hahn, J., C.-J. Kind and K. Steppan
1993 Mesolithische Rentier-Jäger in Südwestdeutschland? Der mittelsteinzeitliche Freilandfundplatz Rottenburg "Siebenlinden I" (Vorbericht). *Fundberichte aus Baden-Württemberg* 18:29-53.
Herrmann, F.
1975 Baiersdorf, ein steinzeitliches Bergbauzentrum—ein Hornsteinhandelsplatz oder eine normale Siedlung? *Acta Albertensa Ratisbonensia* 42:153-189.
1983 Bergbau und Fernhandel im Kelheimer, Abensberger Gebiet. *Archäologische Informationen* 6:22-24.
Housley, R. A., C. S. Gamble, M. Street and P. Pettitt
1997 Radiocarbon evidence for the Lateglacial human recolonisation of northern Europe. *Proceedings of the Prehistoric Society* 63:25-54.
Jaguttis-Emden, M.
1983 Die Radiokarbondatierungen der Ausgrabung Petersfels. In *Naturwissenschaftliche Untersuchungen an Magdalénien-Inventaren vom Petersfels, Grabungen 1974-1976*, edited by G. Albrecht, H. Berke, and G. Poplin, pp. 47-58. Archaeologica Venatoria, Tübingen.
1987 Die Radiokarbondatierung des archäologischen Horizontes IIIb

innerhalb der geologischen Schicht 3b der Magdalénienstation Felsställe bei Ehingen, Ehingen-Mühlen, Alb-Donau-Kreis. In *Das Felsställe. Eine jungpaläolithsch-frühmesolithische Abri-Station bei Ehingen-Mühlen, Alb-Donau-Kreis*, edited by C.-J. Kind, pp. 375-382. Forschungen und Berichte zur Vor- und Frühgeschichte in Baden-Württemberg 23. Konrad Theiss Verlag, Stuttgart.

Jochim, M.
1980 Postglacial adaptations. *Proceedings of the Conference on Northeastern Archaeology* 19:49-70.
1990 *Henauhof-Nordwest—Ein mittelsteinzeitlicher Lagerplatz am Federsee.* Materialhefte zur Vor- und Frühgeschichte 19. Konrad Theiss Verlag, Stuttgart.
1991 Archaeology as long-term ethnography. *American Anthropologist* 93:308-321.
1995 Two Palaeolithic sites on the Federsee, South Germany. *Journal of Field Archaeology* 22:263-273.

Kaiser, K. F. and U. Eicher
1987 Fossil pollen, molluscs, and stable isotopes in the Dättnau valley, Switzerland. *Boreas* 16:293-303.

Kaulich, B.
1983 Das Paläolithikum des Kaufertsberges bei Lierheim (Gem. Appetshofen, Ldkr. Donau-Ries). *Quartär* 33/34:29-97.

Keeley, L.
1982 Hafting and retooling: effects on the archaeological record. *American Antiquity* 47:798-809.

Kelly, R.
1988 The three sides of a biface. *American Antiquity* 53(4):717-734.

Kind, C.-J.
1987 *Das Felsställe. Eine jungpaläolithsch-frühmesolithische Abri-Station bei Ehingen-Mühlen, Alb-Donau-Kreis.* Forschungen und Berichte zur Vor- und Frühgeschichte in Baden-Württemberg 23. Konrad Theiss Verlag, Stuttgart.

Krumbeck, L.
1927 Zur Kenntnis der alten Schotter des nordbayrischen Deckgebirges: Ein Beitrag älteren Flussgeschichte Nordbayerns. *Geologische und Paläontologische Abhandlungen* N.F. 15(3):183-315.

Kuhn, S.
1994 A formal approach to the design and assembly of mobile toolkits. *American Antiquity* 59(3):426-442.
1995 *Mousterian lithic technology : an ecological perspective.* Princeton University Press, Princeton.

Lang, G.
1962 Die spät- und frühpostglaziale Vegetationsentwicklung im Umkreis der Alpen. *Eiszeitalter und Gegenwart* 12:9-17.

Lemcke, K.
1988 *Geologie von Bayern, I: Das bayerische Alpenvorland vor der Eiszeit.* E. Schweizerbart'sche Verlagsbuchhandlung, Stuttgart.

Lipinski, E. and H. Löhr
1987 Vorkommen von Buntsandsteinkarneol im Oberbettinger Triasgebiet nördlich Gerolstein/Eifel. *Archäologische Informationen* 10:137-142.

Lister, A. M.
1988 A 15,000-year isotopic record from Lake Zürich of deglaciation and climatic change in Switzerland. *Quaternary Research* 29:129-141.

Löhr, H.
1987 Achate des Rotliegenden als Rohmaterial geschlagener Steinartefakte. *Archäologische Informationen* 10:142-150.

Löhr, H. and W. Schönweiß
1987 Keuperhornstein und seine natürlichen Vorkommen. *Archäologische Informationen* 10:126-137.

Mall, W.
1968 Die Geologie der Blätter Dettingen am Albuch und Giengen an der Brenz (1:25,000) (Schwäbische Alb). *Arbeit der Geologische-Paläontologishen Institut der Universität Stuttgart,* NF 54.

Malmer, M. P.
1969 Die Mikrolithen in dem Pfeil-Fund von Loshult. *Meddelanden fra Lunds Universitets Historisks Museum* 1966-1968:249-55.

Morrow, T.
1995 Bigger is better: Comments on Kuhn's formal approach to mobile tool kits. *American Antiquity* 61(3):581-590.

Moser, M.
1978 Die vorgeschichtliche Bergbau auf Plattensilex in den Kalkschiefern der Altmühl-Alb und seine Bedeutung im Neolithikum Mitteleuropas. *Archäologische Informationen* 4:45-81.
1981 Katalog der Feuerstein-/Hornstein-Bergwerke, D4 Baiersdorf, Ldkr. Kelheim; D5 Abensberg NO, Ldkr. Kelheim. In *5000 Jahre Feuersteinbergbau*, edited by G. Weißgerber, pp. 446-447. Veröffentlichungen des Detuschen Bergbau Museums Bochum 22, Bochum.

Moss, E. H.
1983 *The Functional Analysis of Flint Implements — Pincevent and Pont d'Ambon: Two Case Studies from the French Final Paleolithic.* British Archaeological Reports, I.S. 177.

Moss, E. H. and M. H. Newcomer
1982 Reconstruction of tool use at Pincevent: Microwear and experiments. In *Tailler! Pour quoi faire: Préhistoire et technologie lithique II/ Recent Progress in Microwear Studies*, edited by D. Cahen, pp. 77-88. Studia Praehistorica Belgica 2.

Müller, S.
1958 Feuersteinlehme und Streuschuttdecken in Ostwürttemberg. *Jahresheft des geologischen Landesamts Baden-Württemberg* 3:241-262.

Myers, A.
1989 Reliable and maintainable technological strategies in the Mesolithic of mainland Britain. In *Time, Energy, and Stone Tools*, edited by R. Torrence, pp. 78-91. Cambridge University Press, Cambridge.

Naber, F. B.
1968 Die "Schrage Wand" im Barental, eine altholozane Abrifundstelle im nordlichen Frankenjura. *Quartär* 19:289-314.
1970 Untersuchungen an Industrien postglazialer Jägerkulturen. *Bayerischer Vorgeschichtsblätter* 35:1-68.
1981 Zur steinzeitlichen Fundstelle Baiersdorf/ Ldkr. Kelheim und zum Plattenkerndepot der Sesselfelsgrotte. *Quartär* 31/32:7-39.

Nelson, M.
1991 The Study of Technological Organization. *Archaeological Method and Theory* 3: 57-100.

Owen, L.
1988 *Blade and Microblade Technology: Selected Assemblages from the North American Arctic and the Upper Paleolithic of Southwest Germany.* British Archaeological Reports, I. S. 441.

Perlès, C.
1992 In search of lithic strategies: A cognitive approach to prehistoric chipped stone assemblages. In *Representations in Archaeology*, edited by J.-C. Gardin and C. S. Peebles, pp. 223-247. Indiana University Press, Bloomington and Indianapolis.

Riek, G.
1957 *Drei jungpaläolithische Stationen am Bruckersberg in Giengen an der Brenz.* Veröffentlichungen der Staatlichen Amtes für Denkmalpflege, Stuttgart, Reihe A. Vor- und Frühgeschichte, Heft 2. Verlag Silberburg, Stuttgart.
1959 Das federmesserführende Magdalénien der Burkhardtshöhle bei Westerheim im Kreis Münsingen (Schwäbische Alb). *Fundberichte aus Schwaben N. F.* 15:9-29.

Rozoy, J.-G.
1989 The revolution of the bowmen in Europe. In *The Mesolithic in Europe*, edited by C. Bonsall, pp. 13-28. John Donald Publishers, Edinburgh.

Rutte, E.
1965 Karst- und Überdeckungsbildungen im Gebiet von Kelheim-Donau. *Quartär* 14:69-81.

Shott, M.
1986 *Settlement mobility and technological organization among Great Lakes Paleo-Indian Foragers.* Ph.D. Dissertation, University of Michigan. University Microfilms, Ann Arbor.
1989 On tool-class use lives and the formation of archaeological assemblages. *American Antiquity* 54(1):9-30.

Straus, L.
1991 Human geography of the Late Upper Paleolithic in Western Europe. *Journal of Anthropological Research* 47:259-278.

Ströbel, R.
1959 Tardenoisspitze in einem Bovidenknochen von Schwanningen am Neckar (Kr. Rottweil). *Fundberichte aus Schwaben* N. F. 15:103-106.

Symens, N.
1986 A functional analysis of selected stone artifacts from the Magdalenian site at Verberie, France. *Journal of Field Archaeology* 13:213-222.

Taute, W.
1971 *Untersuchungen zum Mesolithikum und Spätpaläolithikum im südlichen Mitteleuropa.* Band I: Chronologie Süddeutschlands. (Habilitationsschrift, Unpublished manuscript, Tübingen).
1972 Die spätpaläolithisch-frühmesolithische Schichtenfolge im Zigeunerfels bei Sigmaringen (Vorbericht). *Archäologische Informationen* 2/3:59-66.

Torrence, R.
1983 Time budgeting and hunter-gatherer technology. In *Hunter-Gatherer Economy in Prehistory: A European Perspective*, edited by G. Bailey, pp. 11-22. Cambridge University Press, Cambridge.
1989 Retooling: towards a behavioral theory of stone tools. In *Time, Energy, and Stone Tools,* edited by R. Torrence, pp. 57-66. Cambridge University Press, Cambridge.

Torrence, R., editor
1989 *Time, Energy, and Stone Tools*. Cambridge University Press, Cambridge.

Vaughan, P.
1985 *Use-Wear Analysis of Flaked Stone Tools*. University of Arizona Press, Tucson.
1987 Positive and negative evidence for hafting on flint tools from various periods (Magdalenian through Bronze Age). In *La main et l'outil: Manches et emmanchements préhistoriques*, pp. 135-144. Table Ronde CNRS, Lyon, 26-29 November 1984. Maison de l'Orient Méditerranéen, Lyon.

Vierra, B.
1995 *Subsistence and Stone Tool Technology: An Old World Perspective*. Arizona State University Anthropological Research Papers No. 47. Arizona State University, Tempe.

Weinig, J. G.
1989 Das Rohmaterialspectrum der Region Oberbayern-Nord, Stand 1989. In *Steinzeitliche Kulturen an Donau und Altmühl*, edited by K. H. Rieder, A. Tillmann, and J. Weinig, pp. 217-228. Courier Druckhaus, Ingoldstadt.

Weniger, G-Ch.
1982 *Wildbeuter und ihre Umwelt: Ein Beitrag zum Magdalenien Südwestdeutschlands aus ökologischer und ethnoarchäologischer Sicht*. Archaeologica Venatoria, Tübingen.

Zotz, L. F. and G. Freund
1973 *Die mittelpaläolithische Geröllgeräteindustrie aus der Umgebung von Kronach in Oberfranken*. Materialhefte zur Bayerischen Vorgeschichte 27. Kallmünz.

Zvelebil, M.
1985 Economic intensification and postglacial hunter-gatherers in north temperate Europe. In *The Mesolithic of Europe*, edited by C. Bonsall, pp. 80-88. John Donald Publishers, Edinburgh.

6. Climate and Raw Material Behavior: A Case Study From Late Pleistocene Hunter-Gatherers in the Middle Rhine Area of Germany

Harald Floss

Moving the geographical focus 300 km further up the Rhine valley, this chapter summarizes the results of a major research project. It expands on the empirical work presented in a detailed monograph (Floss 1994) to consider questions about hunter-gatherer mobility in the late glacial in the middle Rhine region. The author conducted extensive surveys to locate outcrops of stone raw materials, and used the resulting petrographic characterizations of stone to investigate systems of lithic raw material acquisition at all of the 40 Paleolithic sites in the middle Rhine area. The investigations document a continual increase in use of transported raw materials over the course of the Upper Paleolithic. The case study presented here is devoted to a comparison of raw material procurement in Magdalenian and final Paleolithic industries. In the Magdalenian inventories, nearly 50% of the lithic raw materials were transported over a distance of 100 km or more, while Late Paleolithic industries consist largely of diverse materials obtainable between 20 and 100 kilometers of the sites. It is argued that this evidence should be interpreted in terms of different scheduling of lithic raw material acquisition in contrasting mobility strategies. Logistically organized Magdalenian hunter-gatherers transported prepared cores and tools over distances greater than 100 kilometers in distinct seasonal migrations, with little collection of raw material along the way. A Late Paleolithic increase in materials collected at intermediate distances is argued to reflect more frequent residential moves, such that raw material was acquired in a less systematic way at a variety of locations in the course of moving camp over intermediate distances. The change in raw material provisioning is seen as related to change in settlement mobility, which is directly tied to late glacial environmental change. – *Editors.*

The area of investigation is situated in the Neuwied Basin, at the center of the Middle Rhine area, about 80 kilometers southeast of Cologne and 100 kilometers northwest of Frankfurt am Main. Excavations in the course of the past 25 years have made known about 30 new sites dating from the Lower to the Final Paleolithic. During the same period, about an equal number of sites known from earlier investigations since the second half of the 19th century have been re-investigated. In addition, some surface inventories can be added to the known Paleolithic sites in the region, making a total of about 70 locations (Figure 1).

Among these Paleolithic sites are included the well-known Magdalenian base camps of Gönnersdorf (Bosinski 1979) and Andernach (Bosinski and Hahn 1972; Veil 1982; Floss and Terberger 1990, in press) as well as some Final Paleolithic ("Azilian" or "Federmesser") sites (Bolus 1992; Stapert and Street 1997; Street and Baales 1997). These Magdalenian and Final Paleolithic sites will be the subject of this article.

Excellent conditions of preservation for Pleistocene sites in the Neuwied Basin are caused by some elements of the local Quaternary geology, such as loess deposits and numerous deposits of volcanic ash from East Eifel volcanism, including the widespread ash layer from the Laacher See eruption at about 11,000 BP (Schmincke 1988).

As part of an extensive program of research instigated by the Institute of Prehistory of the University of Cologne and the division of Paleolithic research of the Römisch-Germanisches Zentralmuseum, Neuwied, the author conducted an analysis of systems of lithic raw material acquisition at all Paleolithic sites in the Middle Rhine area (Floss 1994). As a first step in this project, the distribution of natural lithic resources was investigated, and the materials petrographically characterized. This topic will be summarized here very briefly. The major local materials used in Paleolithic times were Tertiary quartzites (350 outcrops recognized in the present study), chalcedony (50 outcrops) and different types of Rhine river pebbles (quartz, quartzites, lydite, chert, etc.). The most important non-local raw materials transported by Paleolithic hunter-gatherers were different types of Upper Cretaceous flint, mostly originating from Senonian chalk deposits in the Maas river region of Belgium and the Netherlands. These resources are located at a distance of 100 kilometers or more from the Middle Rhine area (Figure 2).

The second phase of this project involved analysis of the kind and origin of the raw materials appearing as lithic artifacts in the Paleolithic sites. In contrast to the Lower Paleolithic represented by sites such as Kärlich and Miesenheim I, which include only local materials, we notice the first evidence of the transport of single flakes and tools made of exogenous raw materials (distance, 100 kilometers) in the inventories of the early Middle Paleolithic (Schweinskopf, Wannen). The first evidence for transport of a larger quantity of non-local flints is found at the site of Remagen-Schwalbenberg, ex-

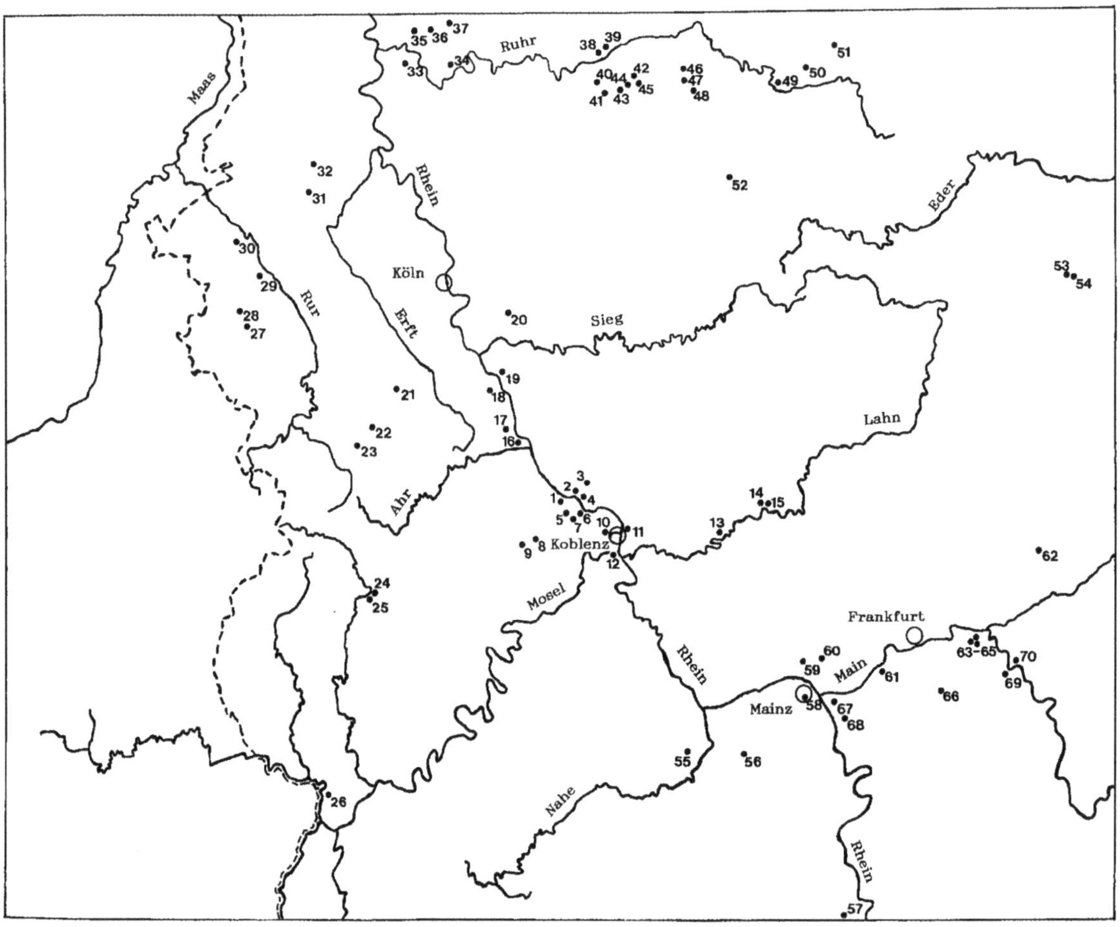

Figure1. Location of Upper and Final Paleolithic sites in the western part of Germany. Dashed lines are national borders. Scale 1 cm = 10 km. T Transition Middle-Upper Paleolithic, A Aurignacian, G Gravettian, M Magdalenian, F Federmesser (Azilian), S Stielspitzen (Ahrensburgian), U Upper Paleolithic not further identified. Capital letters indicate open air sites, small letters cave sites. 1 Andernach-Martinsberg M, F, 2 Gönnersdorf M, 3 Niederbieber F, 4 Irlich M?, 5 Miesenheim II F, 6 Weißenthurm F, 7 Kettig F, 8 Thür F, 9 Mayen M?, 10 Koblenz-Metternich G, 11 Urbar F, 12 Rhens G, 13 Wildweiberlei m, 14 Wildhaus a, 15 Wildscheuer a, g, m, 16 Remagen-Schwalbenberg T, 17 Unkelbach G?, 18 Bonn-Muffendorf A?, G, F?, 19 Bonn-Oberkassel M, 20 Ziegenberg S, 21 Lommersum A, 22 Abri Katzensteine f, 23 Kartstein a, m?, s, 24 Buchenloch a, 25 Magdalenahöhle g, 26 Wintersdorf A, 27 Kinzweiler F, 28 Alsdorf M, 29 Barmen F, 30 Beeck M, 31 Kamphausen M, 32 Korschenbroich F, 33 Duisburg-Kaiserberg S, 34 Essen-Bredeney F?, 35 Oberhausen M?, 36 Essen-Dellwig F?, 37 Bottrop F, 38 Wandhofen F, 39 Schwerte-Ost F, S, 40 Hagen-Herbeck F, S, 41 Holthauser Nordhöhle f, 42 Reingsen S, 43 Burgberg bei Letmathe F, S, 44 Martinshöhle f, s, 45 Grürmannshöhle f, 46 Feldhofhöhle m, f, 47 Honerthöhle u, 48 Balver Höhle a, m?, f, 49 Wennemen (Halloh) S, 50 Bilsteinhöhle f, s, 51 Hohler Stein f, s, 52 Sporker Mulde U, 53 Reutersruh A, 54 Schwalmstadt-Trutzhain F, 55 Heddesheim-Lindengrund U, 56 Sprendlingen G, 57 Fußgönheim M, 58 Mainz-Linsenberg G, 59 Wiesbaden-Adlerquelle G, 60 Wiesbaden-Igstadt A, 61 Kelsterbach A, 62 Büdingen M?, 63 Mühlheim-Dietesheim F, 64 Reuterrain F, 65 Hausen F, 66 Dreieich-Götzenhain M, 67 Rüsselsheim F, 68 Gross-Gerau F, 69 Klein Welzheim F, 70 Karlstein-Dettingen S. (Sources can be found in the following summaries: Bosinski 1982, 1992, Fiedler 1983, Günther 1988, Veil 1978, and in recent publications by members of the Institut für Ur- und Frühgeschichte/University of Cologne, and the Forschungsbereich Altsteinzeit/Römisch-Germanischen Zentralmuseum/Neuwied.)

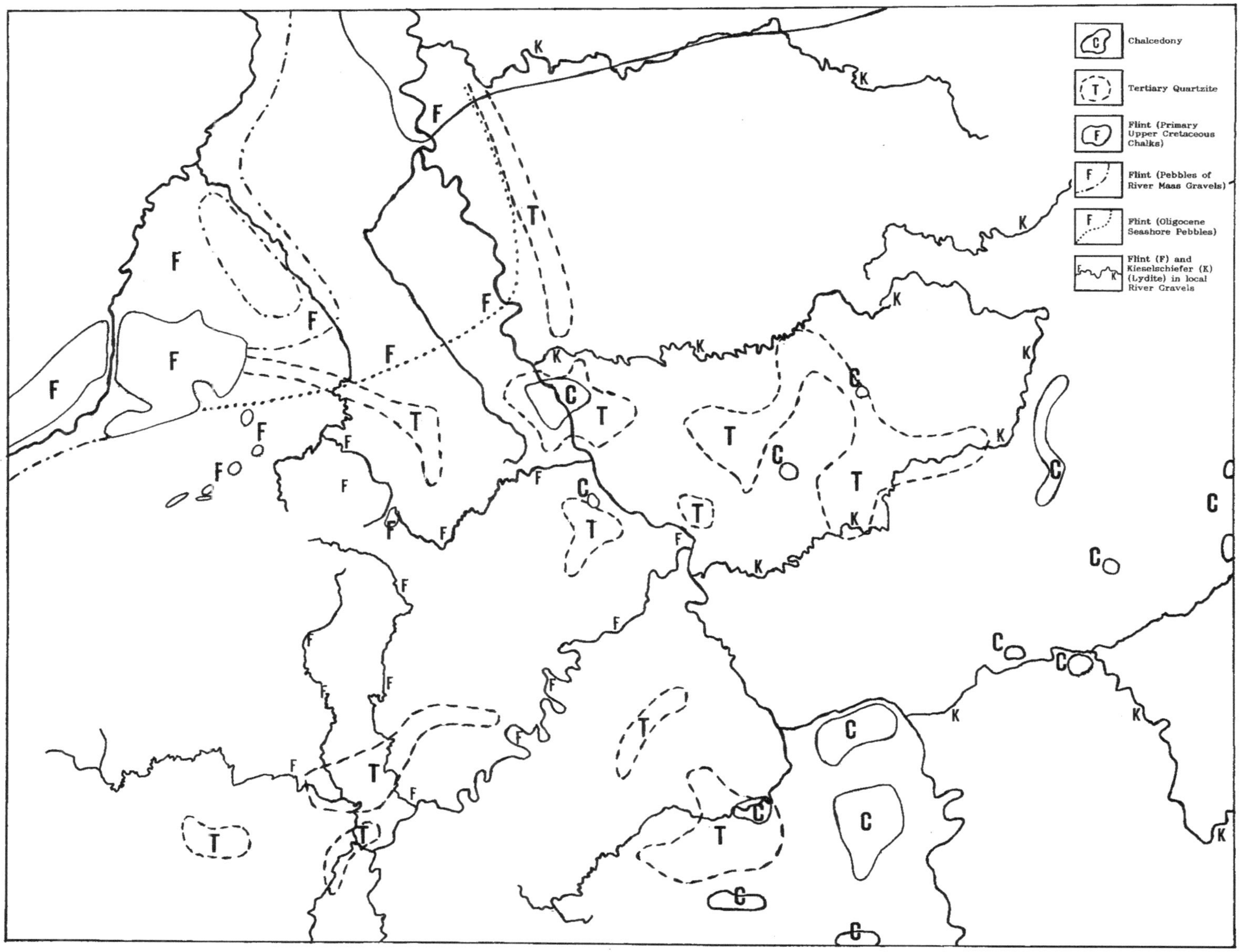

Figure 2. Schematic map showing location of the most important stone raw material resources for the Middle Rhine area of Germany (source Floss 1994). For scale and geographic information see Figure 1.

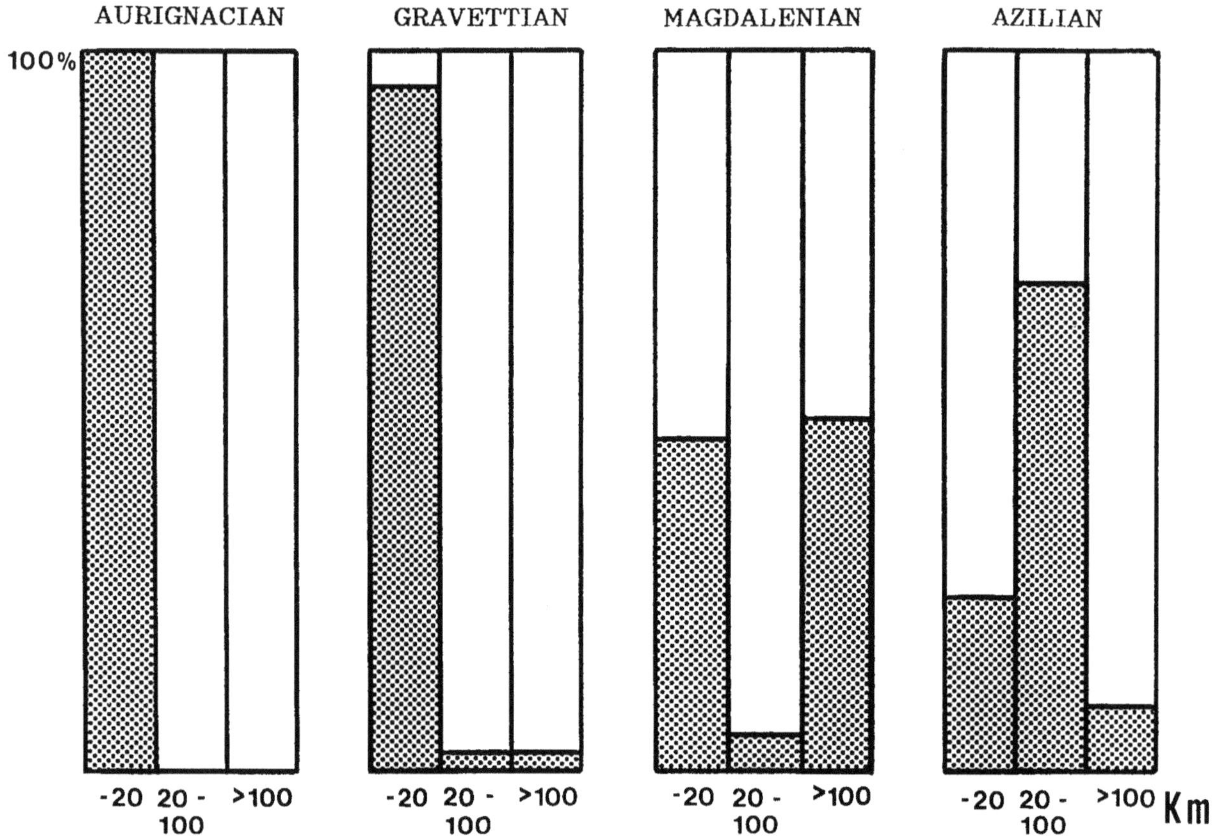

Figure 3. Provenience of lithic raw materials used in Upper and Final Paleolithic sites of the Middle Rhine area (Aurignacian, Gravettian, Magdalenian, Azilian) by distance from source locations to the sites (0-20 kilometers, 20-100 kilometers, >100 kilometers) (source Floss 1994).

cavated by the Institute of Prehistory of the University of Tübingen (App et al 1995). This site is dated to the transition between Middle and Upper Paleolithic times. Over the course of the Upper Paleolithic we can point out an increase in the use of exogenous materials, from the Aurignacian period, characterized by mainly local materials, to the Gravettian, with some increase in use of non-local materials, to the subsequent Magdalenian period, during which nearly 50% of lithic raw materials were transported to the sites over a distance of 100 kilometers or more (Figure 3).

In this paper I will try to explain the differences in raw material procurement between the Magdalenian and the subsequent Final Paleolithic ("Azilian"/"Federmesser"). At Final Paleolithic sites in the region, nearly 70% of lithic raw materials can be found at distances of 20-100 kilometers from the sites (Figure 3), and much less material is transported over longer distances (>100 kilometers). In my view these differences are caused primarily by hunter-gatherer responses to climate and environmental change at the end of the last (Würm/Weichsel) glacial period.

Any consideration of the European late glacial, a period of about 5,000 years from 15,000 to 10,000 BP, must take into account a great deal of recent progress in our knowledge concerning the duration and character of the different climate phases of the late glacial. Previously, the Bølling interstadial that follows the Dryas I cold phase was regarded as a smooth climate oscillation in which tundra vegetation and a cold-climate fauna remained predominant throughout. In the Middle Rhine area, the Magdalenian sites of Gönnersdorf and Andernach previously were dated to the Bølling interstadial. Traditional views then claimed a short-lived deterioration of climate conditions, the Dryas II phase, before the arrival of temperate climate and the spread of forests beginning during the Allerød interstadial before or about 11,800 BP. In the Middle Rhine area, the Federmesser or backed point industries ("Azilian" in the broadest sense) of Niederbieber, Andernach (upper layer), Urbar and other sites have been dated to the Allerød interstadial. After the Allerød, a very marked deterioration of climate referred to as Dryas III took place, with a return of cold conditions, and the appearance of the Ahrensburgian archaeological complex, representing classical reindeer hunters. The Holocene begins at about 10,000 BP.

Analysis of late glacial tree rings (dendrochronology), mollusk shells, pollen, lake marl, and stable isotopes in North America and Central Europe has recently shown that the development of climate in the late glacial was different from this traditional picture (Kaiser 1993). Kaiser argues that the first significant improvement of climate took place during the Bølling interstadial, which was in fact more temperate but less humid than the Allerød interstadial. These two phases are now regarded as a unit (the "big late glacial interstadial") because the supposed intermediate Dryas II cold phase is not very marked, and if it existed at all it was probably only one of several slightly colder climate oscillations during the late glacial. A second very significant recent advance is the extension of 14C calibration from the beginning of the Holocene back to the late glacial interstadial (Street *et al.* 1994). These new results have significant implications for understanding late glacial archaeology. In the Middle Rhine area, for instance, they suggest that the well-known Magdalenian sites of Gönnersdorf and Andernach are to be dated at the end of the Dryas I and not, as previously believed, to the Bølling interstadial. At what time the transition from final Magdalenian to "Azilian" took place, and how the Bølling interstadial, now seemingly not inhabited, is to be interpreted, remains unclear and will be the subject of forthcoming analyses. Some 14C-dates from Middle Rhine "Azilian"/Federmesser sites indicate a non-uniform and partly pre-Allerød age (Street *et al.* 1994), which could mean that the Magdalenian-"Azilian" transition took place in the first part of the "big late glacial interstadial". Generally, as a result of new information about the sequence of late glacial climate change, the whole succession of late glacial "technocomplexes" must be reinterpreted, particularly if we consider, as this article does, that there exists a strong relationship between climate and hunter-gatherer behavior.

In order to compare systems of raw material acquisitions of the Middle Rhine Magdalenian and Federmesser sites, it is necessary to call to mind the general differences between these two complexes (Table 1). Whereas the Magdalenian hunter-gatherers lived in a landscape dominated by steppe conditions, in which the major faunal elements were horse, reindeer, bison, polar fox, saiga, and mammoth, the "Azilian" or "Federmesser" hunter-gatherers found themselves in a completely new environment in which forests of pine, birch and willow increased along with a new fauna represented by elk, deer, beaver, roe, horse, ibex and wild boar. While Magdalenian sites include evidence for large houses (ca. 6-8 meters in diameter) with slate-paved floors, cooking pits and postholes, in the "Azilian" we find mostly small concentrations (ca. 3 meters diameter) with a central fire-place. It is not evident whether these structures represent small tents or, more probably, open air activities. In the Magdalenian bone industry, decoration and art are very frequent, whereas in the "Azilian" decorated objects are nearly absent. Hunting technology also probably changed during these two phases from spear-throwing to archery (see Table 1).

Considerable differences also exist in the acquisition of raw materials, as illustrated by a comparison of the distribution of raw material types from the Magdalenian and "Azilian" layers at Andernach (Figure 4) (see Floss 2000). Both layers (Magdalenian, excavation Veil, 23,166 lithic artifacts, "Azilian", 2,793 artifacts) include both local and exotic raw materials. A closer analysis, though, makes clear that the Magdalenian hunters tended to equip themselves with substantial amounts of high-quality materials, some of which were non-local, such as Cretaceous flints of primary origin. The "Azilian" hunter-gatherers, on the other hand, equipped themselves with single, unsystematically collected pieces often of lesser quality, no matter where and at what distance from the site they were obtained (Figure 4). This results in a uniformity of raw material associations in the Magdalenian, in contrast to a high degree of diversity of raw materials in the "Azilian". In the large Magdalenian inventory from Andernach (ca. 23,000 artifacts), only 5 different raw material types were found, while 12 different raw material types were represented in the relatively small Azilian inventory at the same site (ca. 2,800 artifacts). Magdalenian hunters introduced their raw materials of local origin (0-20 kilometers) in the form of prepared cores and the non-local materials in the form of prepared cores or blades, bladelets and tools. The Azilian pattern is quite different. There the raw materials were imported in nodules or pebbles not subjected to any prior preparation whether they could be obtained close to the site or at distances over 100 kilometers (Figure 4).

In the Magdalenian, there is a high-quality blade industry, while the Azilian lithic technology is quite different, with well-prepared cores and regular blades an exception. The types of retouched tools present vary accordingly. Magdalenian assemblages contain primarily blade tools, while the "Azilian" shows a high percentage of flake tools and unsystematically retouched pieces, and a reduction in tool size.

Without anticipating here the major factors considered responsible for the described changes in raw material behavior, it may be said that changing environment caused a different kind of hunter-gatherer mobility which in turn influenced the ways in which lithic raw materials were handled.

As background for arguments here, I will briefly present some of the history of research on terminal Paleolithic ways of life. Opinions of European archaeologists concerning terminal Paleolithic ("Azilian") ways of life and its material record have changed several times during the last two decades. In the seventies, the "Final or End Paleolithic", de-

Table 1. Comparison of Middle Rhine Magdalenian and Final Paleolithic

	Magdalenian	"Azilian"/Federmesser
Chronology	Dryas I	Allerød (partly before)
Climate	Dry, cold	Humid, temperate
Environment	Steppe/tundra	Increase of forests (pine, birch, willow)
Game	Horse, reindeer, bison, polar fox, mammoth	Elk, deer, beaver, roe, horse, ibex, wild boar
Important sites	Gönnersdorf, Andernach (lower layer)	Niederbieber, Andernach (upper layer) Urbar, Kettig
14C ages BP	12,400 - 13,200	11,300 - 12,300
Houses	Large dwellings (diam. 6-8 m) and small tents, cooking-pits, lamps, postholes, caches, slate-pavedfloors, large numbers of tools, long occupations	Small concentrations (diam. ca. 3 m) with central fire-place (open air activity or tent?), small numbers of tools, relatively short-term occupations
Hunting	Large bone points, harpoons and backed bladelets: spearthrower	Curved backed points, backed bladelets, shaft smoother: archery
Bone, antler and ivory industry	Points, baguettes demirondes, batons percés, barbed points, needles, splinter-technique	Absent (exception: barbed point and antler striker, Kettig, and some unidentified fragments, Niederbieber)
Decoration	Teeth, pigments, fossils, mollusks, wooden beads	Nearly absent
Art	Very frequent: engravings (women, animals,symbols), female ivory figurines, decorated bone tools tradition,	Absent (exception: shaft smoother with stylized female engravings in Magdalenian Niederbieber)
Raw material acquisition	Large quantities of high-quality materials transported as cores and blades, high percentages of non-local materials (max. 95%)	Unsystematically collected single pieces, no testing or preparation outside the sites, non-local materials still frequent.
Lithic technology	Blade industry, prepared cores (grands piéces arqués, talons en éperon), soft percussion (antler)	Poor blade industry, marginal core preparation, direct hard percussion.
Lithic types	Blade tools: burins, scrapers, borers, backed bladelets, splintered pieces	Flake/bladelet tools: short scrapers, backed points, borers, backed bladelets, burins
Mobility	Logistical mobility, seasonal change of base-camps	High residential mobility

fined as the period between Magdalenian and Mesolithic, was regarded as a "decadent Paleolithic" due to an observed decrease in material elements including art, ornaments, and prepared-core blade technology. This view seemed to be reflected in an assumed predominance of local raw materials. In the new wooded environments of the late glacial, human mobility seemed to decrease.

A closer analysis of terminal Paleolithic lithic assemblages (Floss 1994), though, clearly showed that "exotic" raw materials were still very frequent. This suggested to some that the mobility of final Paleolithic groups might have been even greater than in the Magdalenian period, for which large stable dwellings ("base camps") with relatively long occupations imply a certain "settledness". The small, quickly-made lithic tools could be regarded as a pragmatic response to a new biotope in which the abundance of available organic materials, including wood, made careful manufacture of lithic tools such as long blades unnecessary (Bolus 1992). Such views seemed almost to claim that only the Azilian hunter-gatherers had an effective strategy for survival, while

Magdalenian hunter-gatherers were captives of an inflexible concept of lithic technology.

In contrast to this, I will argue that the unsystematic production of terminal Paleolithic stone tools must be understood as a response to the pressure of worsened living conditions with a reduced animal biomass. In my judgment, the rich availability of wood in the Allerød interstadial is only a minor factor and cannot be primarily responsible for changes in the elaboration of lithic technology. The produc-

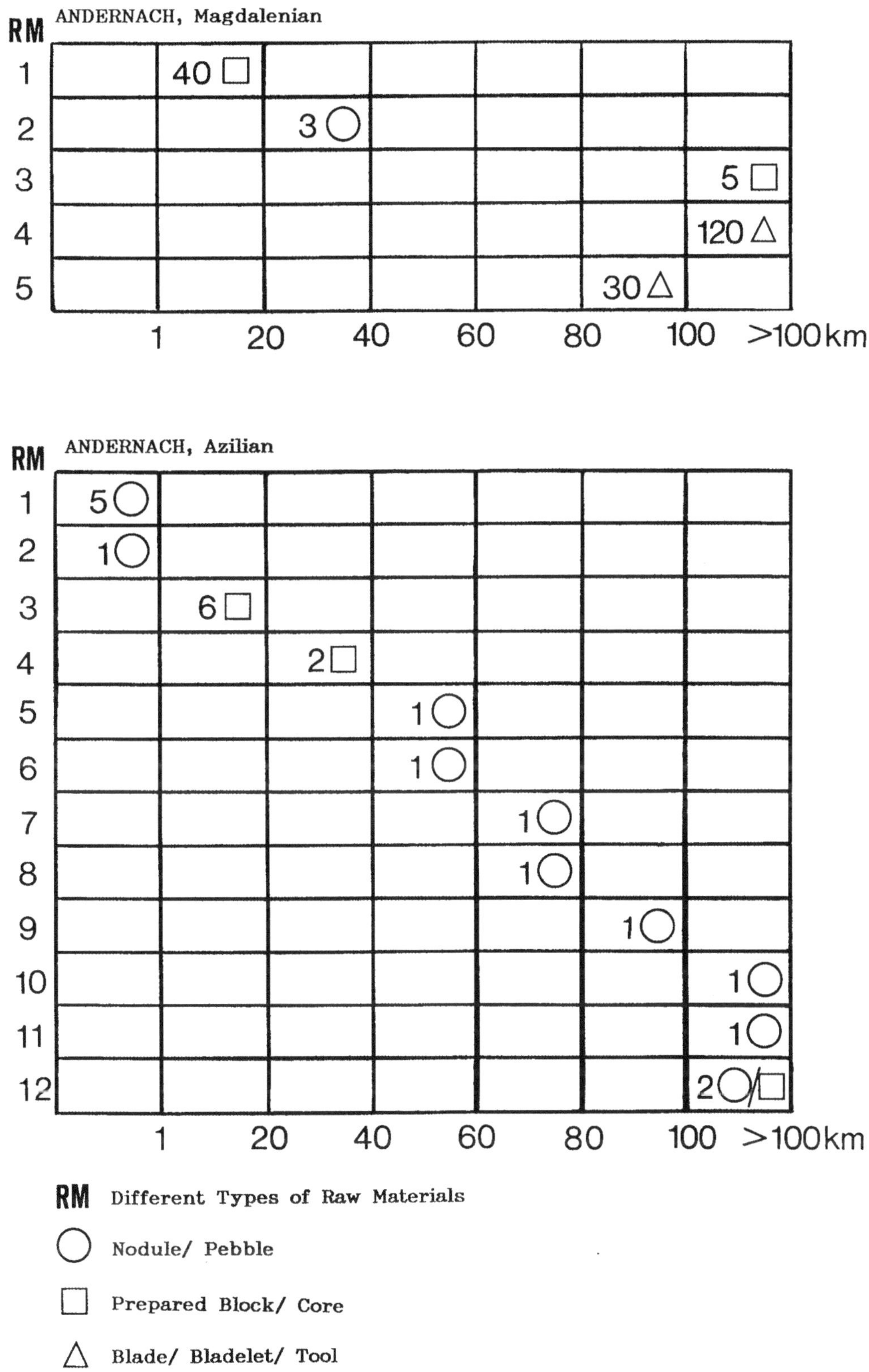

Figure 4. Reconstruction of the form and quantity in which the different types of raw materials (numbered 1-5 and 1-12) were introduced onto the site Andernach-Martinsberg (upper part of figure. shows Magdalenian, lower part "Azilian") in relation to the distance to source locations (source Floss 1994).

tion of antler points and barbed spear throwers that presumably functioned with wooden shafts and spears is a good example of the importance of wood as a material in the Magdalenian as well. Nevertheless, at that time lithic materials were manufactured according to more demanding standards. It is more probable that the terminal Paleolithic ("Azilian") population encountered serious subsistence stress as a result of expanding forests and decrease in animal biomass. Subsistence stress during the Azilian may have ruled out certain kinds of behavior. The decline of art and ornamentation in the Azilian may reflect subsistence stress, since the enormous creative power of art and decoration during the Magdalenian can only have taken place in a milieu allowing sufficient time and leisure for these activities.

If the presence of non-local lithic raw materials in Paleolithic inventories is mainly caused, as I believe, by hunter-gatherer mobility, a change in raw material acquisition must indicate a change in basic hunter-gatherer strategies, including primarily mobility strategies. The presence of various kinds of Upper Cretaceous flint in Middle Rhine "Azilian" inventories (distance to source more than 100 kilometers) shows that materials were transported over distances nearly as great as those in the Magdalenian. Nevertheless, the mechanisms of transport in these two periods appear to be quite different.

Ethnographic research on hunter-gatherers has shown that mobility can be regarded as including two major types, residential and logistical mobility (Binford 1980). A high degree of residential mobility is often linked to hunter-gatherers of the forest, where small and quickly exhausted game may require a frequent change of place. In contrast, hunter-gatherers who live in tundra conditions tend to exhibit a high degree of logistical mobility executed from large central campsites occupied for a longer time. Intensive exploitation of the environment is expressed in the installation of special hunting or acquisition camps established away from the central campsite. In contrast to hunter-gatherers of the forest, in the tundra context residential migrations are relatively rare, over longer distance, and are often of a specific seasonal character.

Applied to the Middle Rhine area, I suggest that these differences in hunter-gatherer settlement patterns and migration are reflected in the analysis of Gravettian/Magdalenian in contrast to the terminal Paleolithic ("Azilian"). Whereas Mainz-Linsenberg (Gravettian), Gönnersdorf and Andernach (Magdalenian) are base camps from which specialized hunting or acquisition camps such as Sprendlingen (Gravettian) or Wildweiberlei (Magdalenian) were maintained, we find in the Azilian context numerous very similar small sites with relatively short occupations. The identification of lithic raw materials shows that the absolute distance to source in both periods was over 100 kilometers for some materials in both cases (Figures 3, 5). In other words, over a year in both periods, hunter-gatherers must have covered a similarly large area, considering that raw materials and single flakes/blades/tools were carried in the course of normal mobility (curation of tools, and embedded procurement of raw materials). Nevertheless, the organization of mobility and raw material acquisition in the periods compared was quite different

Hunter-gatherers of the forest, with high residential mobility, tend over a day or even a year to relatively monotonous, continuous subsistence activities that do not allow systematic procurement of raw materials or retooling. The acquisition of lithic raw materials and the production of tools thus takes place in short episodes. In contrast, food acquisition among tundra hunter-gatherers is, over a year, much more structured and seasonally variable. Food preservation allows some independence from seasonal variation in hunting success.

Hunter-gatherers with a high degree of residential mobility have a low capacity for storage. Because breaks in food acquisition activities are rare, systematic replacement of tool kits is not possible. This is instead carried out opportunistically, such that residentially mobile hunter-gatherers are "constantly at work repairing or making something" (Kuhn 1989:35). These observations lead to a general conclusion: the higher the residential mobility, the less time is available for production of stone tools (Lurie 1989).

In contrast, hunter-gatherers practicing a high degree of logistical mobility with phases of "settledness" in base camps do not have a problem of transport because most of the materials and tools remain in the base camp while exploiting the environment. Specialized task-groups can be expected to need a smaller tool set, which is partly brought back to the central camp. In contrast to forest hunter-gatherers, those living in tundra/steppe conditions should tend toward seasonally structured procurement of raw materials and production of tools. In periods during which food acquisition was less important, or when activities brought groups close to raw material outcrops, these hunter-gatherers would have "geared up", producing a stock of materials and tools in preparation for anticipated needs, when for instance these materials would no longer be available.

Unfortunately, but understandably, since for a long while no other sites were known, research on the Middle Rhine Magdalenian has concentrated primarily on the analysis of large base camps such as Gönnersdorf and Andernach (lower layer). Nevertheless the long occupations of these sites, with large numbers and high diversity of tools, as well as the presence of relatively large amounts of local raw materials, prove without any doubt an intensive logistical mobility and suggest the presence of special activity sites for which we do have some indications, such as the cave sites in the Lahn river valley (Wildweiberlei) (Terberger 1986) or finds of single Magdalenian tools as at Irlich or Mayen (Neuwied Basin) (Bosinski 1992). In Andernach, several accumulations of bones found in artificial pits were interpreted as places for food storage (Eickhoff 1992).

This contrast between different types of hunter-gatherer strategies is, I argue, reflected in the different raw material procurement strategies of Magdalenian and "Azilian" (Federmesser) sites in the Middle Rhine area.

In Magdalenian sites such as Gönnersdorf and Andernach, long-distance transport of prepared cores or tools is well-established (see Floss 1994). The seasonal structure of material acquisition is confirmed by contrasts between raw material assemblages in different parts of the sites (see Figure 3). Some dwellings contain nearly exclusively local raw materials (for instance in concentration I at Gönnersdorf and concentration I at Andernach), probably because their inhabitants had previously occupied a region without good raw material outcrops or where, depending upon the season, other activities were more important than raw material acquisition. In contrast, some Magdalenian dwellings contain almost exclusively non-local, exotic raw materials, primarily Cretaceous flints found in the Maas river region, 100 kilometers or more distant from the Middle Rhine. Examples of concentrations dominated by exotic raw materials include concentration II at Gönnersdorf and concentration II at Andernach (Floss and Terberger 1987). These tested or prepared materials were transported in distinct annual migrations. The almost complete absence of materials of an intermediate origin (see Figures 3 and 4 above) suggest that raw material acquisition was not an important activity during these migrations. There is some indication from faunal analysis that these concentrations, which differ in raw material composition, were inhabited in different seasons of the year, with concentrations including mostly local materials belonging to winter and those with predominantly exotic materials to summer.

In contrast, the "Azilian" context does not show such a systematic procurement strategy, though exotic flints still indicate long-distance movements. I suggest that in the course of a change of residential camps, single pieces of raw material were picked up from a wide variety of locations, and transported to the next residential site without preparation or testing. This leads to a high percentage of materials originating from outcrops in an "intermediate" distance of 20-100 kilometers (see Figures 3 and 4 below), with its maximum at a distance from 20-40 kilometers (Floss 1994:Figures 178, 180), and a considerable diversity of raw materials, as shown for the upper layer at Andernach (Figure 4 below).

In summary, a consideration of hunter-gatherer mobility allows explanation of differences in raw material procurement strategies between Magdalenian and "Azilian" of the Middle Rhine area. In absolute terms (in kilometers), the transport of lithics in the Middle Rhine Upper and Final Paleolithic does not cover distances so large that exchange systems must be invoked to account for them. The presence of non-local lithic raw materials, in my view, is caused by the mobility of the hunter-gatherers themselves. This does not exclude a rare unsystematic exchange of single tools or raw material pieces. If mobility is defined by the frequency of changing a residential camp, the Azilian population was much more mobile than that of the Magdalenian. The Magdalenian hunter-gatherers, in contrast, tended toward fewer seasonal migrations and a higher degree of logistical mobility. The differences in raw material acquisition strategies between the two complexes are caused by a basic change in hunter-gatherer lifeways and mobility that was in turn mainly caused by a fundamental change in climate conditions at the end of the Pleistocene.

References Cited

App, V., B. Auffermann, J. Hahn, C. Pasda and E. Stephan

1995 Die altsteinzeitliche Fundstelle auf dem Schwalbenberg bei Remagen. *Berichte zur Archäologie an Mittelhrein und Mosel* 4:1-137.

Binford, L. R.

1980 Willow smoke and dogs' tails: Hunter-gatherer settlement systems and archaeological site formation. *American Antiquity* 45(1):4-20.

Bolus, M.

1992 *Die Siedlungsbefunde des späteiszeitlichen Fundplatzes Niederbieber (Stadt Neuwied).* Monographien des Römisch-Germanischen Zentralmuseums 22. Rudolf Habelt, Bonn.

Bosinski, G.

1979 Die Ausgrabungen in Gönnersdorf 1968-1976 und die Siedlungsbefunde der Grabung 1968. Der Magdalénien-Fundplatz Gönnersdorf 3. Wiesbaden.

1982 *Das Eiszeitalter im Ruhrland.* Führer des Ruhrlandmuseums Essen 2.

1992 *Eiszeitjäger im Neuwieder Becken.* Archäologie an Mittelrhein und Mosel 1. Koblenz.

Bosinski, G. and J. Hahn

1972 Der Magdalénien-Fundplatz Andernach

(Martinsberg). *Rheinische Ausgrabungen* 11:81-257.

Eickhoff, S.
1992 Wohnplatzstrukturen — Strukturierung eines Wohnplatzes. Zur Besiedlungsgeschichte des Magdalénien-Fundplatzes Andernach. Unpublished Ph.D. Dissertation, University of Cologne.

Fiedler, L.
1983 *Jäger und Sammler der Frühzeit. Alt- und Mittelsteinzeit in Nordhessen.* Vor- und Frühgeschichte im Hessischen Landesmuseum in Kassel, Heft 1.

Floss, H.
1994 *Rohmaterialversorgung im Paläolithikum des Mittelrheingebietes.* Monographien des Römisch-Germanischen Zentralmuseums 21. Rudolf Habelt, Bonn.
2000 La fin du Paléolithique en Rhénanie (Magdalénien, groupes à Federmesser, Ahrensbourgien. L'évolution du choix de matières premières lithiques, reflêt d'un profond changement du climat et du comportement humain. In *L'Europe septentrionale au Tardiglaciaire. Confrontation ds modèles régionaux du peuplement. Colloque Nemours, 13.5.-16.5.1997.* Mém. APRAIF 7:87-96.

Floss, H. and Th. Terberger
1987 Die Konzentration II von Andernach. Ein Beitrag zur Kenntnis der Variationsbreite spätjungpaläolithischer Steinartefaktensembles. *Archäologisches Korrespondenzblatt* 17:287-294.
1990 The Magdalenian of Andernach: Analysis of camp structures by refitting stone artefacts. In *The Big Puzzle,* edited by E. Cziesla, S. Eickhoff, N. Arts and D. Winter, pp. 339-362. Studies in Modern Archaeology 1. HOLOS, Bonn.

Floss, H. and Th. Terberger, eds.
in press *Die Steinartefakte des Magdalénien von Andernach. Die Grabungen 1979-1983.* Monographien des Römisch-Germanischen Zentralmuseums. Rudolf Habelt, Bonn.

Günther, K.
1988 *Alt- und mittelsteinzeitliche Fundplätze in Westfalen, Teil 2.*

Kaiser, K. F.
1993 *Beiträge zur Klimageschichte vom späten Hochglazial bis ins frühe Holozän, rekonstruiert mit Jahrringen und Molluskenschalen aus verschiedenen Vereisungsgebieten.* Eidgenössische Forschungsanstalt für Wald, Schnee und Landschaft, Birmensdorf.

Kuhn, S. L.
1989 Hunter-gatherer foraging organization and strategies of artifact replacement and discard. In *Experiments In Lithic Technology,* edited by D. S. Amick and R. P. Mauldin, pp. 33-47. British Archaeological Reports, International Series 528. BAR Publishing, Oxford.

Lurie, R.
1989 Lithic technology and mobility strategies: The Koster site Middle Archaic. In *Time, Energy, and Stone Tools,* edited by R. Torrence, pp. 46-56. Cambridge University Press, London.

Schmincke, H. U.
1988 *Vulkane im Laacher See-Gebiet. Ihre Entstehung und heutige Bedeutung.* Haltern.

Stapert, D. and M. Street
1997 High resolution or optimum resolution? Spatial analysis of the Federmesser site at Andernach, Germany. *World Archaeology* 29:172-194.

Street, M. and M. Baales
1997 Les groupes à Federmesser de l'Allerød en Rhénanie centrale (Allemagne). *Bulletin de la Societé Préhistorique Française* 94:373-386.

Street, M., M. Baales, and B. Weninger
1994 Absolute Chronologie des späten Paläolithikums und Frühmesolithikums im nördlichen Rheinland. *Archäologisches Korrespondenzblatt* 24:1-28.

Terberger, K.
1986 *Das Lahntal—Paläolithikum.* Unpublished Ph.D. Dissertation, University of Cologne.

Veil, St.
1978 Alt- und mittelsteinzeitliche Fundplätze des Rheinlandes. Kunst und Altertum am Rhein 81. Bonn.
1982 Der späteiszeitliche Fundplatz Andernach-Martinsberg. *Germania* 60:391-424.

7. Lithic Raw Material Utilization in the Final Paleolithic and Mesolithic of Belgium

Marcel Otte, André Gob, Lawrence G. Straus, Eric Teheux and Jean-Marc Léotard

Like the preceding chapter, this paper also offers a long-term perspective on changes in the exploitation of lithic raw materials within a well-defined region. The authors report on the work of a research group that has focused attention over a number of years on excavation and technological studies, including refitting, at a number of Paleolithic and Mesolithic sites in Belgium, as well as recent research on flint sources. Mousterian and early Aurignacian flintknappers here exploited a wide variety of materials, often of local origin. A major shift in procurement strategies is observed in the Gravettian, as flintknappers became very selective in their choice of high quality flint. This is documented by the appearance of large open-air workshop sites at Cretaceous outcrops in the region. In the Gravettian and Magdalenian periods there is also evidence of spatial displacement of technological stages in the production sequence. The authors find quarry-workshop sites, transit sites and residential sites forming a transport network from sources of high-quality flint on the low plateau of Middle Belgium to the caves of Wallonia. A second major shift in procurement strategies occurs in the Allerød. Evidence for long-distance transport of prepared cores is absent, and instead the lithic procurement patterns of the final Paleolithic and Mesolithic are primarily characterized by intensive exploitation of local raw materials. The authors attribute the documented shifts in raw material procurement to a combination of cultural and environmental factors, and consider the role of changes in tool design in determining patterns of raw material procurement. *–Editors.*

Introduction

The present territory of Belgium includes a wide variety of geological formations containing various rocks that are adequate for knapping (Figure 1). Primary deposits in Upper Belgium contain quartzites, quartz and cherts. Middle Belgium has major Upper Cretaceous deposits yielding abundant flints of excellent quality. In addition, detrital materials are distributed throughout the country: in solution clays on the plateaus, in river terraces, and on marine terraces of Tertiary age in Lower Belgium (see Caspar 1982). Human occupation of Belgium began again after the Last Glacial Maximum around 15,000 BP. From then on it was unbroken and intensive. In this field of study, there is evidence for different modes of lithic procurement temporally spread throughout a variety of cultural systems from the Magdalenian through the Mesolithic. Different kinds of lithic exploitation can be studied in order to trace changes in the ways in which materials were dealt with or the impacts of these exploitation modes on the technological processes employed by the different cultural traditions.

Antecedents

Interestingly, the real behavioral break in terms of the exploitation of lithic raw materials does not occur in Belgium with the appearance of anatomically modern humans. Mousterian and at least early Aurignacian sites contain the same heterogeneous range of materials, often of local origin. Territorial opportunism or technological indifference characterizes many of the assemblages of these periods found in the caves of Wallonia, which are far from the main sources of good lithic materials. It was not a question of lack of mobility, since non-local stones do appear along with local ones of equally poor flaking qualities: Campanian flint (from sources 50-80 kilometers away) and Cambrian phtanite (50-60 kilometers) were occasionally used. It seems, however, that the objects found in the cave sites were at the ends of operational chains (bifaces, Levallois flakes, finished endscrapers), isolated from the preparatory reduction stages as would be represented by cores, cortical flakes and other debitage (Otte 1991; Ulrix-Closet 1975).

Recent excavations in le Trou Magrite (Namur) (Otte and Straus, eds., 1995) have demonstrated similarities in raw material exploitation and in technology between the Mousterian and early Aurignacian assemblages (the latter AMS dated to 41,300 ± 1,700 BP), namely the overwhelming use of local limestone, quartzite and poor-quality chert and (consequently?) infrequent production/use of blades. On the other hand, the late Aurignacian (which dates by C-14 to 34,200 ± 1,900 BP) sees a

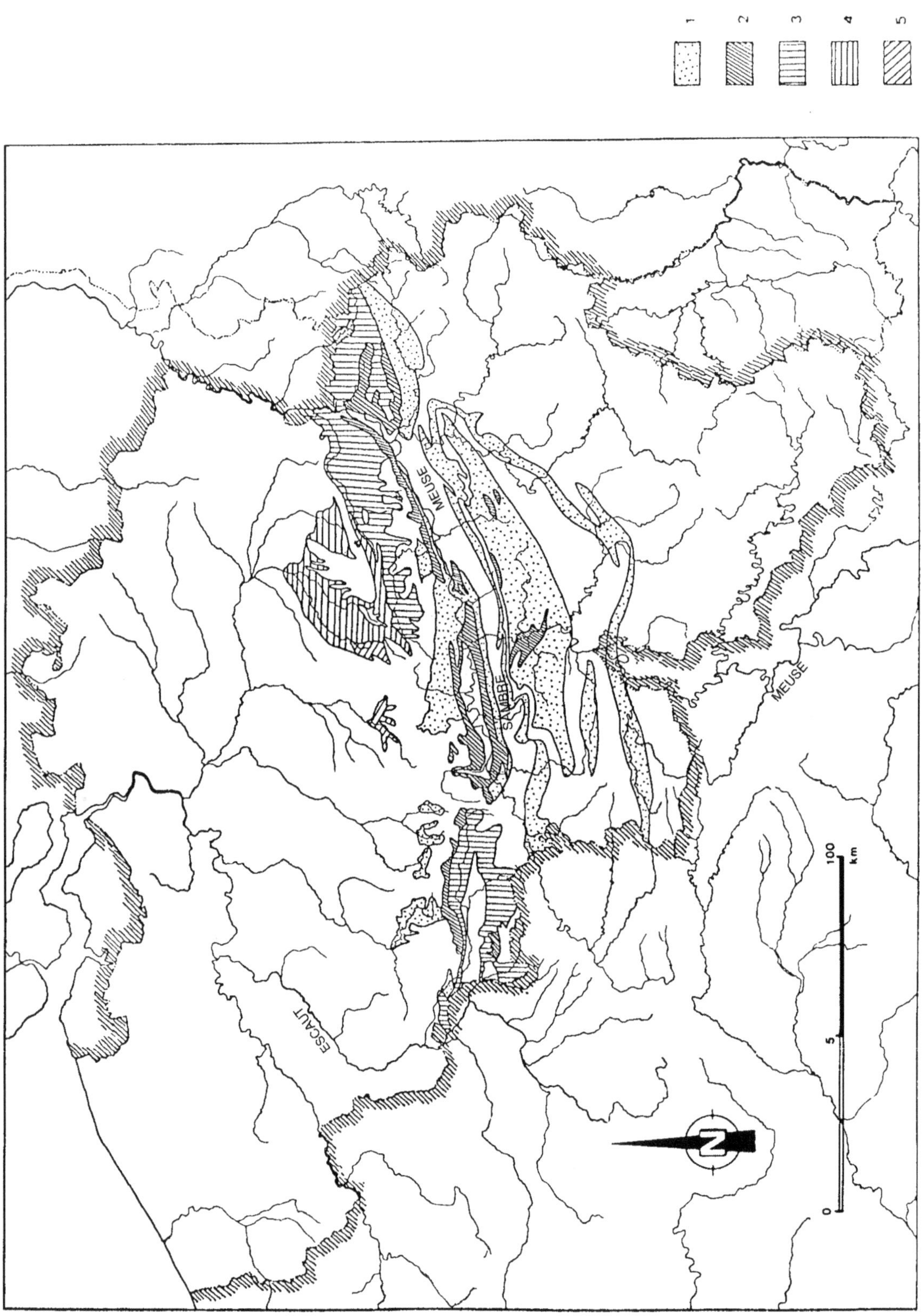

Figure 1. Location of geological units relevant to prehistoric activities in Belgium. 1: limestone outcrops (caves and rockshelters); 2: coal fields (presence of chert); 3: Cretaceous (flint); 4: Landenian sands (quartzite and fossil shells); 5: Cambrian outcrops with phtanite. Source: Otte 1979.

significant increase in the use of good-quality flint, probably from the Cretaceous sources in Hainaut, some 60-70 kilometers to the west. The flint occurs in the forms of blanks (including a relatively high proportion of blades) and finished tools, with few cortical or other debris and virtually no cores. Flint was preferred for the manufacture of Upper Paleolithic tool types. This Trou Magrite assemblage, *typologically* identical to the early Aurignacian one, predates the oldest Belgian Gravettian sites (Maisières-Canal and Huccorgne-Hermitage) by only about 6000 years.

In fact, the radical change in lithic exploitation occurs *within* the Early Upper Paleolithic, especially with the appearance of the Gravettian. The technological expertise required for making weapon tips (Gravette and Font-Robert points, backed bladelets) involved new behaviors relative to the use of raw materials. Humans became very selective in their choice of homogeneous, fine-grained flint. This is indicated by the appearance of massive open-air workshop sites at excellent Cretaceous flint sources such as Maisières (Hainaut) and Huccorgne (Liège) (Haesaerts and Heinzelin 1979; Otte *et al.* 1992, 1993). These sites contain all stages in the operational sequence, which was focused on the elaborate production of narrow, elongated blades from prismatic cores (Martinez and Guilbaud 1993). A few blades were selected for transformation into weapons and other tools, many of which may have been transported away from these sites.

In the Gravettian, blocks and blanks of flint and phtanite that had been roughed out or produced elsewhere were transported to some of the same cave sites that also contained Aurignacian assemblages. In the Gravettian, then, there was spatial displacement of technological stages in the production of blanks that had usually been accomplished at the residential sites themselves in earlier times (in the form of flakes on local raw materials). The elaborate processes of core production in the Gravettian permitted both such physical dissociation of reduction stages and the manufacture of blades whose shape and size were closely governed by their ultimate intended functions in specific tool forms.

The Upper Pleniglacial (ca. 22,000-15,000 BP) saw the apparent abandonment of the northern territories of the Paleolithic world. There is no trace of human occupation temporally equivalent to the Solutrean period in Belgium.

The Wurm Tardiglacial

During the Tardiglacial a recolonization began in the northern plains by means of relatively complex processes (Otte n.d.; Straus 1991). Two principal movements can be detected.

The Magdalenian first appears in Belgium during Dryas I around 14,000 - 15,000 BP, probably originating in the Paris Basin. Several sites of a slightly later phase of this culture are now known that date to the Bølling interstadial (ca. 13,000 - 12,500 BP). Most sites are found in caves and rockshelters in the Meuse Basin (Dupont 1872), but others, probably pertaining to the same period of resettlement, have been found in the open air in Middle Belgium (Vermeersch *et al.* 1985, 1987; Vermeersch and Symens 1988). The second recolonization movement, limited to the Bølling, involves a northerly culture, the Creswellian (Garrod 1926; Danthine 1955-60).

The Magdalenian

Regarding the Magdalenian, the pioneering studies of E. Dupont (1872) insisted on the use of lithic materials from Champagne (the region of Reims) and the Paris Basin, which were also thought to be the source areas of Tertiary shells used as pendants in Belgium (Figure 2, above). Dupont was referring above all to the blond-colored flints and so-called "pyromaque" (strike-a-light) flints found in Meuse and Lesse valley sites. Recent studies (Dewez 1987; Otte 1992; Teheux 1994; Cabboi 1994) have, however, pointed to the importance of *Belgian* sources among Magdalenian assemblages of Wallonia, notably the Cretaceous outcrops in Hainaut (Spiennes, Obourg, etc.). Eric Teheux, in particular, has stressed the presence of flints (and fossil shells) trapped within dolinas formed in the calcareous plateaus and derived from the erosion of Tertiary deposits (Figure 2, below). A wide variety of flints could be obtained by transport to the Lesse-Upper Meuse sites from three ranges of distance, if one takes into account detrital sources that are rarely noted on geological maps, but could have been exploited for raw materials in Magdalenian times. These included:

A. a local zone: poor-quality flint cobbles, phtanite, black limestone, psammite (≤ 20 kilometers);

B. a semi-local zone: more abundant, better quality flint nodules from the area of Thudinie (Hainaut) and Lonzée (Namur) (20-40 kilometers); and

C. a distant zone: major Cretaceous limestone deposits in the Haine Basin (Mons area) and on the Hesbaye Plateau (Province of Liège and Maastricht)(40-80 kilometers); also a few materials of non-local origin (for example, silicified sandstone) that appear along the upper Meuse in France, in the Givet and Charleville-Mézières area, and at the contact with the Paris Basin (Léotard 1988).

Semi-precious or pigment stones (including pyrite, oligist [iron] and fluorine) can also be collected locally.

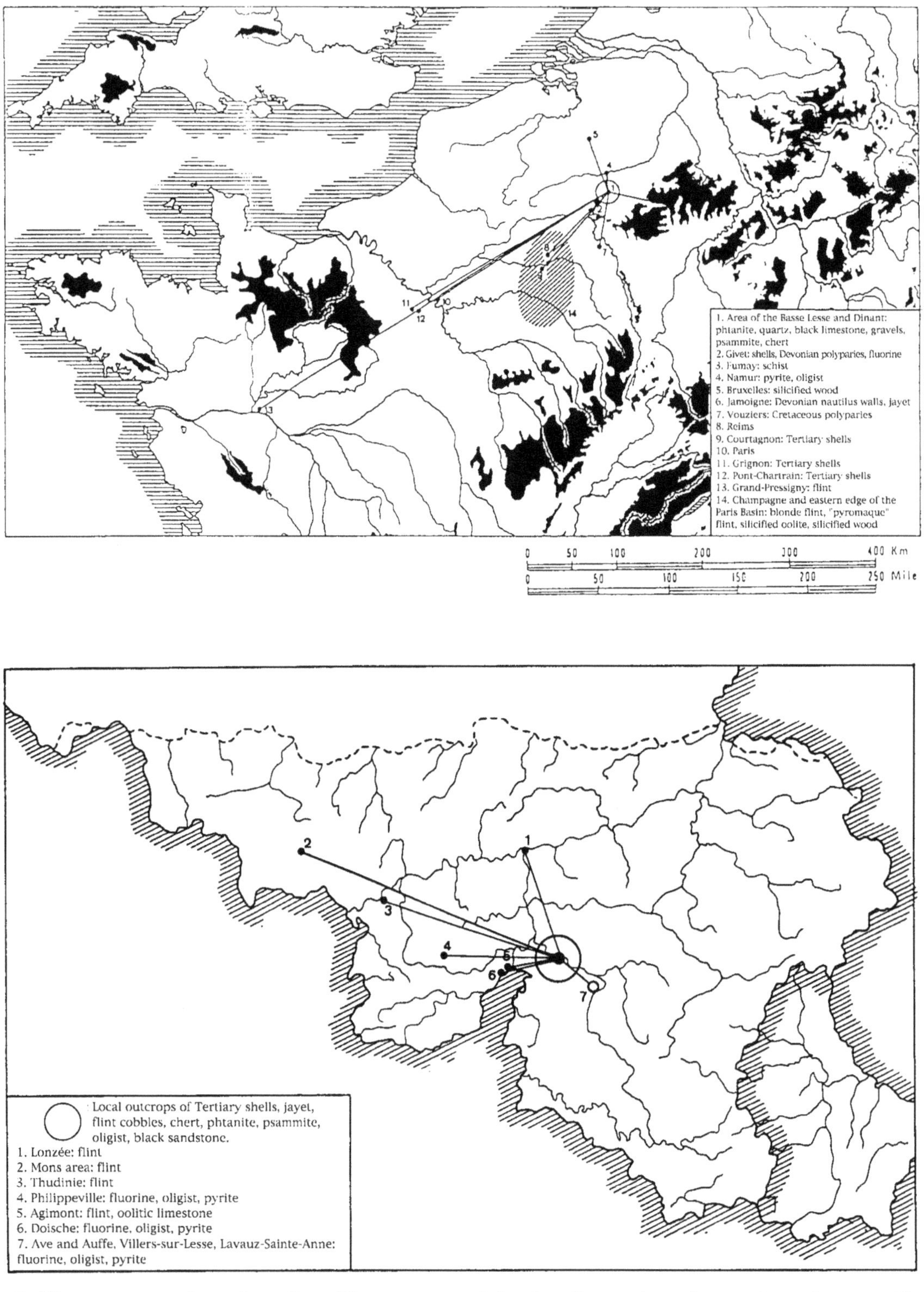

Figure 2. Upper: source locations for siliceous materials, fossils, and various minerals according to E. Dupont. 1. Area of the Basse Lesse and Dinant: phtanite, quartz, black limestone, gravels, psammite, chert. 2. Givet: shells, Devonian polyparies, fluorine.. 3. Fumay: schist. 4. Namur: pyrite, oligist. 5. Bruxelles: silicified wood. 6. Jamoigne: Devonian nautilus walls, jayet. 7. Vouziers: Cretaceous polyparies. 8. Reims. 9. Courtagnon: Tertiary shells. 10. Paris. 11. Grignon: Tertiary shells. 12. Pont-Chartrain: Tertiary shells. 13. Grand-Pressigny: flint. 14. Champagne and eastern edge of the Paris Basin: blonde flint, "pyromaque" flint, silicified oolite, silicified wood. Lower: new hypothesis concerning sources of siliceous materials and Tertiary fossils and shells (after Teheux 1994). Circle: Local outcrops of Tertiary shells, jayet, flint cobbles, chert, phtanite, psammite, oligist, black sandstone. 1, Lonzée: flint; 2, Mons area: flint; 3, Thudinie: flint; 4, Philippeville: fluorine, oligist, pyrite; 5, Agimont: flint, oolitic limestone; 6, Doische: fluorine, oligist, pyrite; 7, Ave and Auffe, Villers-sur-Lesse, Lavauz-Sainte-Anne: fluorine, oligist, pyrite.

Unfortunately these interpretations are imprecise due to the lack of conclusive petrographic or micro-paleontological identifications, in which regard we are not much further advanced than in Dupont's time (mid-late 19th century!), despite several efforts at source determination.

In terms of modes of procurement, two types of situations have been observed among the Magdalenian cave sites of the Upper Meuse Basin. Some sites (Trou des Nutons, Trou Reuviau, Grotte du Bois Laiterie), which were limited-function loci, include primarily end products, the last phases of operational chains: blanks, tools and broken weapon tips. On-going excavations of Bois Laiterie in Namur Province (Otte *et al.* 1994), a small, cold, uncomfortable, but strategic (hunting?) gorgeside site dated to 12,600 BP, have so far yielded only one core, virtually no cortical debitage, very little microdebitage of the sort found at knapping locations and no certain local raw materials. Raw materials at Bois Laiterie consist primarily of good-quality Cretaceous flint probably from either Hainaut or Hesbaye. The lithic assemblage is dominated by blades, bladelets, finished tools (endscrapers, burins, perforators) and especially backed bladelets (as well as antler points). There are also several fossil shells, some perforated. Their source(s), which remain to be positively identified, *might* be in the Paris Basin.

In contrast, other, larger (residential?) sites (Chaleux, Trou du Frontal) (Léotard and Cauwe 1986) contain all stages of the operational sequence, from initial core reduction, blank production and tool manufacture to tool use and discard. The nodules that had been selected were transported unmodified to these habitation sites. Thus, within a single cultural entity and during a brief time period, different types of behavior relative to lithic transport and manufacture were adopted in relation to the different functions and circumstances of particular sites.

In Middle Belgium and the southeastern part of the Netherlands there are, from the same period, several quarry-workshop sites at rich outcrops of excellent Cretaceous flint: Orp, Kanne, Sweikhuizen and Mesch (Rensink 1991). The outcrops, covered with recent loess, were systematically exploited for the production of huge blades, sometimes transformed into burins, perforators, etc. (Figure 3). This distinctive Magdalenian "facies" relates clearly to workshop situations as well to the performance of other specialized activities, such as antler working, though evidence of the latter has unfortunately disappeared at these open-air sites due to bad preservation. Warm-season (?) ambush hunting may also have been conducted at some of these sites, as at Huccorgne and Maisières.

The appearance of "southern" materials (such as fossils and other exotics) in Belgian sites has often been used to argue for a French origin of this regional Magdalenian concentration. Although there is little clear evidence of southern contacts, a general northeastwardly human expansion seems all the more likely in light of the presence of Belgian Meuse Basin flints in sites of the German Middle Rhineland (Gönnersdorf, Andernach) (Bosinski 1988; Floss, this volume) — continuing the same directional pattern of movement.

The Creswellian

The Creswellian, of Britannic affinity, shows evidence of lithic raw material exploitation that is quite similar to that of the Magdalenian. Most of the materials used seem to have come from the Campanian outcrops in Hainaut: abundant, excellent quality black flint (Obourg) (Letocart 1970). But as in the Magdalenian sites, the same materials considered to be of non-local origin amount to about 20% of raw materials (for example at the recently studied site of Presle) (Otte and Léotard 1985; Léotard and Otte 1988). This system of exploitation, analogous in the two cultures, suggests two hypotheses: either contacts existed between the most southern Creswellian people and the Magdalenian people, or, more probably, the mass and the variety of the available materials were identical in the confines of the territories used by these two populations (see Djindjian 1988; Fagnart 1992).

As to the mode of procurement, we can observe, since most of the material probably comes from the Mons area, an exploitation of cortical blocks of good quality and shape. For the non-local materials, two behaviors can be distinguished. For the "worst" of them all the phases of the operational sequence are represented. The "best" (silicified sandstone) seem, by contrast, to be brought in the form of finished products.

Otherwise, intermediate ("transit") sites along the operational sequence, such as Trou Jadot (Liège), show evidence of distinctive behavior: debris from blade production is present, but not the blades themselves, since they had been taken away from the sites (Figure 3) (Toussaint *et al.* 1993). Transport distances of blocks of excellent material, carefully decorticated and preformed at the quarries, are considerable (ca. 80 kilometers).

The Allerød and Dryas III

This whole picture changed again radically during the temperate Allerød oscillation, when human occupation of Belgium seems to have been more widespread than before. Sites are particularly numerous on the sandy plains of northern Belgium, the best-known of which are the Tjongerian localities at Meer (Van Noten 1978; Otte 1981; Nijs 1990). Small, local pebbles of marine origin were exhaustively used for all aspects of technological opera-

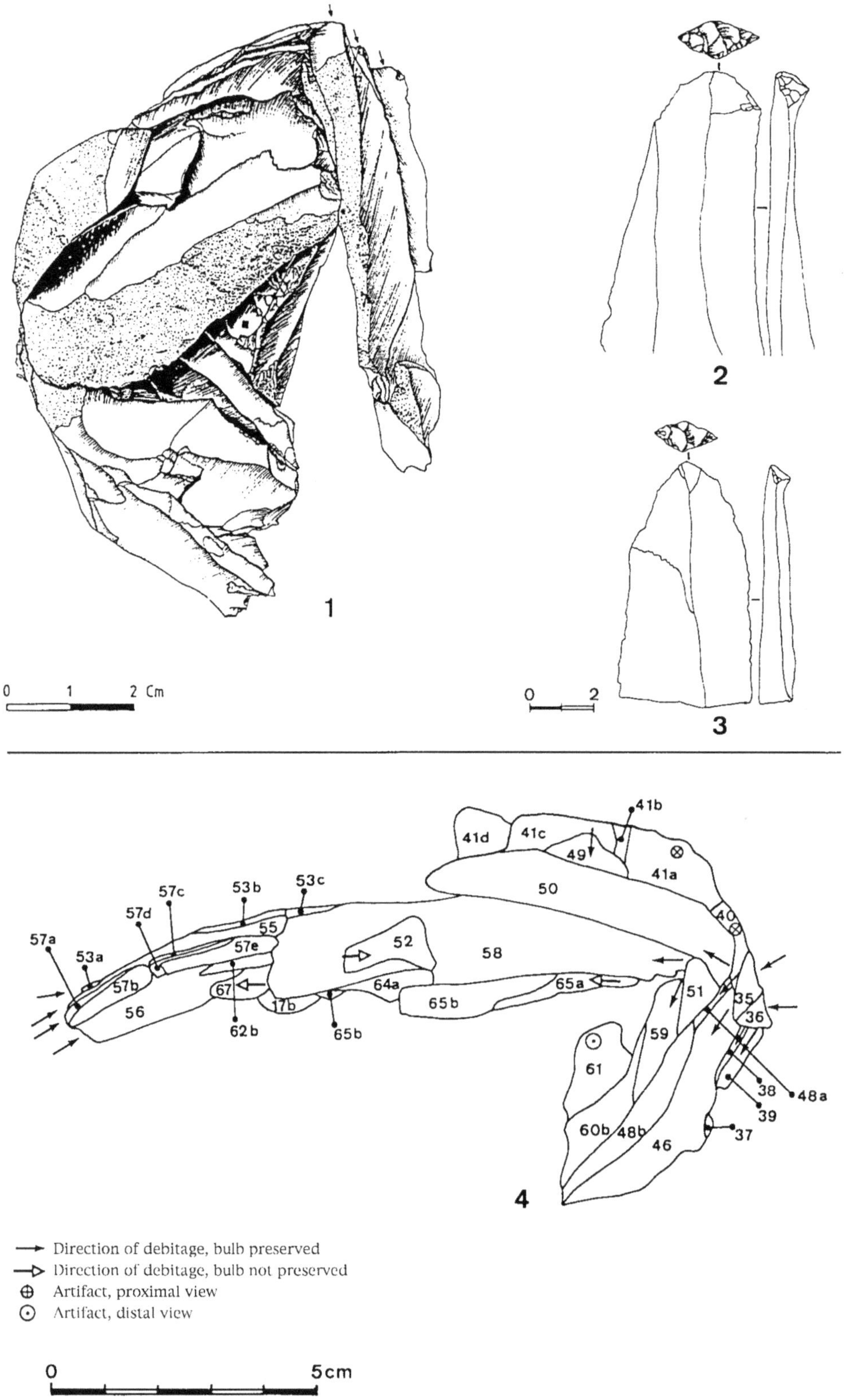

Figure 3. Examples of core technology from Magdalenian (Kanne) and Creswellian (Trou Jadot) sites. 1: refitted core, Kanne; 2,3: faceted butts, Kanne; 4: refitted core, Trou Jadot. Sources: 1-3 after Vermeersch *et al.* 1985, 4 after Toussaint *et al.* 1993.

tions. The domestic tool kits are composed of small elements (e.g., circular and thumbnail endscrapers) and the curved backed points on bladelets required no elaborate preparation (Figure 4). Although the geographic extent of such industries within the same technological tradition is very broad, there is no longer any evidence of long-distance transport of large preformed lithic blocks. The system of distribution via functionally diverse "way-station" or transit sites had come to an end, replaced by intense exploitation of local detrital materials and cobbles in secondary context.

The climatic interruption of Dryas III saw a retreat into Belgium of northerly groups which had tanged and truncated points (the Ahrensburgian). Discovered both in caves (e.g., Remouchamps, Coléoptère) (Dewez *et al.* 1974; Dewez 1975) and in the open air, sites of this culture show that humans exploited local raw materials in ways analogous to those of the Tjongerian during the Allerød (Figure 4) (Dewez 1977).

The Holocene

During the Early and Middle Mesolithic (Preboreal-Boreal), lithic procurement continued to be mainly local, namely flints extracted from veins within Cretaceous deposits, often weathered by solution. Remnant flint blocks left over in clay deposits can also be found on plateaus in the Meuse Basin (particularly above the valleys of the Ourthe and Amblève) (Gob 1981). These deposits are often not shown on the geological maps. Other flints can be found in secondary context in river beds or terraces. These cobbles are brought back to the sites in an unaltered state. Thus all reduction stages are represented at the sites, from decortication to tool production. The case of the Mesolithic sites at la Place Saint-Lambert in Liège (Gustin *et al.* 1994) is particularly typical in this regard, as are the sites of la Roche-aux-Faucons and the Leduc open-air location at Remouchamps (Gob 1976).

In direct contrast to this general type of behavior, one does observe preferential use of a few very specific materials from highly localized sources that were very widely disseminated in very small quantity (Figure 5). Wommersom (Brabant) quartzitic sandstone is the best-known case of this (Gendel 1982; Ophoven *et al.* 1948). It was transported in the form of bladelets and weapon tips and has an extremely broad geographic distribution throughout Belgium. This popularity was probably linked to its fine grained characteristics and aptitude for being pressure flaked, a technique which was developed toward the end of this period.

Two other kinds of raw material with similarly very localized sources also appear in the archeological record in the Mesolithic: Vetchauer flint from the area of Aachen (near Liège, but in Germany) and phtanite from Ottignies near Brussels (Caspar 1982; Cumont 1904). These are never found in large quantities, but were distributed over long distances, apparently for specific functions.

In the recently excavated Middle Mesolithic levels (dated to 8,800 and 7,800 BP uncalibrated) at the Abri du Pape (in the Meuse gorge just north of the French border), a variety of both poor quality local materials and some possibly imported flints have been found (Otte *et al.* 1994). All the lithics are very small, suggesting total exhaustion, especially of the few good flint cores, which are miniature. Besides abundant flakes, both tiny narrow bladelets and short, thick, wide, poorly-made bladelets (or blades) were produced. Rarely were these transformed into formal tools or weapons. Among retouched tools, only three armatures, some retouched, notched, truncated or denticulated blades/flakes, and a very few formal endscrapers, perforators or burins were found at this site. These characteristics are typical of the Ardennian Mesolithic (Rozoy 1990). Access to large blocks of good-quality raw material in this upland region seems to have been difficult, especially under the heavily wooded conditions of Boreal times, so all lithics were economized.

Conclusions

The study of changes in the systems of lithic raw material exploitation in Belgium is favored by the existence of a wealth of potential sources and a long history of human occupation during the Paleolithic and Mesolithic. Despite deficiencies in petrographic sourcing and in geological mapping of surface detrital deposits, cobble sheets and solution pockets that contain remnant flint nodules, one can discern some significant behavioral tendencies in the record.

During the Mousterian and (early) Aurignacian a very wide variety of heterogeneous materials — often very local — were exploited. In the Gravettian and Magdalenian periods, elaborate operational sequences were broken up into stages across space (i.e., among functionally different types of sites). There were quarry-workshop sites, transit sites and residential base camps forming transport chains from sources of rich, high-quality flint on the low plateaus of middle Belgium to the caves of Wallonia. The open, steppic, glacial landscapes and a highly mobile set of lifeways that led humans to cover large areas during their annual rounds, fit well with this sort of long-distance movement of lithics, as well as with the specialized nature of the technologies, particularly focused on the production, transformation and use of large, standardized blades.

Everything changed with the onset of the environmental amelioration of Allerød—even before the development of the Mesolithic *per se*. Technology

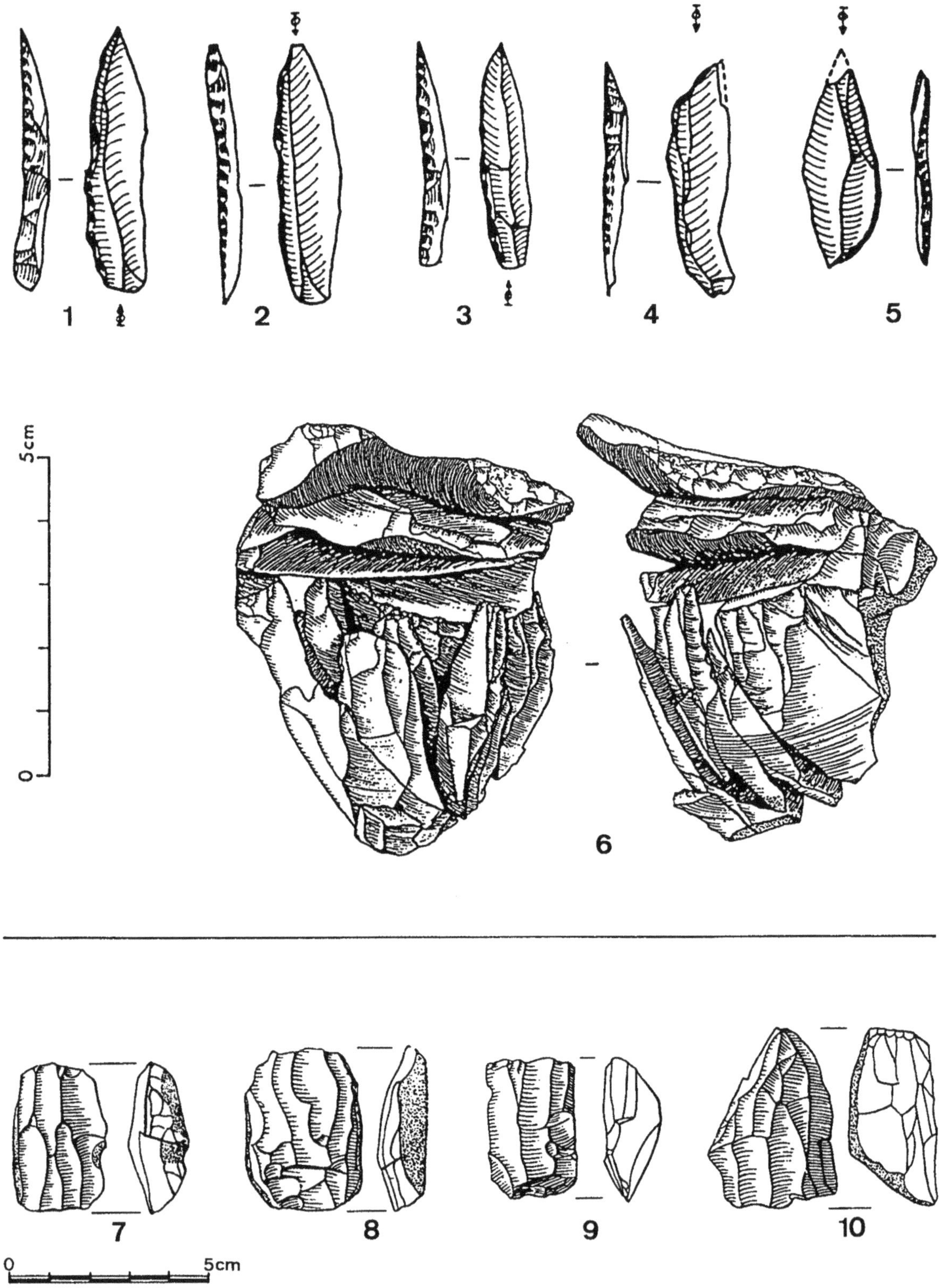

Figure 4. Examples of core technology and backed points from Tjongerian (Meer) and Ahrensburgian (Remouchamps) sites. 1-5: arched backed points, Meer (after Verbeek and Vermeersch 1993); 6: refitted core, Meer (after Nijs 1990); 7-10 cores, Remouchamps (after Dewez *et al.* 1974).

and typology had to change in concert with a newly wooded landscape, maximally using smaller, often more widely scattered stones — often to exhaustion. A few "noble" materials (phtanite, quartzitic sandstone) were exceptions to the rule of highly local lithic procurement, and were distributed widely, but as shaped or retouched items for very specific functions.

The result is a combination of cultural and environmental factors that came into play in determining the variety of types of lithic raw material exploitation over time. Different adaptive choices

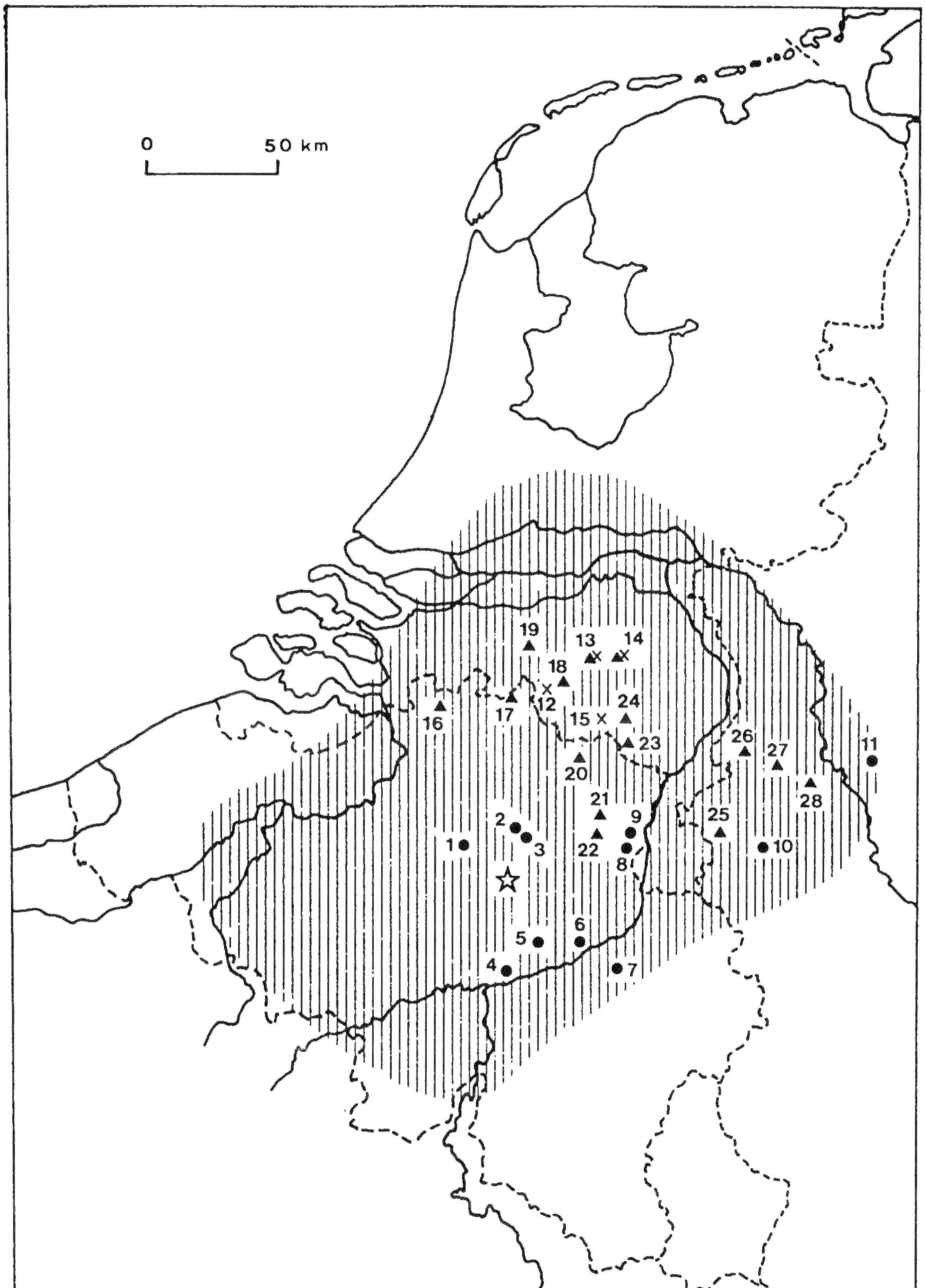

Figure 5. Approximate archaeological distribution of Wommerson quartzite (shaded) and location of sites (circles: Early Mesolithic; crosses: Middle Mesolithic; triangles: Late Mesolithic; star: outcrop near the Wommerson village) (after Gendel 1982).

were made according to traditions and the requisites of new technological inventions, but always in harmony with environmental conditions and the reigning lifeways.

References Cited

Bosinski, G.
1988 Upper and final Paleolithic settlement patterns in the Rhineland, West Germany. In *Upper Pleistocene Prehistory of Western Eurasia*, edited by H. Dibble and A. Montet-White, pp. 375-384. University Museum Monograph 54. University of Pennsylvania, Philadelphia.

Cabboi, S.
1994 Les fouilles récentes. In *Le Magdalénien du Trou de Chaleux*, edited by M. Otte, pp. 47-159. Etudes et Recherches Archéologiques de l'Université de Liège 60. Université de Liège, Liège.

Caspar, J-P.
1982 L'exploitation du phtanite d'Ottignies et Mousty et sa distribution. *Notae Praehistoricae* 2:63-82.

Cumont, G.
1904 Utilisation du phtanite cambrien des environs d'Ottignies et du grès tertiaire bruxellien par l'homme préhistorique. *Bulletin de la Société d'Anthropologie de Bruxelles* 23: 53-63.

Danthine, H.
1955-60 Fouilles dans un gisement préhistorique du domaine de Presle. *Documents et Rapports de la Société Royale d'Archéologie et de Paléontologie de l'Arrondissement de Charleroi* 50(1):3-25.

Dewez, M.
1975 Nouvelles recherches à la Grotte du Coléoptère à Bomal-sur-Ourthe (Province du Luxembourg): rapport provisoire de la première campagne de fouille. *Helinium* 15:105-133.
1977 Problématique de l'étude des groupes culturels du Paléolithique Final en Belgique. In *La Fin des temps glaciaires en Europe: Chronostratigraphie et écologie des cultures du paléolithique final*, edited by D. de Sonneville-Bordes, pp. 791-796. Colloques Internationaux du CNRS 271. CNRS, Paris.
1987 *Le Paléolithique Supérieur Récent dans les Grottes de Belgique*. Publications d'Histoire de l'Art et d'Archéologie de l'Université Catholique de Louvain 57. Oleffe, Court St. Etienne.

Dewez, M., M, Brabant, J., Bouchud, M. Callut, F. Damblon, M. Degerbøl, C. Ek, H. Frere, and E. Gilot
1974 Nouvelles recherches à la Grotte de Remouchamps. *Bulletin de la Société Royale Belge d'Anthropologie et de Préhistoire* 85:5-161.

Djindjian, F.
1988 Les rapports entre les industries magdaléniennes, creswelliennes et hambourgiennes du nord de l'Europe. In *De la Loire à L'Oder. Les Civilisations du Paléolithique Final dans le Nord-Ouest Européen, Vol. 2,* edited by M. Otte, pp. 683-705. British Archaeological Reports, S 444, Etudes et Recherches Archéologiques de l'Université de Liège 25. BAR Publishing, Oxford.

Dupont, E.
1872 *Les Temps Préhistoriques en Belgique. L'homme Pendant les Âges de la Pierre dans les Environs de Dinant-Sur-Meuse.* Muquardt, Bruxelles.

Fagnart, J.-P.
1992 Le Paléolithique final dans le nord de la France : le Magdalénien et les cultures septentrionales. In *Le Peuplement Magdalénien. Paléogéographie Physique et Humaine,* edited by J.-Ph. Rigaud, H. Laville, and B. Vandermeersch, pp. 247-258. Actes du colloque international de Chancelade (Périgueux 1988). C.T.H.S., Paris.

Garrod, D.
1926 *The Upper Palaeolithic Age in Britain.* Oxford University Press, Oxford.

Gendel, P.
1982 The distribution and utilization of Wommersom Quartzite during the Mesolithic. In *Le Mésolithique entre Rhin et Meuse*, edited by A. Gob and F. Spier, pp. 21-50. Société Préhistorique Luxembourgeoise, Luxembourg.

Gob, A.
1976 Le Mésolithique du gisement inférieur de la Roche-aux-Faucons (Plainevaux). *Bulletin de la Société Royale Belge d'Anthropologie et de Préhistoire* 87:45-76.
1981 *Le Mésolithique dans le Bassin de l'Ourthe. Mémoire* 3. Société Wallonne de Palethnologie, Liège.

Gustin, M., J.-M. Léotard, and M. Otte
1994 Le Mésolithique de Liège: Campagne 1994. *Notae Praehistoricae* 14:113-123.

Haesaerts, P. and J. de. Heinzelin
1979 *Le Site Paléolithique de Maisières-Canal.* Dissertationes Archaeologicae Gandenses 19. De Tempel, Bruges.

Léotard, J.-M.
1988 Occupation magdalénienne au Trou de Somme, Massif de Roche-al-Rue (Waulsort). *Notae Praehistoricae* 8:17-23.

Leotard, J.-M. and N. Cauwe
1986 Fouilles 1986 au Trou du Frontal à Furfooz. *Notae Praehistoricae* 6:83-93.

Léotard, J-M. and M. Otte
1988 Occupation paléolithique final aux Grottes de Presle. In *De la Loire A l'Oder. . Les Civilisations du Paléolithique Final dans le Nord-Ouest Européen, Vol. 1*, edited by M. Otte, pp. 189-215. British Archaeological Reports, S 444, Etudes et Recherches Archéologiques de l'Université de Liège 25. BAR, Oxford.

Letocart, L.
1970 Un gisement du Paléolithique final à Obourg Saint-Macaire (Hainaut). *Hannonia Praehistorica* 3:353-361.

Martinez, A. and M. Guilbaud
1993 Remontages d'un nucléus à lames gravettien à Huccorgne: aspects d'une chaine opératoire. *Préhistoire Européenne* 5:146-160.

Nijs, K.
1990 A Tjonger and a Mesolithic Site at Meer, Belgium. In *The Big Puzzle. International Symposium on Refitting Stone Artefacts*, edited by E. Cziesla, S. Eickhoff, N. Arts and D. Winter, pp. 493-506. Studies in Modern Archaeology 1. HOLOS, Bonn.

Ophoven, M., E. Saccasyn Della Santa, and J. Hamal-Nandrin
1948 *Utilisation à l'Âge de la Pierre (Mésolithique) du Grès-Quartzite dit de Wommersom.* Liège.

Otte, M.
1979 *Le Paléolithique Supérieur Ancien en Belgique.* Monographies d'Archéologie Nationale 5. Musées Royaux d'Art et d'Histoire, Bruxelles.
1981 L'industrie Tjongérienne de Meer IV. *Notae Praehistoricae* 1:42-44.
1991 Evolution in the relationships between raw materials and cultural tradition in the European Paleolithic. In *Raw Material Economies Among Prehistoric Hunter-Gatherers*, edited by A. Montet-White and S. Holen. Lawrence, pp. 161-167. University of Kansas Publications in Anthropology 19. University of Kansas, Lawrence.
1992 Processus de diffusion à long terme au Magdalénien. In *Le Peuplement Magdalénien. Paléogéographie Physique et Humaine,* edited by J.-Ph. Rigaud, H. Laville, and B. Vandermeersch, pp. 399-416. Actes du colloque international de Chancelade (Périgueux 1988). C.T.H.S., Paris.
n.d. Paléolithique final du nord-ouest. Migrations et saisons. *Colloque d'Amiens* (1994).

Otte, M. and J.-M. Léotard
1985 Fouilles au site creswellien de Presle (Hainaut). *Notae Praehistoricae* 5:33-35.

Otte, M. and L. G. Straus, editors
1995 *Le Trou Magrite.* Etudes et Recherches Archéologiques de l'Université de Liège 69. Université de Liège, Liège.

Otte, M., L. G. Straus, A. Gautier, and P. Haesaerts
1992 Fouilles dans le Paléolithique moyen et supérieur de Belgique Méridionale. *Notae Praehistoricae* 11:3-28.

Otte, M., L. G. Straus, A. Gautier, A. Martinez, M. Newman, Cl. Schutz, P. Haesaerts, P. Noiret, and J.-M. Léotard
1993 Fouilles paléolithiques 1992 en Belgique. *Notae Praehistoricae* 12:15-31.

Otte, M., L. G. Straus, Ph. Lacroix, A. Martinez, P. Noiret, J.-M. Léotard, V. Ancion, and I. López-Báyon
1994 Fouilles 1994 à l'Abri du Pape et à la Grotte du Bois Laiterie. *Notae Praehistoricae* 14:45-68.

Rensink, E.
1991 L'observation du gibier et le débitage des nucléus: un poste de guet du Magdalénien à Mesch. *Helinium* 31:5-59.

Rensink, E., J. Kolen, and A. Spieksma
1991 Patterns of raw material distribution in the Upper Pleistocene of northwestern and central Europe. In *Raw Material Economies Among Prehistoric Hunter-Gatherers,* edited by A. Montet-White and S. Holen, pp. 141-159. University of Kansas Publication in Anthropology 19. University of Kansas, Lawrence.

Rozoy, J-G.
1990 La Roche-à-Fépin et la limite entre l'Ardennien et le Tardenoisien. In *Contributions to the Mesolithic in Europe*, edited by P. Vermeersch and P. VanPeer, pp. 413-422. Leuven University Press, Leuven.

Straus, L. G.
1991 Human geography of the late Upper Paleolithic in western Europe. *Journal of Anthropological Research* 47:259-278.

Teheux, E.
1994 *Le Magdalénien de la Vallée de la Lesse.* Unpublished Mémoire de Licence, Université de Liège.

Toussaint, M., A. Becker, M. Burhenne, J.-M. Cordy, E. Gilot, J. Heim, and E. Juvigne
1993 Le Trou Jadot à Comblain-au-Pont (Province de Liège, Belgique). Paléoécologie et archéologie d'un site du Paléolithique supérieur récent. Etudes et Recherches

Archéologiques de l'Université de Liège 58. Université de Liège, Liège.

Ulrix-Closset, M.
1975 Le Paléolithique Moyen dans le Bassin Mosan en Belgique. Universa, Wetteren.

Van Noten, F.
1978 *Les Chasseurs de Meer.* Dissertationes Archaeologicae Gandenses 18. De Tempel, Bruges.

Verbeek, C., and P. Vermeersch
1993 Epipaleolithicum en Mesolithicum te Weelde-Eindegoorheide 2. *Notae Praehistoricae* 12:77-81.

Vermeersch, P., R. Lauwers, and Ph. Van Peer
1985 Un site Magdalénien à Kanne (Limbourg). *Archaeologia Belgica* 1:17-54.

Vermeersch, P. and N. Symens
1988 Le Magdalénien de plein air en Belgique. In *De la Loire A l'Oder. . Les Civilisations du Paléolithique Final dans le Nord-Ouest Européen, Vol. 1*, edited by M. Otte, pp. 243-258. British Archaeological Reports, S 444, Etudes et Recherches Archéologiques de l'Université de Liège 25. BAR Publishing, Oxford.

Vermeersch, P., N. Symens, P. Vynckier, G. Gijselings, and R. Lauwers
1987 Orp, site Magdalénien de plein air (commune Orp-Jauche). *Archaeologica Belgica* 3:7-56.

8. Etiolles: A Blade Production Site?

Monique Olive and Yvette Taborin

The chapter presents a case study featuring the pivotal Magdalenian site of Etiolles (Esonne, France). Following a description of the site, the archaeological remains, and the geomorphological environment of the Paris Basin the authors proceed to evaluate a series of questions concerning the possibility that Magdalenian occupation at Etiolles was to some extent specialized for blade production. Many years of excavation, refitting, and analysis at Etiolles have produced an extraordinary record of prehistoric behavior at this site, including evidence for movements of material between different dwelling and activity areas and a partial internal chronology of dwelling units on the site. The authors use these data to explore a series of questions relating to a possible functional specialization of the site. The lithic raw material acquisition system at Etiolles is distinguished by the exploitation of nearby, abundant, and exceptional flint deposits. A dynamic analytical approach and systematic use of refitting allows the authors to thoroughly document the ways in which lithic raw material was exploited at here. Flexibility in the socioeconomic management of the flint industry can be observed in the variation in structure, spatial organization of debitage, lithic production, and shaped tools between dwelling units. The authors emphasize the large number of different factors that could affect intrasite variations in organization of debitage, including group size and duration of occupation, aspects of social organization, and flint-knapping skill. They propose a dynamic model of occupation at the site, in which related groups coordinate a wide variety of activities, including blade production, hunting, and a range of domestic activities. The limits of the territory visited by the Magdalenian people at Etiolles, as well as their movements within this area, are assessed through provenience studies of the fossil shells they collected and the few lithic tools brought to the site from other raw material sources. Etiolles seems to belong to a larger network of Magdalenian period mobility in the Paris Basin. *–Editors.*

Discovered by chance by plowing farmers, the Etiolles site consists of a series of fluvial deposits at the junction of a rivulet called Les Hauldres and the river Seine. Fortunately, the deposits were preserved and found almost intact in the Magdalenian levels. The whole site extends over less than 10,000 square meters, 1,000 square meters of which have been excavated. Other Magdalenian sites are known in the area, particularly on the opposite bank of the river Seine, where two settlements, "Les Tarterets 1" (Schmider 1975) and "Les Tarterets 2" (Brézillon 1971), share a number of similarities with Etiolles, especially in the use of the same characteristic flint. During the glacial period, the river Seine must have been so low that it could be crossed on foot.

In this area, the Seine valley follows an outcrop of Tertiary limestone extending from the junction with the river Marne, in the northeast, about 50 kilometers to the southwest. The promontory that stands out at Etiolles was formed by a fast-flowing rivulet which eroded the limestone down to the Ludien beds (Figure 1). In the Ludien-age limestones, blocks and sometimes enormous nodules of flint formed within great silicified deposits. Some of these nodules have been moved downslope by water and freeze-thaw effects, making a good-quality, homogeneous, fresh-water flint available to knappers. This erosion process, placing otherwise inaccessible flint nodules within arm's reach, has affected various places at the foot of the outcrop, probably wherever a powerful stream gushed down the hill. We know that this flint source was extensively used in the surrounding region. We were able to collect flaked flint and cores from the surface around the site thanks to nearby road construction.

In addition to yielding fresh flint, the site area is inviting because of its location in a valley with a southwestern exposure, sheltered from the north and the east by the plateau. Its situation at the confluence of the rivulet and the Seine reminds us of prehistoric Ukrainian settlements located on promontories. At Etiolles, the promontory, which is an alluvial cone, is low but sufficient to insulate the site from the humidity of the confluence area. The discovery of remains of horse, reindeer, buffalo, and mammoth at the site indicate the existence of a varied fauna in the valley. Faunal remains are few, due to the poor preservation of organic matter in these water-transported sediments, but the number of backed bladelets probably repaired around

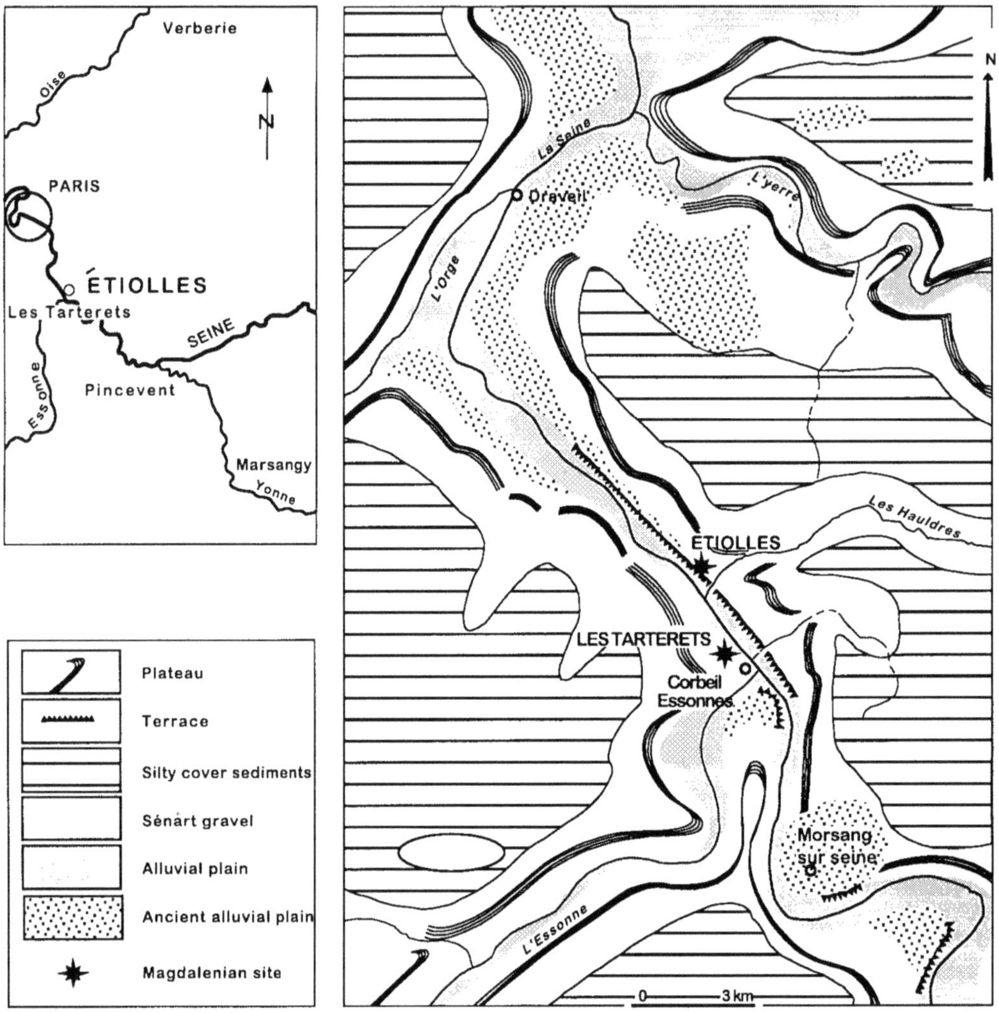

Figure 1. Location of Etiolles.

the hearth provides ample proof of hunting (Coudret *et al.* 1994).

This location beside the rivulet that meandered from its course before joining the river Seine provided green banks and flint nodules, while herds of large mammals wandered the Seine valley, and other species such as deer, birds, and fish could easily live nearby. No wonder the place was frequently visited!

Interpretive Problems of the Settlement

In 1972, during the first period of excavations, unit W 11 yielded many large and compact cores, piles of flakes and long blades. The unusual aspect of the blade debitage and the scarcity of retouched tools led to the impression, before any preliminary analysis was done, that it was a flintknapping area such as those we can see around the flint mines of a later period. Nevertheless, the circle of stone slabs that demarcated an inner space and the big central hearth made clear that the flintknapping work was accomplished during a stay which was long enough to justify such an arrangement of the place. Further research has allowed continued excavation of this structure, which obviously was one of the most beautiful dwellings in the Upper Paleolithic world of western Europe. The dwelling was structured to shelter the versatile activities of a human group for a given period. Among these activities, the flintknapping industry was carefully organized and showed the existence of several levels of skill (Pigeot 1987).

The subsequent discovery of the nearby dwelling unit P15 revealed a more simply organized dwelling, far poorer in flint, notably in the production of blades. This cast doubt on the assumption of an occupation motivated only by blade production. This idea had occurred to us while refitting cores which clearly showed the absence of some beautiful blades. We had come to think that high quality blades were carried away by people leaving the camp. More recent analysis has led us to question this view. Continued excavation has confirmed the importance of the space occupied by each dwelling and has revealed the fact that some of them were united in the sort of community we would call a camp. Evidence for this appears in the circulation of the blade blanks between dwellings. The interpretation we now suggest is that several human groups that were united enough to help each other must have camped together at Etiolles. Repeated occupation of the site proves that the place was well known, probably because of its flint. Magdalenian occupants took advantage of the flint resource, but they engaged in other activities during their stays at Etiolles, particularly the maintenance of social networks indicated by the very idea of the camp.[1]

The Magdalenian Camps at Etiolles

Magdalenian groups came to establish their camps at Etiolles on many occasions, as demonstrated by the superposition of living floors at the site. The progressive enlargement of the excavated area gives an idea of the organization of the camps, composed of several occupied units.

Because of the ambiguous character of the prehistoric dwelling structures, their interpretation demands strict argumentation. Without going into the details of this demonstration, we thought it might be useful to recall here in a few words the methodology on which the analysis of the space in inhabited areas is based[2]. A necessary condition for this analysis is obviously the particular dynamics of fluvial deposits that must bury the remains without transporting them. Whenever the deposition process allows it, as is the case at Etiolles, observation of the general organization of the remains reflecting the state of the place as it was abandoned provides a first approach: the hearths, the accumulations of flint, and particular aspects of artifact distributions (e.g., "*effets de paroi*") provide the clearest evidence of the way people arranged space on the site. Furthermore, the systematic use of the refitting technique applied to hearthstones and to numerous flint blocks knapped on the site yields a particularly rich source of data allowing an analysis of behavior within the camp. This methodology reveals both how they dealt with the space for their activities and how they disposed of waste material. In addition, refitting of cores has shown that relationships of contemporaneity existed between several dwelling units and has enlarged the framework of the analysis. The subject matter therefore becomes the camp inhabited by several social units[3]. The way is opened now for research on the social organization of nomadic groups that have settled down for a time at Etiolles. In addition, the very existence of several floors of inhabited areas provides an opportunity for diachronic analysis.

The Organization of the Camps

At the current state of research, we are aware of 20 inhabited units arranged around hearths scattered over about 1,000 square meters and distributed on at least 6 levels. The units have not all been studied, but refitting already gives an idea of patterns within the settlement.

Dynamic spatial analysis enables us to differentiate two types of structures: dwellings, and auxiliary hearths that do not seem to have been covered by a shelter. These auxiliary hearths were associated with the dwellings in a complementary context (Olive 1993). The most complete analyses concern the single level that corresponds to the inhabited units U5 and P15, locations which have already been the subject of monographs (Pigeot 1987, Olive 1989). This level, which now includes two contemporaneous dwellings[4] and three auxiliary hearths, shows an adjustment to the local environment — the two dwellings or tents were set up on the upper part of a bank, whereas the auxiliary hearths were located along the slope of a former channel of a small tributary of the river Seine (the rivulet of Hauldres) (Figure 2). We must point out immediately that the total surface of these inhabited areas is and will remain unknown since it has been disturbed in its northwest part by road construction. Coring carried out on the other side of the road produced no further archaeological remains, and we can assume that the occupied surface did not extend beyond the disturbed area. However, we still do not know either the number of the dwellings in the camp or the number of auxiliary hearths that may have been present[5]. Excavation and on-going research on other living floors reveal a similar organization of space, depending on changes in the local environment linked to the meandering of the stream.

[1]This camp was discovered in Autumn 1971. Excavations began in June 1972 and have taken place in June and July of every year since that time. The settlement is excavated by a team of students from the University of Paris I, including both confirmed researchers and beginners from France and from other countries. Students receive training courses in refitting, interpreting stratigraphic sections, and excavation methods. The same people have been in charge of the settlement since 1972: M. Olive (CNRS), N. Pigeot and Y. Taborin (professors at l'Université de Paris I). Funds for the excavation are granted by the Ministry of Culture, the National Research department and in large part by the council of Essonne.

[2]For a more detailed argumentation about the Etiolles settlement, see the following papers: Pigeot, 1987, Olive, 1993.

[3]Of course, Etiolles is not the only settlement to raise such issues. The application of a similar methodology to all the Magdalenian and Late Glacial sites of the Paris Basin provides some comparative space/time data on a regional scale.

[4]In fact, the refitting of knapped flints only reveals a partial contemporaneity, since dwelling U5 was occupied for a longer period.

[5]Thanks to the general concentration of inhabited areas discovered until now we suspect that the number of dwellings could not have been much bigger . We may hope that the extension of the excavation to an area further off the road will enable us to judge the extent of the camps and to estimate thus the importance of the groups who have come to settle at Etiolles.

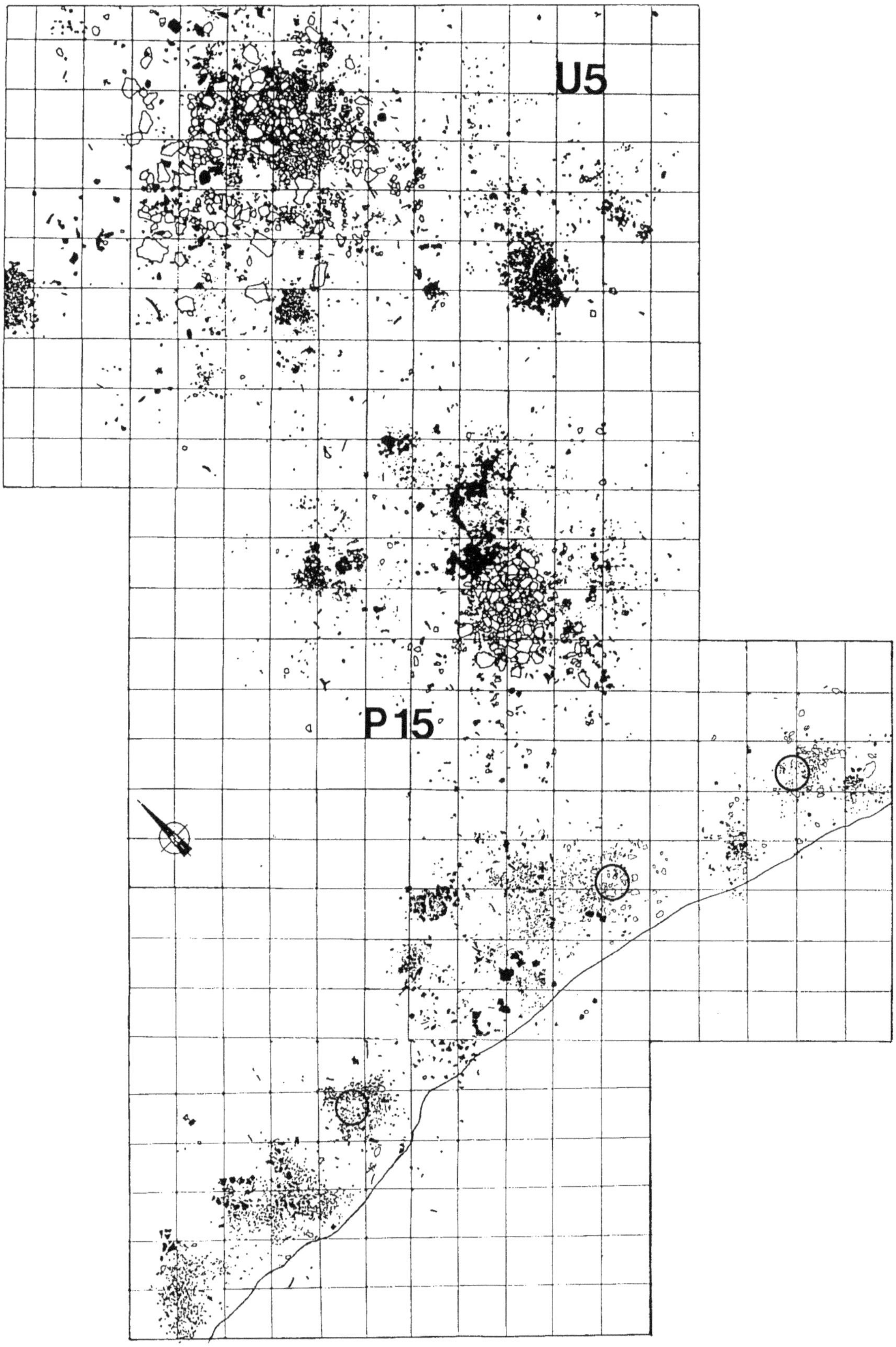

Figure 2. The living floor including contemporaneous dwelling units U5 and P15.

The Variability of the Arrangements

The dwellings at Etiolles are characterized by some recurrent features such as the central location of the hearth in the shelter, and by a certain similarity in the way the space was used, with activities taking place around the hearth inside the shelter as well as outside. Waste from flintknapping was thrown out of the protected area in order to keep the area surrounding the hearth clean, though it is constantly occupied for various activities.

Beyond this similarity, an impression of diversity among structures prevails (Taborin 1983). This diversity is apparent in the presence or absence of particular features. The dwelling that shows the heaviest structure is unit U5, whose limits are defined by a circle of stone slabs and whose inner space is divided by two lines of stone (Figure 3). Two dwellings are simply surrounded by stone slabs, while four others have no stone structures (Figure 4).

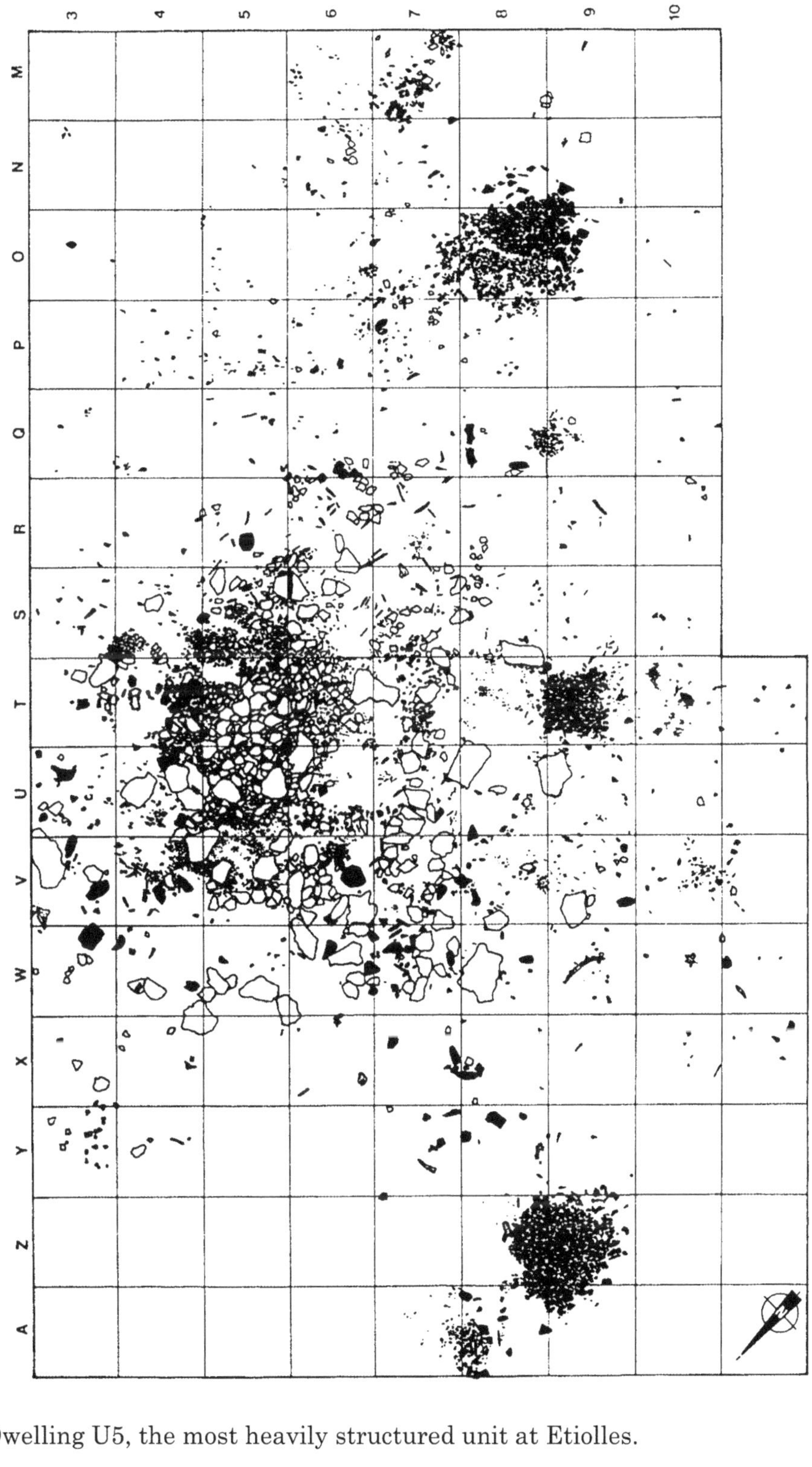

Figure 3. Dwelling U5, the most heavily structured unit at Etiolles.

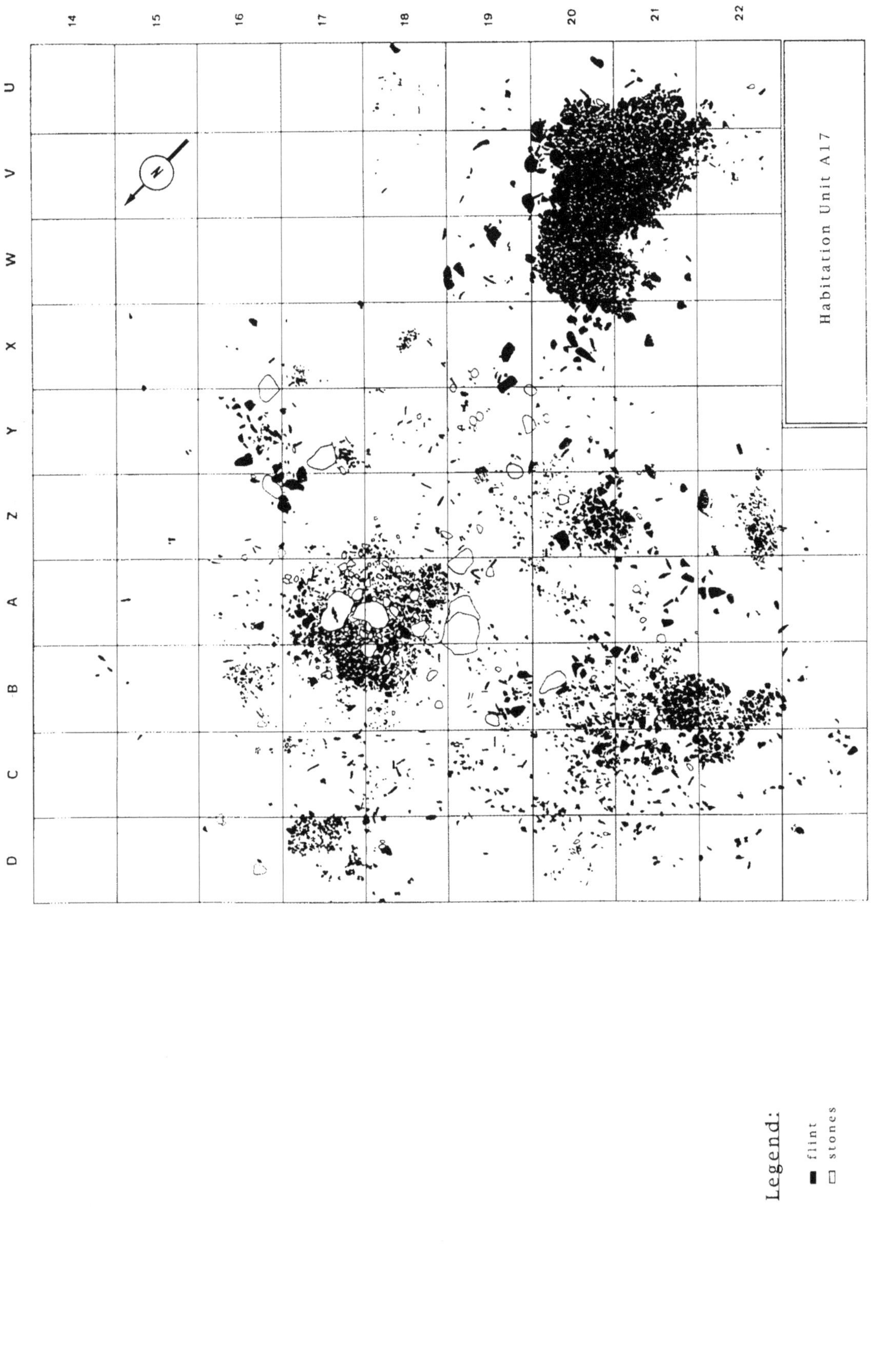

Figure 4. Dwelling Unit A17, lacking stone structures.

Hearths also vary in form, and no single model strictly prevails, either inside or outside the dwellings. Inside the domestic structures, hearths either covered or surrounded by stones can be found. The auxiliary hearths, although generally comprising fewer stones, also show some diversity, including types with or without stones, and with or without a slab lining.

This variety among dwelling units may be due in part to variations in duration of occupation, but can also be related to the spatial organization of the camp. The contemporaneous dwellings U5 and P15 illustrate this lack of standardization in the huts. U5 is clearly structured while P15 lacks any stone structure. We are still unable to find an explanation (or explanations) for these differences, but we can propose different hypotheses based on the function of structures, the duration of occupation of different structures, the evolution of behavior, or just a certain freedom of choice in the organization of dwellings. It seems that we cannot reduce this variation to a univocal diachronic explanation.

Activities in the Camps

In addition to the analysis of dwelling structures, the artifact inventory and the evaluation of activities performed on the site provide other means by which to examine site function. Still, the analysis encounters many problems, including differential preservation of remains. In addition, the areal extent of the site, the number of its inhabitants, and the length of their stay at the site, all parameters likely to determine site structure, remain difficult to estimate. The 100,000 flint objects found at Etiolles to date indicate the importance of flintknapping in the camp. Nevertheless, this impressive wealth must not divert us from questioning the actual role flintknapping played during the Magdalenian occupations at Etiolles and more generally in the whole cycle of their activities.

Multiple Activities

Living in a given space implies that the occupants spent a certain lapse of time at the site and that they must have carried out a lot of various domestic and technical activities. As we have said before, flintknapping left numerous traces at Etiolles. Magdalenian occupants made the most of good local material (from 70 to 80 cores at the richest dwellings to 20 at the poorest ones). This material occurs in the form of unusually large blocks (Mauger 1994). Nodules 20 cm long are common at Etiolles, while those 40 cm long are quite numerous and some very large nodules measure as much as a meter in maximum dimension[6] (Figure 5). The debitage of such material obviously demanded great skill and a very specific process to produce long regular blades in series (Pigeot 1987). Although flintknapping behavior might differ from one place to another, all the stages of the operational sequence which produced flint tools are represented at the site, from core shaping to the shaping and resharpening of tools[7]. The exceptional size of the blocks, the very special care required for their shaping, and the large number of flaked cores in each unit have combined to produce a sizable volume of flint waste accumulated into dense layers and piles (Figure 6).

Compared to the large number of flintknapping by-products, shaped artifacts are relatively rare at Etiolles. For example, 503 shaped tools represent only 2.32% of the total count of lithic material in unit U5[8]. In two other units where the debitage assemblage was just as large, Q31 and A17, the percentage of retouched tools was less than 1%. This is of course very little compared with the figures yielded by some of the richest units at the Magdalenian site of Pincevent (for instance, 8.8% for T112 and 5.5% for V105).

Those figures could make us think that among the thousands of blades produced at Etiolles, few of them were eventually used. Yet the number of tools, in the classic sense of the term, represents only part of the number of flintknapping products actually used. A better understanding of artifact use must take into account the blanks showing partial, irregular retouch and above all the numerous blades that were selected and removed during production. Among these blanks are included what we call the "missing blades" not found during refitting and separated out, often in large numbers, to be carried away from the knapping area. Blades make up the majority of products selected in this way. Consideration of these blades, whether they were actual or possible tools, significantly enlarges the number of useful products, and leads us to correct our impression of a very moderate use of flintknapping products. For example, in unit U5, 50% to 80% of blades from each debitage have been used or selected in a special way (Pigeot 1987).

The presence of such tools, sometimes found in the camp, thus provides proof that activities other than flintknapping were carried out. The nature of these activities can be determined through microwear analysis. The microwear analysis H. Plisson performed on some of the tools shaped in

[6]That is the reason why the number of cores appears to be a more relevent criterion than their weights if we are to compare the relative importance of the knapping activity between the different sites of the Paris Basin.

[7]A general behavior observed in the surveyed units must have consisted in bringing the cores to the site after having knapping them to test their quality.

[8]In this number we include about 40 tools made on exogeneous material that were brought onto the site as utilitarian products.

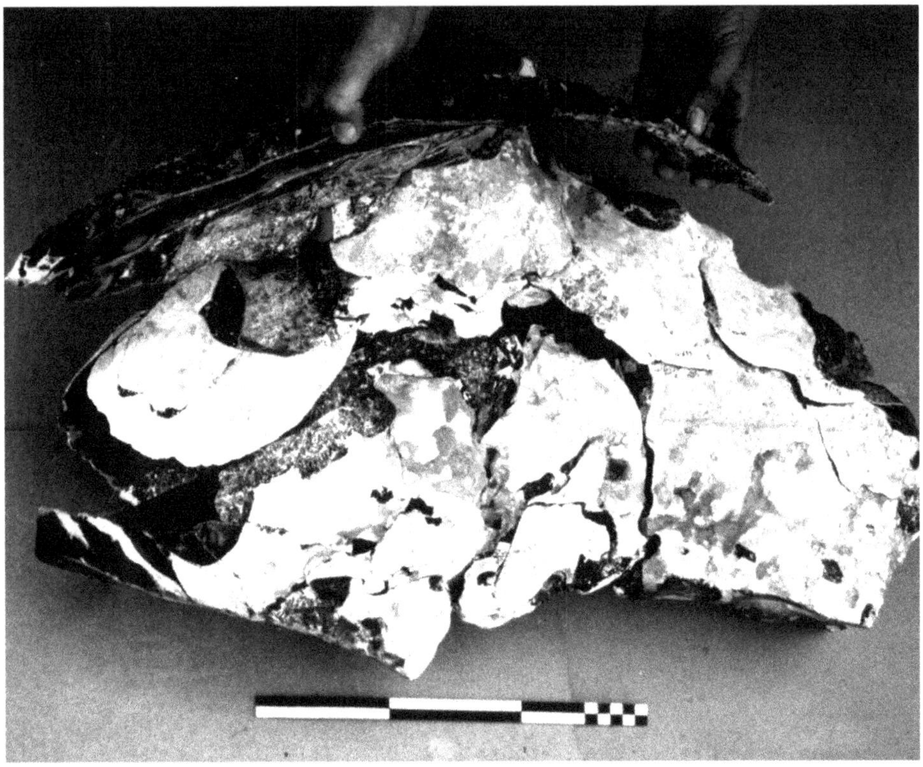

Figure 5. The exceptional raw material found at Etiolles.

Figure 6. A large concentration of flint waste.

dwelling unit A17 (34 out of 77 studied tools yielded positive results) reveals several types of activities, such as butchering, skin-working (mainly on dry skins), and work on wood and bone (Larrière-Cabiran 1993). The presence of ochre around the domestic hearths and a few antlers (one of them with grooves) confirms our assumption. The usual range of activities performed on living floors can be found at Etiolles (Plisson 1985).

Compared to the sites of Pincevent and Verberie, evidence of hunting is definitely less abundant at Etiolles. Animal remains are few compared to the rich lithic material. Several species were present, including buffalo, horse, mammoth, and reindeer, but on the whole it is reindeer that predominate (Poplin 1994)[9]. The numerous backed bladelets found near the hearths testify to hunting activities in an indirect way. As in most of the Paris Basin sites, the number of backed bladelets found varies from one unit to the next (over 45% of the shaped tools in U5, 25% in A17, and 17% in P15).

Nothing is surprising in the fact that the Magdalenian occupants of Etiolles did not only make blades. The tasks they accomplished in the camps include all the necessary daily ones such as supplying and consuming food and maintaining fires. The question is whether or not these activities actually played a secondary role compared to flintknapping. Answering this is of course tricky, if not impossible. However, we would like to point out two facts that might account for a possible functional specialization of the site. First of all, the image of Etiolles as a "bone desert" (quoting F. Poplin) must be corrected in view of the fact that bones were badly preserved. Even if the Etiolles fauna was not as rich as that at Pincevent or Verberie, the fact remains that the filter of time has been drastic (Poplin 1994). The second point deals with the microwear observations on the utilitarian artifacts. To appreciate properly the importance of the activities performed on the site, it would be interesting to analyze all the tools (in the broad sense of the term) including the selected blades that were used on the site. However, this kind of work encounters two major difficulties: first, it is a long term concern, and second, microwear traces are not easy to decipher on material from Etiolles because of a soil polish, particularly on the oldest levels[10] (Christensen 1991).

The Spatial Organization of Activities

Current studies show that activities performed at Etiolles took place in a space that is not limited to the domestic area inside the dwelling. The whole grouping of contemporaneous units, including dwellings and satellite areas, must be used for the analysis of spatial organization of activities.

Domestic hearths have always represented privileged concentration points, since we can find there both numerous products and various tools, ochre, animal remains, and even a few shell ornaments. Some are characterized by a bipartite spatial organization on both sides of the central hearth, depending on the length of occupation or on the nature of the activity. Units U5, P15 and A17 display such a bipartite division of the sheltered space (Pigeot 1987, Olive 1989, Larrière-Cabiran 1993) (Figure 7).

Spatial organization can also be viewed within the framework of the single activity of debitage. Though flintknappers at Etiolles did produce exceptional series of long blades, not all the cores are equivalent and not all debitage sequences show the same quality. N. Pigeot's analysis of the debitage carried out in the dwelling unit U5 reveals that this variability in quality is connected to the complexity of the debitage operation (elaborate or opportunistic debitage) and also to the technical skill of the knappers (Pigeot 1987). Some debitage sequences in particular, showing manifest signs of clumsiness, are attributed to beginning knappers who were probably young individuals[11]. Within the dwelling, space appears to be organized relative to the hearth, such that the best knappers, who work the biggest cores, work near it, while less skilled knappers work far from it.

Activities were also carried out outside the dwellings, around the tents or dwellings and around the auxiliary hearths. Near auxiliary hearths, activities were less diversified than those performed around domestic hearths. Products of debitage done on the spot, a few remains of animals and also blade blanks and shaped tools were discovered near them. In addition, some activities that demanded a lot of space seem to have been carried out in isolated places. For example, an aggregate of bones representing some remains of three horses was discovered quite far from any hearth.

The current analysis of the occupied floors where dwellings U5 and P15 were integrated shows how some cores and blade blanks were removed on both sides of the knapping space. We can infer from this that interaction between the two dwelling units was complex (Figure 8). For example, several knapped blades produced in U5 were left in the dwelling P 15 or near an auxiliary hearth. Further refitting may fill the gaps in the blade series and help to find some missing blades. This circulation of blade-blanks requires researchers to think about the extent of the occupied surfaces, either at the level of the whole site, which is difficult if not impossible, or at least a part of the site extended enough to allow some assessment of the scale of on-going activities.

Variability in Activities

Since flintknapping was particularly well developed at Etiolles, analysis can show the ways in

[9]Only two bones (a shoulder-blade and a fragment of hip-bone) indicate the presence of mammoth. The possibility that these parts might have been cut off of carcasses of animals that died a natural death may be considered (Poplin 1994).

[10]This kind of analysis would obviously raise a problem of profitability. Limiting the analysis to a single level of the site might represent an intermediate solution. To date, only 3 units have been throughly tested or surveyed by H. Plisson and M. Christensen.

[11]The reality of the knapping apprenticeship is based on a documented survey that mixes technical, economical and spatial criteria (Pigeot, 1987).

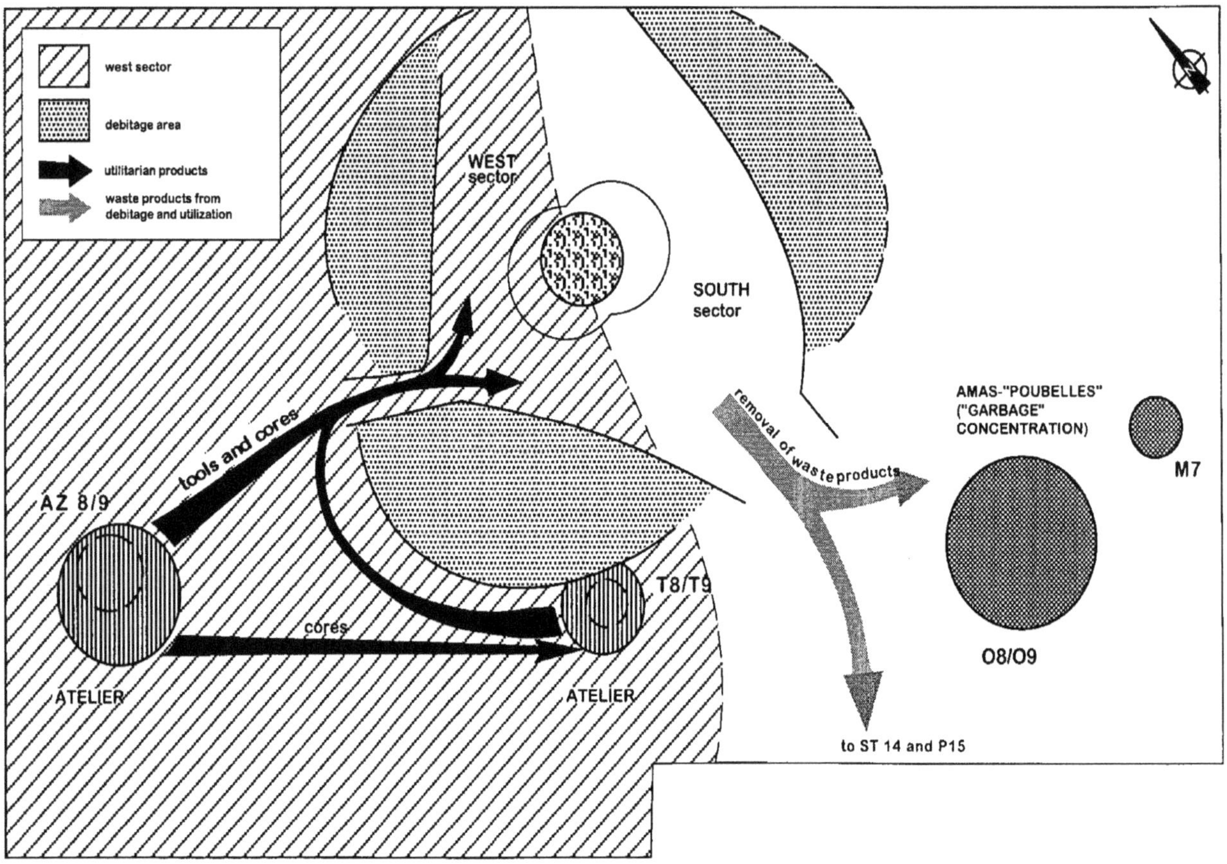

Figure 7. Spatial organization of dwelling unit U5.

which lithic raw materials were exploited, and demonstrates that this activity does not conform to a single, constant model. The flexibility of the socio-economic management of the flint industry can be observed in the scale and spatial organization of the debitage.

No two dwellings at Etiolles are alike in evidence of flintknapping activities. As mentioned above, the number of knapped cores and the proportion of shaped tools, especially of retouched bladelets, differ significantly from one unit to another. Dwelling units U5 and W11 are both surrounded by a circle of stones and dedicated to knapping activities (about 70 cores knapped in each), but the former yielded 500 shaped tools, the latter only a dozen.

Since numerous parameters are likely to be involved, these variations are difficult to account for. Length of occupation, though only one possible parameter, might be an explanatory factor, as suggested by connections established through refitting between contemporaneous dwellings U5 and P15. The habitation U5, which shows the richest artifact assemblage and the most structured dwelling, was occupied during two main phases which show an evolution in the spatial organization of debitage. Habitation P15 was occupied for a shorter time, corresponding to the first period of occupation of U5 (Pigeot 1987, Olive 1988).

The spatial organization of knapping activities related to knapping expertise and the quality of cores shown in U5 is not found in all the dwellings. The contemporaneous unit P15 and the more recent Q31, for instance, do not exhibit this kind of spatial patterning. In both of these units, a different socio-economic management of the raw material related to a lesser quality of the blocks produced entails an organization of space that is less strict. Various explanations can be proposed to account for this difference in the exploitation of lithic resources. First, we might propose a certain social complexity, involving acquisition of tools by some groups from others in the camp. In P15 , a lack of large cores providing long series of blades has led occupants of this dwelling to borrow beautiful blade blanks produced in the next dwelling, U5 (Olive and Pigeot 1992). Second, an alternative explanation focuses on changes due to flint availability. The knappers working in Q31, one of the latest in the settlement, generally produced smaller and less regularly shaped blocks. If the best beds of flint were difficult to reach later in the Magdalenian occupation of Etiolles, this might account for this phenomenon.

The Choice of the Site and its Function

How should we understand the choice of Etiolles as a location for these activities, and what role did

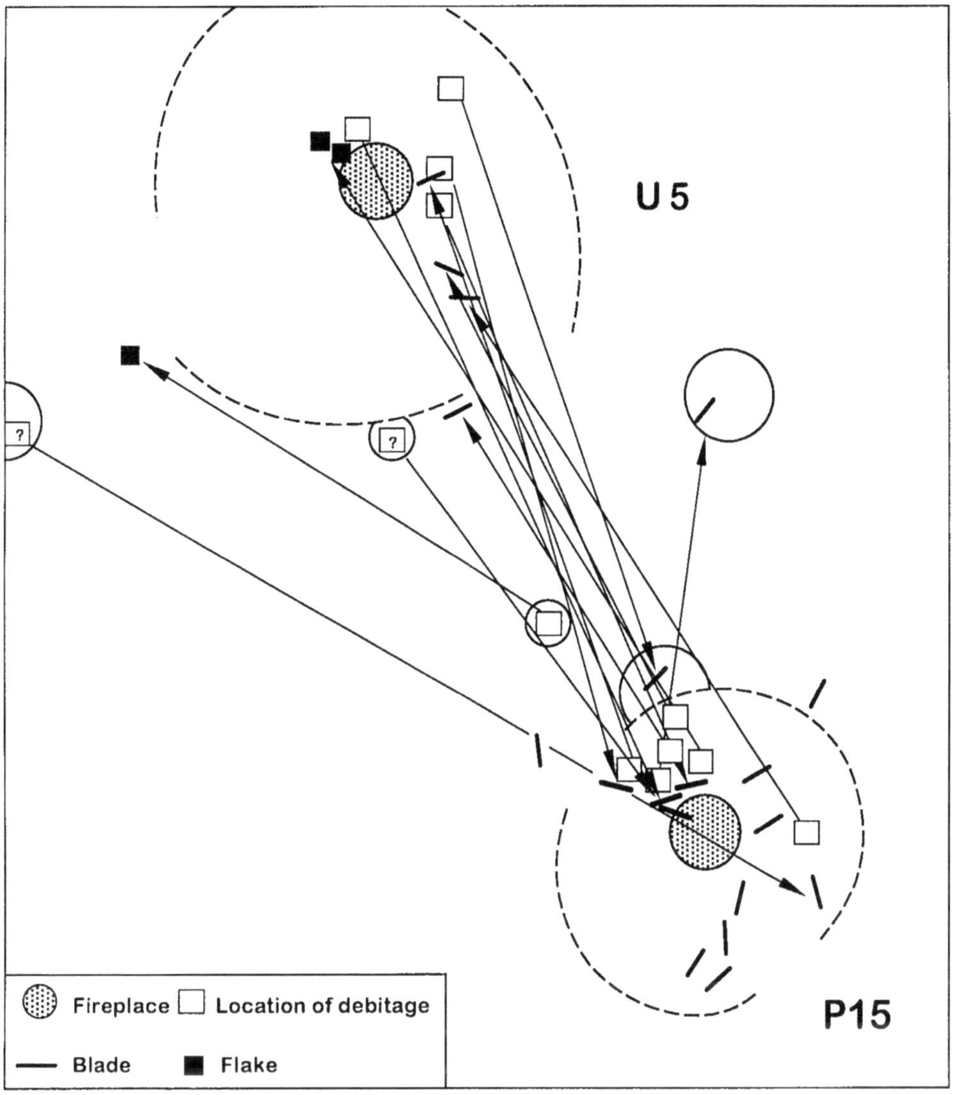

Figure 8. The movements of blade and flake blanks between dwellings U5 and P15. Shaded circles indicate location of central fireplaces, dashed lines show boundaries of dwellings, solid arrows show movement of products of debitage based on results of refitting, with arrow pointing from area of production to discard location. Symbols indicate the type of product: black lines are blades, black squares are flakes.

Etiolles play in the whole cycle of activities? We may wonder why Magdalenian groups came to Etiolles and, first of all, whether they made any preliminary plans or found the place again by chance as they wandered the region. We might expect that site location involved intentional planning by analogy with the seasonal migrations of arctic populations (Binford 1980) and because we suppose that the precariousness of the hunters' economics must have been abated by a rational organization of their needs. Dependence on hunting must have required a degree of planning to allow adjustment to the habits of prey species. On the other hand, if we assume that rich pastures and a stable ecosystem in this environment ensured dense populations of prey species that did not move over long distances, we might suppose that Magdalenian hunters did not plan journeys in set periods, but that they wandered their territories according to their immediate and various needs.

How can we favor one or the other view? If they did plan their journeys, the periodic return of the same group must appear in our field data. If they settled at this location opportunistically, analysis of structures and artifacts should reveal a diversity caused by occupation of the site by different groups. The question of the choice of locality is therefore to be tackled in two ways: either this choice was carefully considered and planned in advance, and thus related to other stopovers and needs, or it was an immediate choice meeting their needs independently of any advance plan, in which case the choice of location was totally disconnected from time, space or production patterns.

The distinction criteria between the two types of choices concern the field of tool and weapon technology and morphology and the spatial organization of dwellings, as well as the range of activities performed. This analysis is difficult because the range of activities performed is a matter of individual behavior, whose diversity, routine and randomness make research of possible planning uncertain.

The idea of a planned choice must be demonstrated by the identity of the group who returned to the site and by repetition of the primary on-going activities during the occupation. The identity of the group is based on technique, debitage, morphology and tool-making, the mode of occupation, the organization of the dwelling and the nature of the prevalent activity. We admit, though, that the proximity of flint beds has always entailed the same kind of activity, flintknapping, in a most intensive way whatever the motivation of the occupation. Technical organization on a large scale seems to be a good criterion in favor of a planned visit. Alternatively, the idea of an expedient choice of location by different groups could be supported by variability in the techno-morphology of implements and in spatial organization of dwelling structures, and from diversity of prevalent activities.

What evidence along these lines does Etiolles bring to this argumentation? Many indications, including the technique of blade debitage, the composition of tool assemblages, and the organization of activities inside and outside dwellings, argue for the general cultural identity of the groups coming to Etiolles. However, the dwelling structures do not all indicate the same degree of planning in the choice of location. Several dwellings share a characteristic, intensive production of blades from large cores. In these dwellings, the occupied surface is extended, the surface reserved for flintknapping is clearly delimited, and accumulations of refuse are significant. Dwellings at the site vary considerably, though, in the presence or absence of structural elements, size and construction of hearths, and the organization of space.

A common point stands out—in most of the known dwellings, the number of cores proves that debitage was an important activity (Pigeot *et al.* 1991). The exceptional sizes of some nodules and consequently the dramatic aspect of the refuse clusters have imposed the idea that there were flintknapping areas, more precisely areas where blade blanks were produced for use outside the knapping area. Etiolles might have been famous for supplying blade blanks, thus entailing the periodic return of groups! Let us try to examine the arguments supporting this hypothesis. Debitage must have played an important part in the life of the groups coming to Etiolles. The collective memory was probably reminded of its location thanks to the quality of raw materials there, the large dimensions of the nodules, and the isolation of the flint-beds in the center of the Paris Basin. All these things made Etiolles a very attractive geographic location. Yet flint was not rare in the Paris Basin, since rivers transported it extensively. Debitage was more or less intensively carried out in dwellings at Etiolles, and was more important in large dwellings than in smaller ones. We have then to think of two new parameters which cannot be determined: the number of occupants and the duration of their occupations at Etiolles. The number of occupants can be the product of the wildest imagination! Who can say whether the people of Magdalenian times lived in their shelters squeezed together in a narrow space like the Patagonians or dispersed around like the Tuaregs! It is interesting, though, to note that the analysis of technological knowledge has enabled us to provide evidence of different stages of training for the debitage of cores, probably performed by adolescents or even children (Pigeot 1987). This analysis suggests the presence of whole families, or at least allows us to dismiss the idea that only a small group of skilled knappers came to Etiolles in order to supply the rest of the community with blade-blanks.

The duration of stays at Etiolles was limited by the rhythms of the Seine floods annually linked to the thaw. We are thus in a position to admit that undoubtedly the people could not have camped at Etiolles in February or March. But throughout the rest of the year, occupation could last 10 months or a few weeks, or a period of occupation could be divided into several stays separated by periods of absence. The occupation layers are separated by layers of fluvial deposits that distinguish them from one another. In some cases the dynamic study of the remains reveals, as in dwelling U5, changes in the interior organization of the dwelling that are significant enough to suggest re-settlement within a structure similar to the one existing before the flood.

As a matter of fact, the duration of occupation is calculated according to patterns of spatial organization. This is easy as long as occupations of the site are short in duration and tends to become more confused as daily refuse clutters up the space. At Etiolles, evidence that fireplaces were cleared out by dumping ashes and that lithic refuse was moved out of activity areas and toward debris concentrations provides some indication of length of occupation. Nevertheless, the fact that we do not know the number of occupants makes such inferences weak. In addition, the example of the relationships established between the dwelling units U5 and P15 shows that the occupation of these two units, however strongly linked, partly overlaps: P15 is contemporary with the first occupation of U5 (Olive 1988). This is enough to show how little we can rely on any single proposed model of camp occupation. Neither the composition of the camp nor the intended duration or purpose of the occupation responded to strict rules!

On the other hand, the current analysis demonstrates that there were mutual relationships closely linking the occupants of several units. The concept of a settlement established by family clusters or related groups begins to emerge from this analysis. Each dwelling is only a unit among the whole functional camp of a group. We might suggest that the activities located in a particular dwelling were profitable to everybody. For example, while some individuals flaked flint, others could hunt, gather plants or wood, or work around hearths. This view enlarges the narrow link uniting the prevalent activity to the intended purpose of the occupation, for the activity that seems to be increased tenfold might have been an answer to the immediate needs of several families. Families had a continual need for blades and flint tools, and those needs were probably reactivated by the presence of large nodules.

As a result, Etiolles flintknapping debitage cannot be used as evidence that this activity belonged to a planned scheme or deliberate choice of place with periodic returns. Nor does it imply a narrow restriction of activities, such that the occupants of each hut were compelled to flake flint. The camp and the exchange of goods among the occupants provide ample proof of the fact that some of them, probably the most skilled knappers, must have flaked flint more skillfully and in a larger proportion than the others.

The view we now have of the techno-economic life of the Magdalenian groups at Etiolles is different from the one originally suggested. This view is based on extensive excavations that indicated the relationships between the occupants of different units. The diversity of the structures and the variability in lithic production and in shaped tools can be understood by considering that each dwelling, with its activities, represents only a part of the larger dwelling unit of a human group whose economic balance cannot be understood without taking into account all the other units of the camp (Conkey 1980).

The Hunting Territory of the Magdalenian of Etiolles

It seems that in spite of some variation in flintknapping skills, the morphological homogeneity of the tools is too obvious to dismiss the idea of a cultural identity of the group coming to Etiolles. The dates when people came to the site are roughly framed a little before and during the Bølling, and spread over a few centuries. So, the question of these groups' familiar territory is naturally stated in rather specific temporal and geographical terms (White 1985).

Several different means are available to estimate the extent of the group's territory. A thorough analysis of the sources of non-local shell used for ornamental purposes and of ochre found at Etiolles, where there is no trace of art, is under way. Also, since the flint yielded by Cretaceous beds near Etiolles is easily identified, we can expect to find this material deposited elsewhere in the form of tools at locations visited after leaving Etiolles. These tools are always recovered at the bottom of occupation debris, which proves they had been used during the occupation. The research done by M. Mauger (1994) on the silicified fossils has shown that they came from a variety of locations within the Paris Basin. Similarly, all the shells so far identified are fossils derived from the numerous Eocene deposits of the Paris Basin. Ochre, on the other hand, probably came from the region near Burgundy.

Specific facts have been revealed about that first level of territory estimation, limiting it to the Paris Basin. The non-local flint comes from different sites along the rivers (Eure, Loing, Basse-Marne and Oise), which suggests travel routes revolving around Etiolles (Figure 9). Did Magdalenian groups move from the outskirts of the Paris Basin to the center to settle at Etiolles at the right period from the beginning of spring until the wintertime? Or did the wanderings of each group lead them to chance upon the Tertiary flint beds and to stop at Etiolles, one site among others such as Les Tarterets, or even unknown? Stopping at Etiolles could enable them to increase their usual rhythm of flintknapping (though they may not necessarily choose to), while performing their numerous domestic tasks. As the introduced tools were made of lithic materials coming from different places, we can suppose that there was some mobility, but the possible routes are many, and the longest distances do not exceed 200 kilometers, which corresponds to a normal exploitation territory for large animal hunters.

Within the territory, some areas may have been busier, better known or more famous than others, especially those where some shells and non-local flint can be found. Examples might include the Basse-Marne region or the Basse-Oise valley (Taborin 1994). But the comparison cannot go beyond this observation because, unlike flint tools, the shell ornaments were probably kept by their owners for a long time. There is no evidence of work on shell objects during the occupation at Etiolles. As a result, the origin of flint from distant places fairly marks off the region through which groups traveled before settling at Etiolles. Source areas of the shells, on the other hand, merely shows that the inhabitants of Etiolles had knowledge about more distant territories or had social relationships with groups living in these places. If we extend this question of distant territories to include data from Pincevent, Verberie and Marsangy, this interpretation is reinforced. The shells found in each of these multiple-unit settlements come from the same places as those found at Etiolles. That is to say they originate from the Lutetian or the Auversian levels (Taborin 1994). The currently known fossil deposits define a territory of about 50 to 150 kilometers around each Magdalenian settlement (Figure 10).

Developing the premise of a concentration in a well-defined territory yields a wealth of data for understanding the inner evolution of the Magdalenian individual living north of the river Loire. This stability seems unusual compared to other regions. In the Perigord and in the Pyrénées ornamental shells come from various and faraway places such as the Atlantic Ocean. The shells picked up from the beaches and shell-marl deposits of the

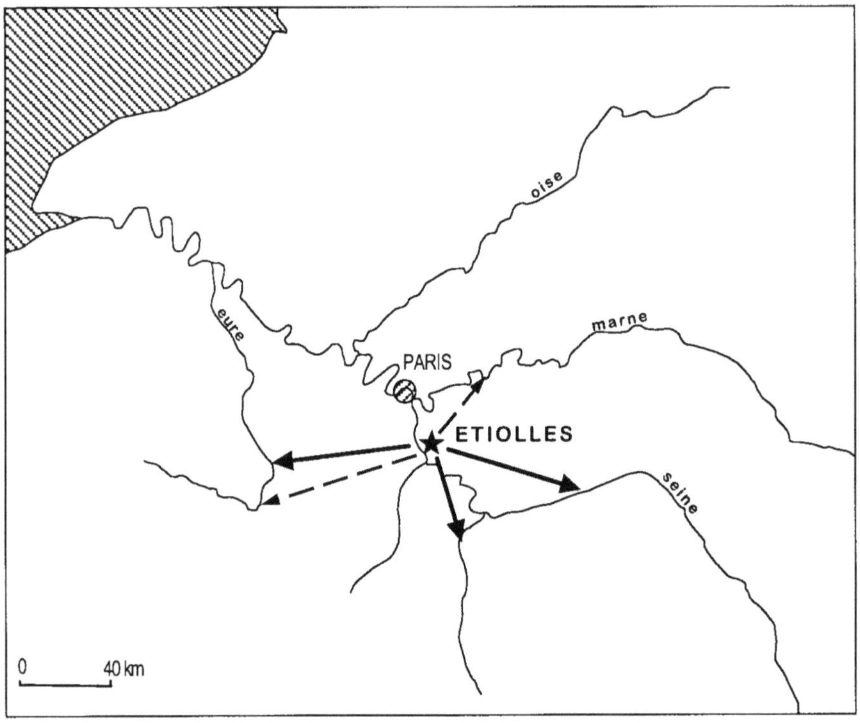

Figure 9. Source areas of non-local flint found at Etiolles.

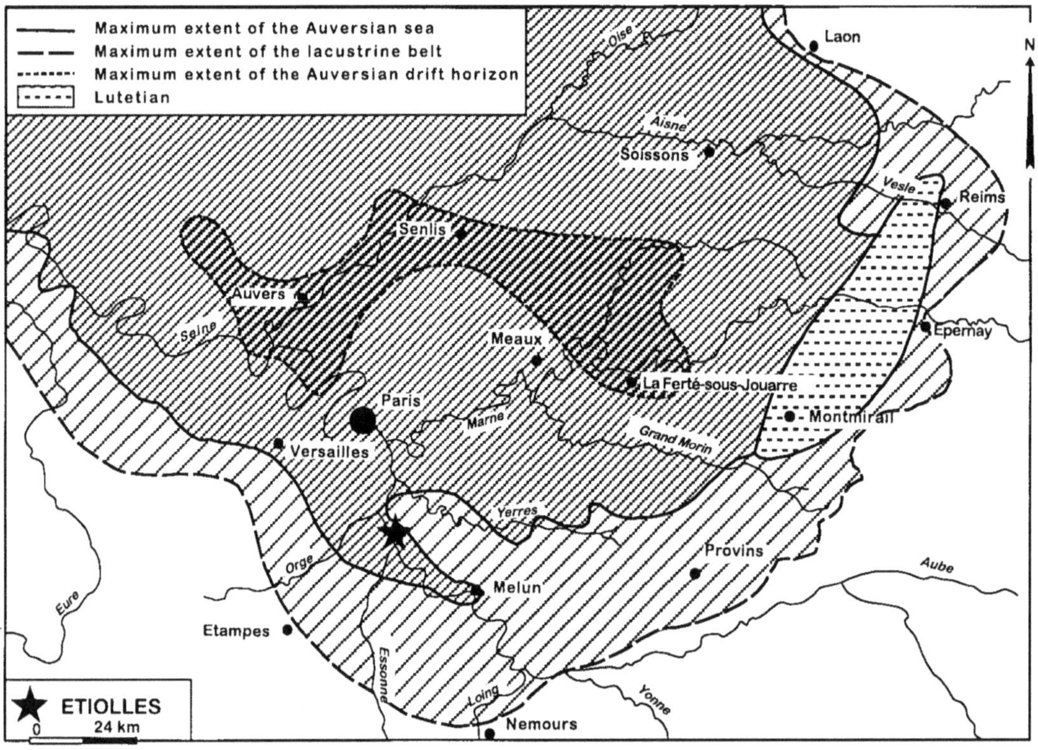

Figure 10. The boundaries of the Lutetian and Auversian levels in the Paris Basin.

Bordeaux region come from the Mediterranean sea, whereas the fossilized shells come from Mediterranean Pliocene deposits. Nevertheless it would be interesting to compare the Paris Basin with another regional concentration of Magdalenian sites, in the Cantabrian region for example, where Magdalenian shell ornaments seem to come from sources in the local area. In Cantabria, though, the source areas of lithic raw materials used still remain to be determined. All these arguments tend to evoke a time (around 13 000-12 000 B. P.) when the demography was balanced, the availability of resources stable, the way of life harmonious and sufficient. Magdalenian groups lived in the Paris Basin during an exceptional time!

References Cited

Binford, L. R.
1980 Willow smoke and dogs' tails: Hunter-gatherer settlement systems and archaeological site formation. *American Antiquity* 45:4-20

Brézillon, M.
1971 Les Tarterets II, site paléolithique de plein air à Corbeil-Essonnes (Essonne). *Gallia-Préhistoire* XIV(1):3-40.

Christensen, M.
1991 *Les burins d'Etiolles. Etude fonctionnelle des burins de l'unité O16.* Mémoire de DEA, Université de Paris I.

Conkey, M.
1980 The identification of prehistoric hunter-gatherer aggregation sites: The case of Altamira. *Current Anthropology* 21:609-630.

Coudret, P, M. Larriere-Cabiran, M. Olive, N. Pigeot, and Y. Taborin
1994 Etiolles. In *L'Environnement et l'habitat magdalénien au centre du Bassin parisien,* edited by Y. Taborin, pp. 132-146. Documents d'Archéologie Française 43. Editions de la Maison des Sciences de l'Homme, Paris.

Larrière-Cabiran, M.
1993 Organisation et utilisation de l'espace dans une unité d'occupation magdalénienne à Etiolles (Essonne, France): l'habitation A17. Thèse de doctorat, Université de Paris I.

Mauger, M.
1994 L'approvisionnement en matériaux siliceux au Paléolithique supérieur. In *L'Environnement et l'habitat magdalénien au centre du Bassin parisien,* edited by Y. Taborin, pp. 78-93. Documents d'Archéologie Française 43. Editions de la Maison des Sciences de l'Homme, Paris.

Olive, M.
1988 *Une habitation magdalénienne d'Etiolles. L'unité P15.* Mémoire de la Société Préhistorique Française 20. Société Préhistorique Française, Paris.
1989 Etiolles: quels foyers pour quels usages?. In *Nature et fonction des foyers préhistoriques: Actes du colloque international de Nemours 1987,* edited by M. Olive and Y. Taborin, pp. 197-207. Mémoire du Musée de Préhistoire d'Ile de France 2. Association pour la promotion de la recherche archéologique en Ile-de-France, Nemours.
1993 En marge des unités d'habitation d'Etiolles: Les foyers d'activité satellites. *Gallia-Préhistoire* 34:85-140.

Olive, M. and N. Pigeot
1992 Les tailleurs de silex magdaléniens d'Etiolles: Vers l'identification d'une organisation sociale complexe. In *La pierre préhistorique,* edited by M. Menu and P. Walter, pp. 173-185. Ministére de la Culture, Paris.

Pigeot, N.
1987 Magdaléniens d'Etiolles: Économie de débitage et organisation sociale. *Gallia-Préhistoire, supplément XXV.* CNRS, Paris.

Pigeot, N. , M. Philippe, G. Le Licon, and M. Morgenstern
1991 Systèmes techniques et essai de technologie culturelle à Etiolles: Nouvelles perspectives. In *25 ans d'études technologiques en préhistoire. Bilan et perspectives,* pp. 169-185. XIèmes Rencontres internationales d'archéologie et d'histoire d'Antibes. Éditions de l'Association pour la promotion et la diffusion des connaissances archéologiques, Antibes.

Plisson, H.
1985 Etude fonctionnelle d'outillages lithiques préhistoriques par l'analyse des micro-usures: Recherche méthodologique et archéologique. Thèse de doctorat, Université de Paris I.

Poplin, F.
1994 La faune d'Etiolles: milieu animal, milieu taphonomique, milieu humain, In *L'Environnement et l'habitat magdalénien au centre du Bassin parisien,* edited by Y. Taborin, pp. 94-104. Documents d'Archéologie Française 43. Editions de la Maison des Sciences de l'Homme, Paris.

Schmider, B.
1975 Le gisement paléolithique supérieur des Tarterets I, à Corbeil-Essonnes (Essonne). Stratigraphie, outillage lithique,

organisation des vestiges. *Gallia-Préhistoire* 18(2):315-340.

Taborin Y.

1983 La configuration des sols d'occupation à Etiolles. *Cahiers du Centre de Recherches Préhistoriques* 9:33-44.

1994 Les coquillages marins. In *L'Environnement et l'habitat magdalénien au centre du Bassin parisien*, edited by Y. Taborin, pp. 70-77. Documents d'Archéologie Française 43. Editions de la Maison des Sciences de l'Homme, Paris.

White, R.

1985 Upper Paleolithic Land Use in the Perigord. A Topographical Approach to Subsistence and Settlement. British Archaeological Reports, International Series 253. BAR Publishing, Oxford.

9. Lithic Raw Material Economy in the Late Glacial of the Paris Basin: Case Studies from Magdalenian, Federmesser and Long Blade Technology Hunter-Gatherers

Boris Valentin, Michéle Julien and Pièrre Bodu

Picking up the thread from the preceding chapter, this paper presents a complementary view on lithic raw material economy through time in the late glacial Paris Basin. Through comparative analysis of core technology on several Magdalenian, Federmesser and Long Blade Technology sites, they present a discussion of general trends in lithic production over time, as well as intersite variability within each time period. The authors compare three Magdalenian sites (Pincevent, Ville-Saint-Jacques, and Marolles), in the southern Paris Basin, and point out consistent features of raw material acquisition and use. Long-distance transport is indicated at all the sites, but the quantities are small, and most raw material is locally acquired. The evidence suggests movement of Magdalenian groups in repetitive cycles over distances on the order of 50 to 100 kilometers. Use of local materials indicates a variable focus at each site on production of standardized blanks not intended for immediate use and simplified production of blades for immediate use. Variations between the sites are attributed to adjustments to seasonally varying activities, particularly seasonal changes in scheduling of hunting and need for advance preparation of tools. Discussion of the later time periods are based on summaries of evidence from several recently discovered Federmesser and Long Blade Technology sites. Though there seem to be no radical changes in lithic raw material acquisition in the Allerød, lithic production nevertheless undergoes considerable change, resulting in reduced productivity and standardization of blank production. The authors relate these changes to changed economic and settlement patterns in late glacial environments. By contrast, Long Blade Technology sites dating to the Preboreal show a uniform focus on standardized production of blades at what are apparently specialized workshop sites. Southern Paris Basin sites belonging to the Long Blade Technology complex can only be understood in the context of larger regional patterns of group movement and anticipatory production. *–Editors.*

The region of the Paris basin considered here is the southeastern Ile-de-France, around the confluences of the rivers Yonne and Loing with the river Seine (Figure 1).

Comparison of Some Magdalenian Sites

Around the confluence of the river Yonne with Seine, about 80 km southeast of Paris, several Magdalenian sites have been discovered in the vicinity of Pincevent. We present here a comparison of three of them that have yielded, along with a great quantity of lithic material, a large number of animal remains. Pincevent is situated in the valley along the left bank of the Seine (Leroi-Gourhan and Brezillon 1966, 1972). Just above Pincevent, about one kilometer away on the edge of a plateau rising above the valley, there is another large site. This site, Ville-Saint-Jacques, is known through intensive surface collections made over the past century (Degros *et al.* 1994). In the valley between the Seine and Yonne, two other major sites have recently been discovered in the construction of an autoroute and a gravel quarry (Alix *et al.* 1993). These two sites, "Le Grand Canton" and "Le Tureau des Gardes", are close together and it is possible they might be directly linked. Both together will be referred to as Marolles in this paper. All the radiocarbon dates for Ville-Saint-Jacques and Marolles fall within the phase of occupation at Pincevent, and they might have been contemporary (around 12,500 BP).

As is now well established, the site of Pincevent was repeatedly occupied in autumn by groups focused on reindeer acquisition and subsequent processing of animal carcasses. Reindeer remains represent between 97 and 99% of all the fauna. Other animal remains include some hare, very few horse bones, and wolf (David and Enloe 1992). At Marolles, on the other hand, (especially at Le Grand Canton where faunal analysis is further advanced at present) horse comprises the majority of the fauna, making up about 98% of identified specimens (Bridault and Bérnilli 1999). The remaining 2% includes occasional reindeer and some wolf. According to archaeozoological data, animal acquisition at Marolles is not restricted to a particular season. Excavation data indicate that these very extensive sites were probably not

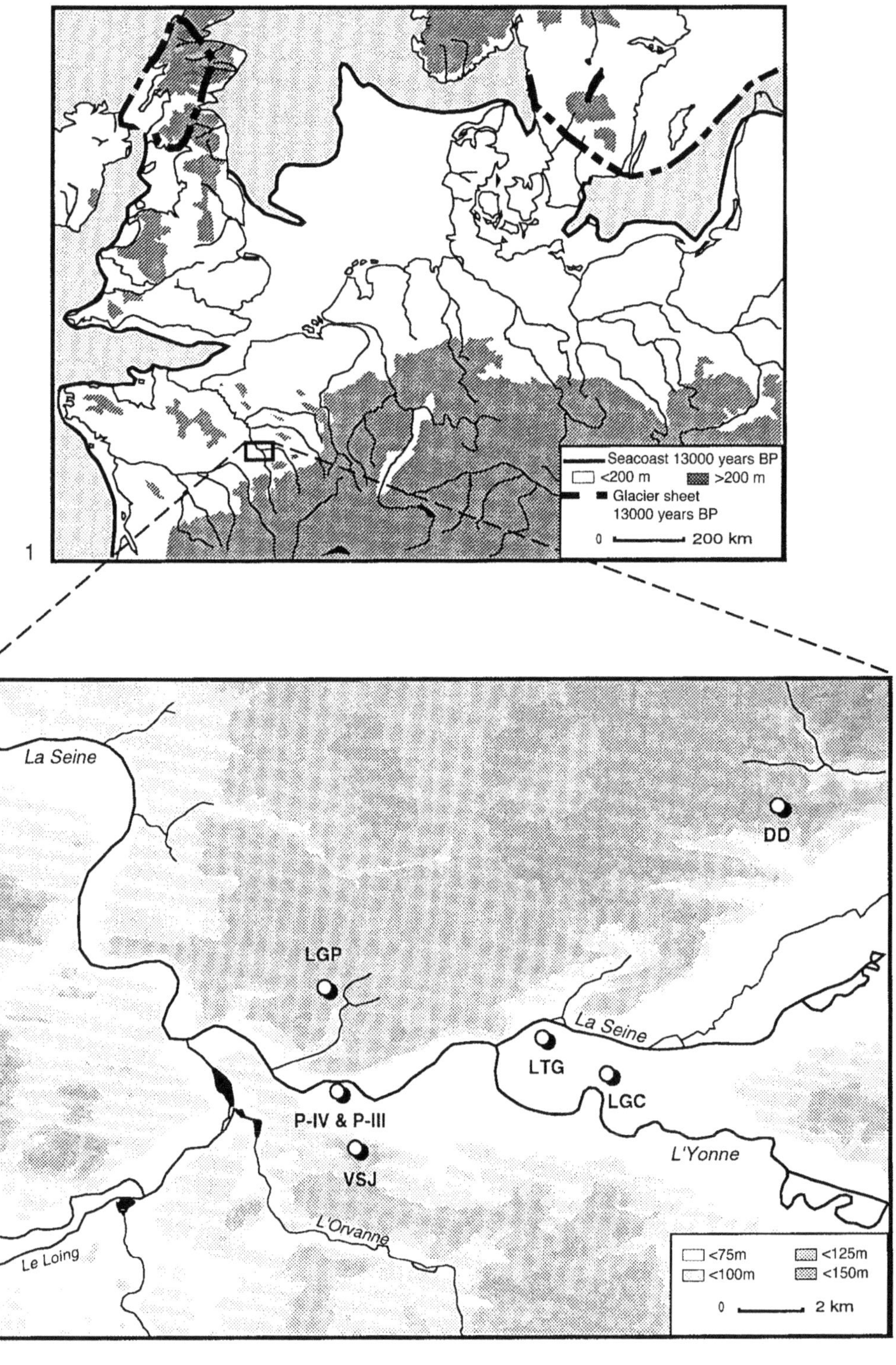

Figure 1. Map of the study area. 1: Location of the study area in northwestern Europe during the Late Glacial (after Burdukiewicz 1986). 2: Location of sites mentioned in the text. P-IV, Pincevent, Level IV (Magdalenian), VSJ: Ville-Saint-Jacques (Magdalenian), LGC: Marolles - "Le Grand Canton" (Magdalenian), LTG: Marolles - "Le Tureau des Gardes" (Magdalenian), P-III: Pincevent, Level III (Federmesser industry), LGP: La Grand Paroisse (Federmesser industry), DD: Donnemarie-Dontilly ("Long Blade Technology").

occupied all around the year but rather through various successive short periods, over a long time interval. Activities were diverse around each of the numerous hearths at Marolles, with no evidence for the kind of activity specialization often seen at Pincevent hearths. In the third site, Ville-Saint-Jacques, where the dwelling structures have been destroyed by colluvial phenomena and plowing, faunal remains are more diverse: reindeer bones make up the majority, but horse is of nearly equal importance, and some others species such as wolf, brown bear and polecat are present. The season of occupation at Ville-Saint-Jacques is still unknown, but based on the thousands of lithic tools collected and the presence of numerous heated stones, it seems clear that this site was occupied repeatedly over quite a long period.

Consequently, the subsequent comparison will be focused on two specialized subsistence settlements and another one that shows a more diversified strategy of animal acquisition.

Raw Material Procurement

Local Flint

The sites of this region are all located on the southern Cretaceous fringe of the Paris Basin, where flint is abundant and easily available in the stream alluvium and in slope solifluction deposits, where they are in a secondary position, or sometimes in primary position in calcareous beds (Mauger 1985). Time investment in raw material procurement could then be very minor depending on the location of the settlements.

At Pincevent, the origin of the nodules used is mainly fluviatile and water flow has often broken their natural irregular protuberances (Bodu 1993; Karlin 1991; Ploux *et al.* 1991) (Figure 2). The cortex of these nodules has a washed appearance, and is generally lacking fragments of the embedding chalk. Magdalenian flintknappers at

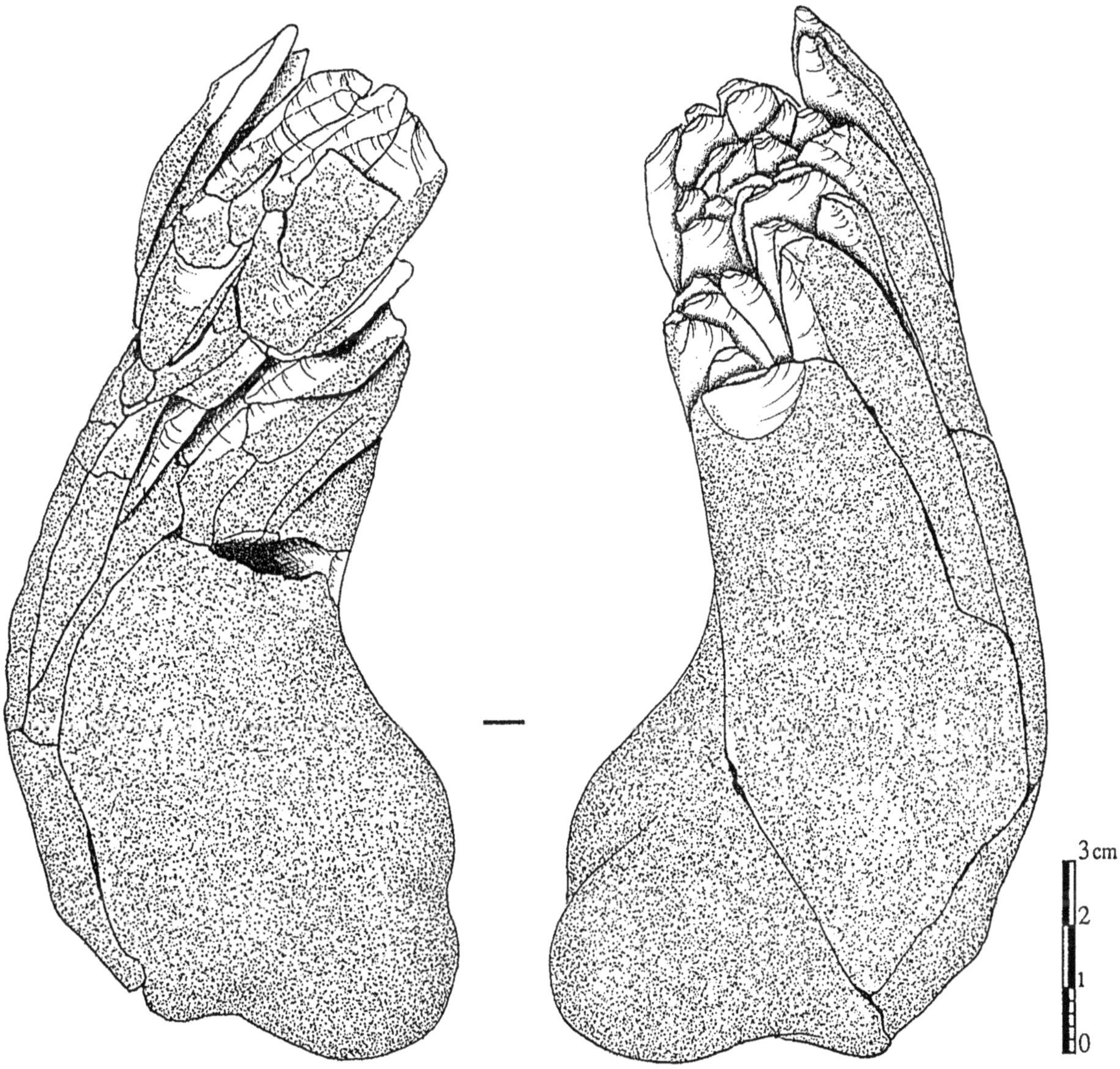

Figure 2. Bladelet core from Pincevent, Level IV-20 (Magdalenian). Pieces absent in the refitted nodule are blanks for backed bladelets. Drawings by D. Molez.

Pincevent usually selected ovoid forms with narrow cross-sections that present natural edges or convexities well suited to rapid core preparation. Rarely, some small slabs were also collected. The dimensions of selected blocks are rather small, mainly between 100 and 200 mm long, and the overall quality is fair (mediocre or more or less good).

At Ville-Saint-Jacques, the nodules, collected from secondary deposits, have the same geological origin as those exploited at Pincevent, but they present external surfaces less rolled (Degros *et al.* 1994). The overall quality seems better, and the nodules, often longer, must have come from a slightly different source.

At Marolles, the morphology and quality of the nodules suggests at least two sources in alluvial and colluvial deposits (Alix *et al.* 1993; Hantai 1993; Valentin *et al.* n.d.). Numerous small nodules of mediocre quality might have been collected from the surface directly at the settlement. Some of these show internal flaws caused by freezing. Others, bigger and of better quality, could come from the Seine and Yonne banks, and tend to be much more regular and of larger dimensions than nodules used at Pincevent (Figure 3, no. 1). This heterogeneity of raw material, greater here than in the two previous sites, may be the consequence of a diversity of sources or of the absence of preliminary quality control at the collecting location. Among various possible explanations, two at least can be proposed for the small nodules: either good material was in short supply, or these mediocre local materials reflect opportunistic acquisition of raw material for immediate use. In the latter case, the occupants of the site may have preferred to use immediately available but mediocre material if the objective of lithic production was only a few blanks for some urgent activity. This opportunistic strategy seems to contrast with the situation at Pincevent or Ville-Saint-Jacques where, within a supply of mediocre to good quality, a particular nodule morphology was selected and nodules were usually tested for homogeneity before being transported to the site. This suggests more time investment for greater future productivity.

Distant Flint Sources

In the three sites (see summary in Table 1), flint from sources other than the locally available Senonian flint has been found in small quantities. At Ville-Saint-Jacques, the flints are too patinated for identification of provenience. If we consider the total number of retouched tools, non-local raw materials make up only about 1.7 to 2% of this total at Marolles, but are more common at Pincevent, where non-local material amounts to 4% in Habitation 1, and 8% in the upper level of the IV.20 campsite. At both Ville-Saint-Jacques and Pincevent, this non-Senonian flint comes from Tertiary deposits. Although the analysis is not advanced enough at Marolles to determine their origin, at least five varieties of non-Senonian materials could be distinguished. At Pincevent, only one distant source seems to have been utilized during the entire phase of occupation. This material has been identified as coming from Ludian deposits in the center of Ile-de-France, near the confluence of the Marne and the Seine, about 50 kilometers north/northwest of Pincevent (Mauger 1985). It is interesting to note that these deposits extend to the region of the Magdalenian site at Etiolles, where the same raw material was extensively used some centuries before (Olive and Pigeot 1992; Pigeot 1987).In the three sites, the non-local flint is generally present in the form of laminar or lamellar blanks and of tools, with very few cores (two at Pincevent, one at Marolles). Since the various tools or blanks of Tertiary flint do not seem to have been used for special tasks and show the same range of functions as the local ones, it may be assumed that this non-local flint was not obtained through exchange but that it was probably brought from a previous halting-place by the groups occupying these sites, in anticipation of the first needs at the new camp.

Tool Production

In all these Magdalenian sites, as in the Upper Magdalenian in general, core reduction was aimed at production of standardized blades and bladelets that could be either used without further shaping

Table 1. Comparison of Raw Material Use

Raw Material	Pincevent Hab. 1 + IV.20	Marolles	Ville-Saint-Jacques
Local sources	Cretaceous flint	Cretaceous flint	Cretaceous flint
Origin	alluvium	alluv./colluvium	alluvium
Quality	fair	bad/good	good
Size	small to medium	small/large	medium
Distant sources	Tertiary flint	Tertiary flint	Undetermined
% of assemblage	4 to 8%	1.7 to 2%	too patinated

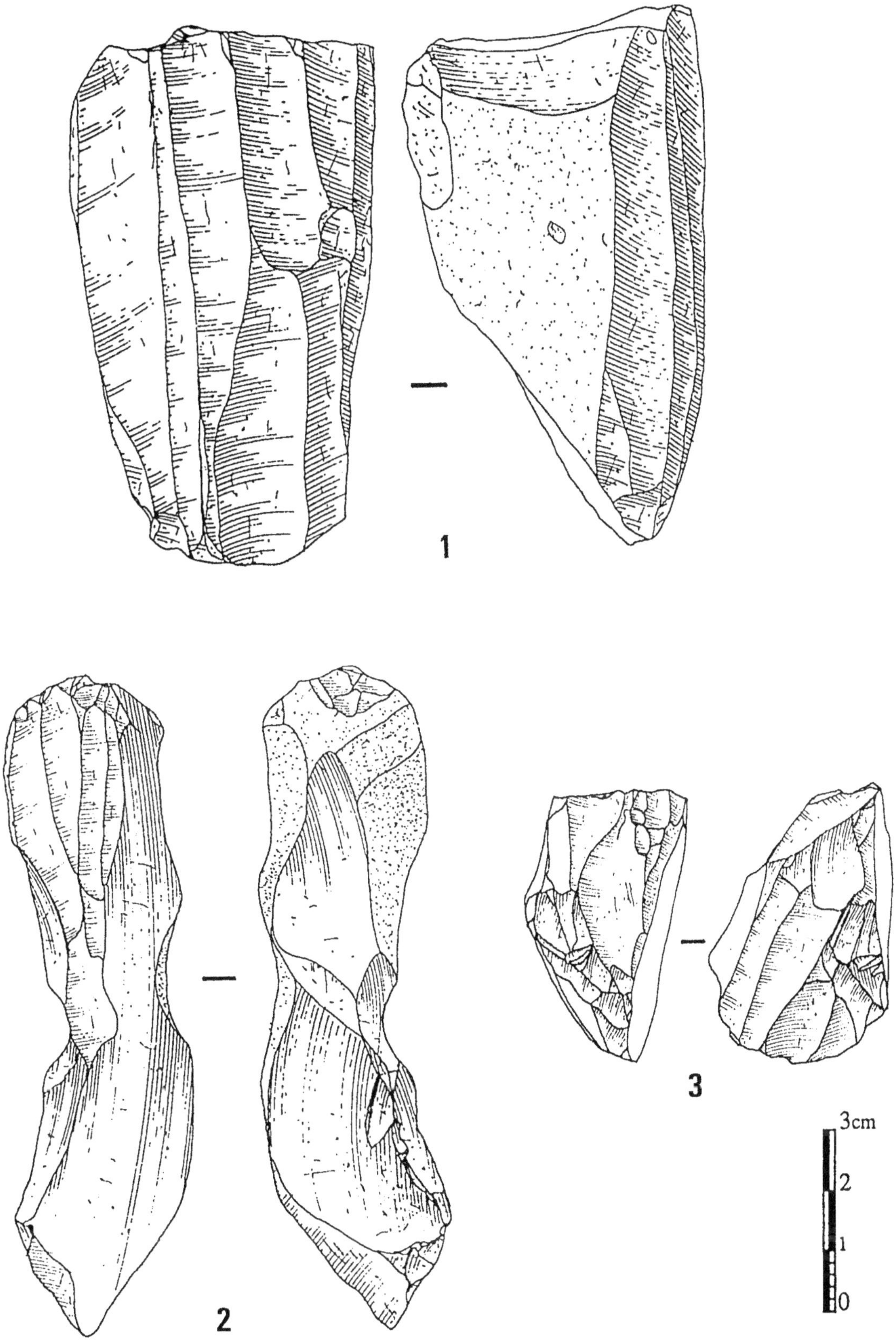

Figure 3. Le Grand Canton (Magdalenian). 1: a productive blade core, 2: a less productive bladelet core on a flake, 3: a bladelet core on which several faces were worked in succession. Drawings by P. Alix.

as knives or transformed by retouch into a variety of tools such as burins, end-scrapers, borers, backed bladelets, etc. (Figure 4). The core technology is characterized by an emphasis on high productivity, systematic use of soft hammers (probably made of cervid antler) for the extraction of laminar blanks, and careful preparation and maintenance of the volumes to be knapped (Audouze *et al.* 1988; Bodu and Valentin 1993a; Karlin 1991; Karlin *et al.* 1993; Olive and Pigeot 1992; Pigeot 1987; Ploux *et al.* 1991; Valentin 1995).

Though the extent of excavated areas and methods of data collection differ from one site to another (surface collection at Ville-Saint-Jacques, rapid exploration over several thousand square meters at Marolles), or even from one level to another (as at Pincevent, where the IV.20 campsite has been excavated over an area of 4,500 square meters while the area surrounding the older Habitation 1, regarded as a whole unit of 70 square meters, remains unexplored), a preliminary comparison must be attempted if we want to go further in this comparison of raw material strategies.

Comparison of the proportions of backed bladelets and laminar tools (Table 2) shows large discrepancies between Pincevent level IV.20, where backed bladelets make up 61% of the total number of retouched pieces, and all the other assemblages. At Marolles, only 19% of retouched tools are backed bladelets, similar to Habitation 1 of Pincevent (20%). The proportion is known with less certainty for Ville-Saint-Jacques, where the tool assemblage comes from a surface collection. For a broader comparison, we also provide data from the Magdalenian site of Marsangy, located about 30 kilometers south of Marolles and Pincevent, on the bank of the Yonne river (Schmider 1993). It can be seen that the ratio of backed bladelets is about the same as at Marolles. In these two sites, in addition, there are a few convex and angular backed points and, at Marsangy, some shouldered points. At Marsangy, shouldered or backed points represent around 7% of the tool assemblage and seem to have been abandoned in the course of their manufacture, while at Marolles these make up less than 5% of tools. This does not radically change the discrepancy in numbers of backed pieces with level IV.20 of Pincevent.

Table 2. Laminar Tools* and Backed Bladelets as a Percentage of Total Retouched Tools

SITES	% laminar tools	% backed bladelets
Pincevent IV.20	39%	61%
Pincevent Hab. 1	80%	20%
Marolles	81%	19%
Marsangy	82%	18%

*Burins, scrapers, piercers and truncations. Note that in these assemblages, flake tools always make up less than 1% of total retouched tools.

Tool Production Strategies

Coming back to the tool production strategies, it is then possible to find some variation among the three sites. At Pincevent, in level IV.20, it seems that, in a majority of cases, bladelets were the primary aim of lithic production, though other kinds of blanks, by-products of bladelet production, were often recovered for fabrication tools (Figure 2). However, at Pincevent IV.20, there is also evidence of opportunistic production of blades for immediate use in domestic tasks, with blanks retouched and probably used on the spot. Core reduction is in this case often simplified, since the objective is only to obtain a few blades of more or less predetermined width or thickness. In some other cases, good-quality, larger nodules were selected for skillful production of long regular blades, taken into various areas of the campsite for use or, in the case of some elements not recovered on the excavated surface, possibly reserved for use outside the site.

In Habitation 1 of Pincevent, as at Marolles, the main objective seems to be intensive laminar production to obtain series of regular laminar blanks. A special technical skill has been applied to the pre-shaping of the nodule, the preparation of butts (often by "en eperon" technique), and a careful maintenance of core shape. At Marolles, the best nodules were selected for this intensive laminar production, whereas small nodules of mediocre quality, or frequently by-products of blade production (shaping flakes, core fragments), were used for an opportunistic and very simplified production of bladelets. It appears that time investment in core preparation and reduction varied quite a bit depending on varying needs. The rapid procurement of bladelets, at lower cost in time and energy, does not show a concern for high productivity, rather, simply an intent to obtain a limited number of blanks. The more time-consuming selection of better and bigger nodules not found on the site itself reflects an intention of obtaining good blades for other tools. This variation indicates a certain flexibility in selection of blocks and in reduction strategies, evidently adapted to local needs.

Socioeconomic Implications

The subtle differentiation that appears through this comparison of tool production strategies allows an estimation of certain priorities in the activities performed at the three sites. At Marolles, the most important tendency is the exploitation of good raw material for producing series of regular blades intended for processing and maintenance activities (Figure 4, no. 1-9). The secondary tendency shows more opportunistic strategies of bladelet production (Figure 3, no. 2-3). It seems that, here,

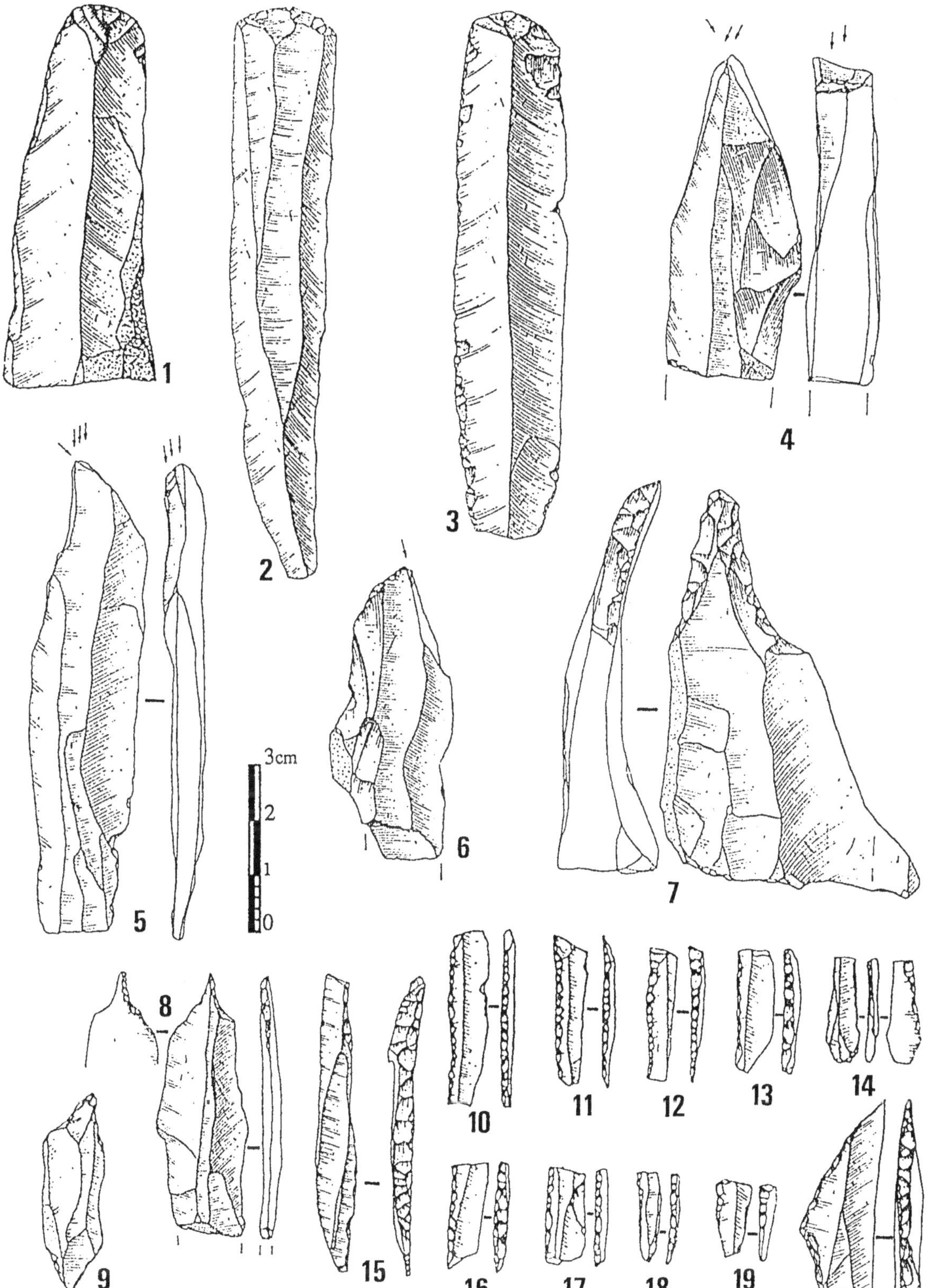

Figure 4. Le Grand Canton (Magdalenian). 1-3: endscrapers, 4-6: burins, 7: borer, 8-9: piercers, 10-19: backed bladelets, 20: shouldered point. Drawings by P. Alix.

bladelet production is not the primary production activity, but was performed for occasional repair of hunting weapons. The qualitative priority would appear here to be accorded to the production of tools related to transformation activities, beyond and after the operational sequences related directly to resource acquisition.

In level IV.20 of Pincevent, on the other hand, the large quantity of backed bladelets, either discarded after use or broken in the course of their fabrication, and the numerous absent blanks in refitted cores indicate that use of barbed points for hunting weapons was important (Figure 2). Although this production activity seems to have been privileged, and carefully prepared for in raw material selection, it must not be forgotten that transformation activities were also performed at the site, either by selecting by-products of bladelet cores, or by obtaining blades for immediate use through intentional knapping. This would suggest that hunting was performed in the course of the occupation and that processing of animal materials took place as soon as the carcasses were brought back from the nearby kill site.

It is possible that these two strategies of raw material procurement and tool production correspond to two functional contexts: activities performed at Pincevent are concentrated around highly seasonal collective hunting, with subsequent carcass sharing between dwelling units, followed by more or less collective animal processing (Enloe and David 1989). Since reindeer hunting was expected to be intensive, optimal conditions had been ensured by collecting selected nodules and preparing weapons. The Magdalenian group here was probably a large one, with several families gathered at the same time for this important seasonal activity.

At Marolles, we know that horse hunting was, instead, a repeated short-term activity over the course of the year. In consequence, it demanded only a smaller group of hunters tracking small horse bands along the stream, and subsequent animal processing was undertaken very near the kill site. This anticipated pattern of activity, involving a short, uninterrupted operational sequence from killing to processing a small number of animals, could require preparation of weapons before arriving on the spot as well as collection of appropriate flint raw materials for butchering and for on-site preparation of the tools needed for the processing animal materials. If, in the course of this process, some other horse bands appeared, they could, with what they found at hand, rapidly repair their weapons and kill more animals. Preparation was thus at short range. It can be proposed that this type of short range hunting was carried out by people who had divided into small groups outside of the season of collective hunting.

In these two examples, relating faunal data to flint raw material procurement and processing might indicate some of the various differences between strategies that move resources to people and those moving people to resources. Returning to consider tool production at Habitation 1 and Ville-Saint-Jacques, the importance of blade tools for transformation activities and the relative scarcity of backed bladelets would tend to suggest a function similar to Marolles. But there are some differences. In Pincevent Habitation 1, reindeer hunting is evidently seasonal, though this remains to be established by re-analysis of the faunal assemblage according to new methods for determining the age of killed animals. One of the main contrasts with level IV.20 is the large number of costal and vertebral remains in Habitation 1, elements that are extremely rare in the upper level. Since backed bladelets are much less numerous than in the upper level IV.20, one explanation could be that hunting took place over a very short time, and not over a period of weeks. Still, part of the Marolles model may be proposed: hunters arrived with prepared weapons and settled at Pincevent Habitation 1 just after the hunting session to process the killed animals and probably consume them more or less entirely on the spot; in the course of the occupation, they may have repaired some of their hunting weapons. At Ville-Saint-Jacques, in contrast, repeated occupations over a long period are related to diversified animal procurement activities, mainly focused on reindeer and horse. This may suggest a kind of foraging hunting during the time of the occupations, but apart from this, the diversity of the tool-kit suggests varied activities, perhaps not only directly related to animal acquisition. Was Ville-Saint-Jacques a base-camp for the more specialized hunting and butchering location of Pincevent? Was it occupied in another period of the year by the same group, or does it correspond to another phase of Late Magdalenian? Answers to these questions remain conjectural. All the questions raised here deserve further investigation.

Some years ago, we proposed a regional model of site function and occupation for the Paris Basin based on the relations between animal or raw material procurement and transformation activities in the course of a hypothetical year-round subsistence cycle. We suggested that economic activities at Pincevent and Verberie were specialized in seasonal reindeer hunting and primary animal processing, while the main economic trend, though probably not the only one, at Etiolles and Marsangy seemed more closely related to transformation activities. These activities included intensive blade production on an excellent flint at Etiolles and diversified tool production at Marsangy, probably corresponding to large-scale work on bone or cervid antler in anticipation of activities during the next season (Julien 1989). If these general economic tendencies have not thoroughly changed and have simply been made more precise (Audouze 1992), the discovery of new sites and the advancement of lithic technology and archaeozoological studies indicate that there existed, in the Paris basin, a lot of other types of settlements, varying with the availability of seasonal or year-round resources, hunting strategies in relation to behavior of the different animals, etc., and, what is more elusive but very attractive, the social habits of the various Magdalenian communities in the course of at least one millennium. Even at Pincevent, it appears that the aim of the activities may have evolved from the lower levels (IV.40, IV.30 and Habitation 1) to the upper ones. Research on Magdalenian economy is just beginning, and every new site discovered, clarifying overall Magdalenian economic behavior, will add a new facet to the general synthesis we are trying to construct.

These preliminary results, that must now be compared to observations made on the Magdalenian of other regions (Rensink *et al.* 1991), point out a certain regularity of flint acquisition strategies in the southern part of the Paris Basin. In this region, Magdalenian hunter-gatherers often exploited local Cretaceous raw material of an adequate but never exceptional quality. Resources found in secondary position (alluvium or slope deposits) appear to have been clearly preferred to those in primary context, the quality of which was however much better. The recorded qualitative differences in raw material procurement between neighboring settlements might be related to slight adaptations of the knapping objectives to seasonal variations in activity. In all these sites, flint acquisition and production are almost totally integrated with other activities. However, evidence is present in each case for local production of standardized blanks for non-immediate use, but this production is often limited and its intended end-point remains difficult to determine (nearby use, for instance in killing sites, or use in other settlements?). Though transport of flint from somewhat more distance sources is everywhere attested, the transported quantities are always small and the quality of the products is rarely better than the best locally produced ones. The origins of these foreign materials are not very diverse and give some evidence of movement in repetitive circuits over rather limited distances (50 to 100 kilometers to Ile-de-France, perhaps a little farther to the south of the Loire river). During the Magdalenian, the southern Paris Basin thus appears to have yielded enough diversified and predictable resources to engender a system of exploitation that, though flexible, is nevertheless rather constant in its general rules.

Evolution of Raw Material Acquisition in Post-Magdalenian Settlements

In the same region, data from recently discovered sites of Allerød and Pleistocene/Holocene transition periods (Bodu and Valentin 1993a, Valentin 1995) allow some consideration of diachronic developments in the evolution of strategies related to acquisition, transformation and management of lithic raw materials at the end of late glacial times.

Federmesser Sites

During the Allerød period of climatic amelioration, Magdalenian industries in the Paris Basin are replaced by Federmesser industries, in which tools bear the mark of a general Azilian trend (Fagnart 1988, 1992a, 1993; Valentin 1995). Insofar as it is possible to determine from the two sites known in the region (Figure 1), raw material procurement strategies do not change radically in this period. Cretaceous materials collected here are of small dimensions and appear to be of a very variable quality depending on the location, while use of non-local flint is non-existent (Bodu 1996; Valentin 1995, 1996). The greatest contrast with Magdalenian is in the transformation of lithic reduction techniques: soft-hammer percussion (with cervid antler) is abandoned, and blade production is always achieved with hammer stones (Figure 5). This requires continuous maintenance of flaking faces, and gives a characteristic rhythm to the knapping sequence, formed by alternating production of elongated flakes and regular blades. Frequent recourse to a second striking platform produces inversions in the course of debitage. The consequences of this flexibility, observed at every step of the knapping process, for raw material economy include low productivity of standardized blanks relative to the quantity of less standardized by-products, and an evolution in the conception of blank production: Federmesser blank production shows a very strict selection of blanks for hunting weapons, and much more flexible selection of blanks for other tools (Figure 6).

Change in environmental factors has often been proposed as an explanation for the transformations of these late Paleolithic groups and, especially, of their techniques. In other northern and eastern regions, several prehistorians noted that the quality of flint raw material was less good than in the Magdalenian (David 1992; Fagnart 1993; Floss 1992; Lauwers 1988), but this contrast is not so strong in the south Paris Basin. Here, the systematic use of stone hammers might be related to the variability of raw materials, since it allows exploitation of their diverse qualities through the flexible application of hard-hammer percussion. If this shift in technique is related in part to ecological

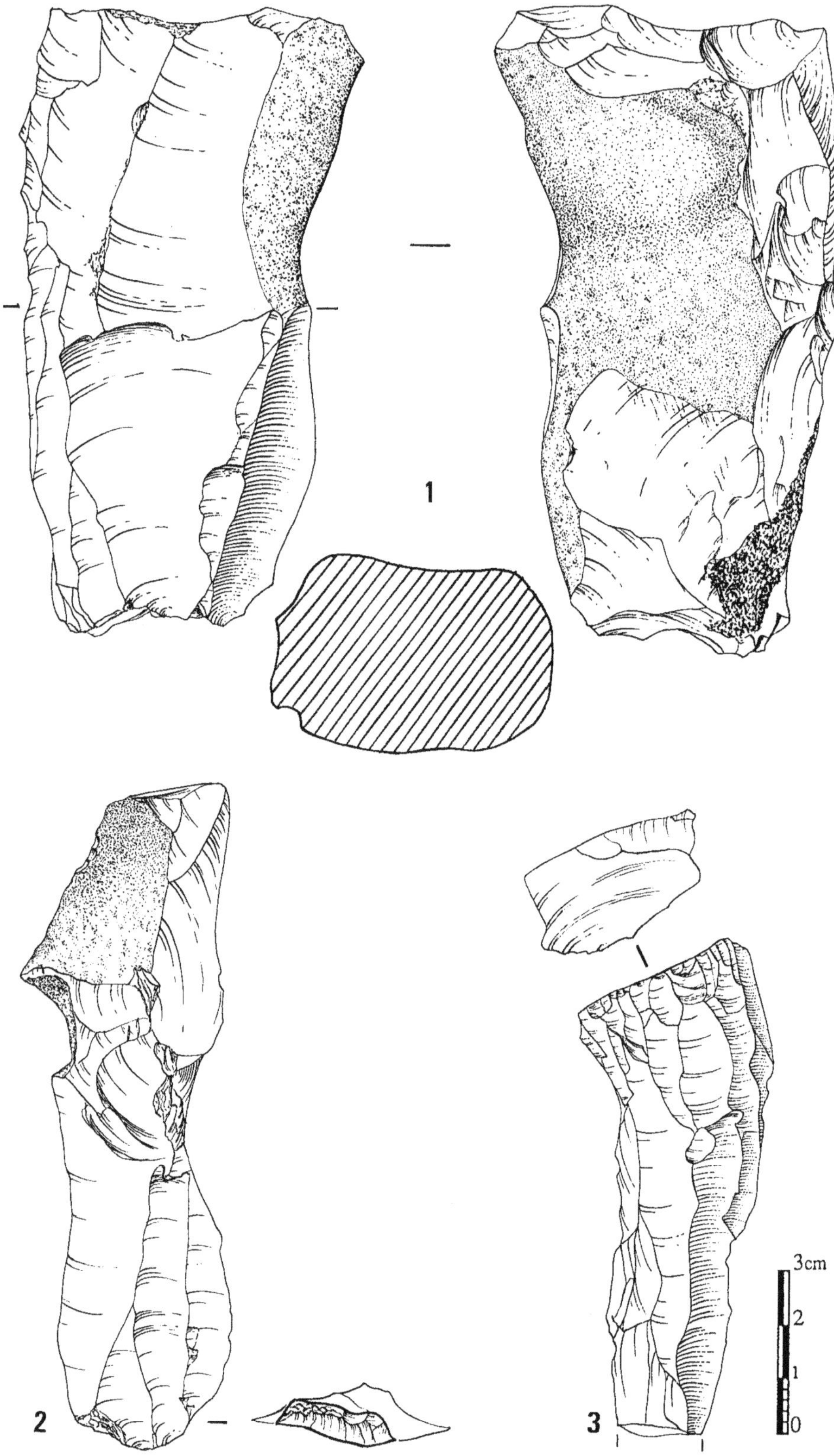

Figure 5. La Grande Paroisse (Federmesser). 1: a core worked with a sandstone hammer, 2: a blade extracted during the first stages of the core reduction process, 3: a rejuvenation blade struck from an opposite striking platform. Drawings by D. Molez.

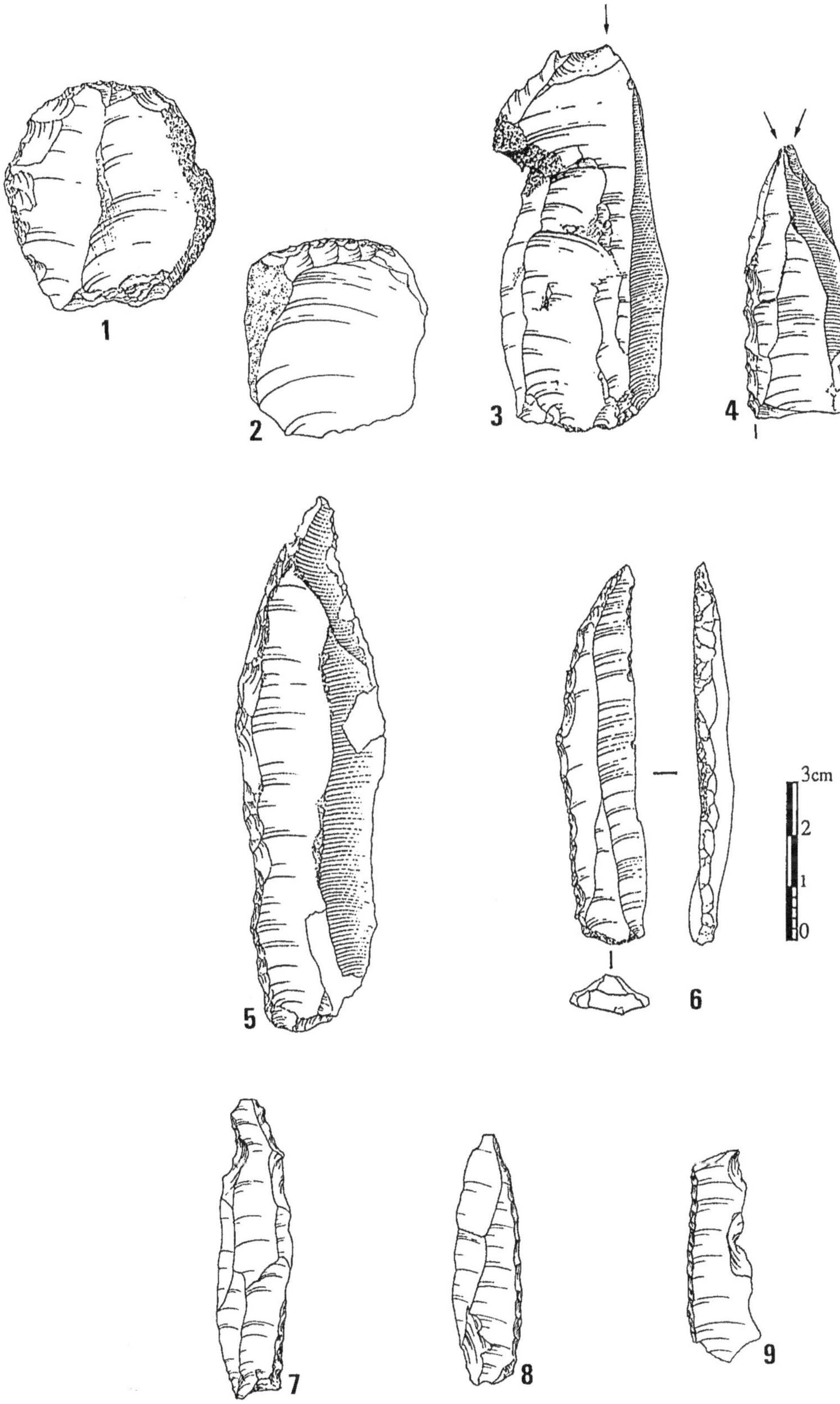

Figure 6. La Grande Paroisse (Federmesser). 1-2: scrapers on flakes, 3-4: burins, 5-6: backed blades, 7-8: backed points, 9: fragment of a backed point. Drawings by D. Molez.

change, it could be because the development of vegetation made raw material procurement less predictable.

In any case, as observed above, this evolution in percussion technique accompanies a transformation in the conception of lithic production: less productivity, less standardization and more flexibility in the selection of blanks for retouch or reshaping. These changes certainly derive from more fundamental economic changes that we are not able to identify in this region at the present time, but certainly including: length of occupation and nature of the settlements, kind of activities, and relative importance of the lithic component in the material culture (Bolus 1992; Bosinski 1987; Lauwers 1988; Van Noten 1978).

Long Blade Technology Sites

Only a few years ago nothing was known about Late Paleolithic occupations after Allerød times in the Paris Basin. Recently, several sites have yielded industries corresponding to what is known as *Long Blade Technology*, first identified in the Thames Basin in England (Barton 1989, 1991; Gob 1991) and in the French Somme basin (Fagnart 1988, 1992a, 1992b, 1993). In the southern Paris Basin, as elsewhere, these sites are always located close to a good raw material source, the exploitation of which implies thorough knowledge of the geology of the region. These industries, dating to the beginning of the Preboreal period, are considered by some prehistorians to be a specialized southern facies of Ahrensburgian groups.

The site of Donnemarie-Dontilly, excavated since 1991, is the best known of this tradition in the Seine-Yonne region (Bodu and Valentin 1992, 1993b). It is located on a cuesta border, above the Seine-Yonne confluence, where geological deposits contain Tertiary flint blocks of very good quality and size (up to 50 cm long). Several localized, low-density knapped flint concentrations suggest that occupation of the site was rather brief. Core reduction modalities, achieved entirely with "soft" sandstone hammers and a careful preliminary shaping of large volumes for knapping, indicate a high degree of technical skill (Figure 7). The objective of the debitage is production of well-calibrated blades of various lengths. Reduction of these large volumes by percussion with a sandstone hammer required management of very rectilinear debitage, using a specific knapping rhythm: rapid alternating use of two striking platforms established at the beginning of reduction. Debitage concentrations found on the site yield only knapping by-products. A clear quantitative deficit, confirmed by multiple absent blanks in refit cores, indicates that a large proportion of the laminar products, especially small well-calibrated blades, were taken

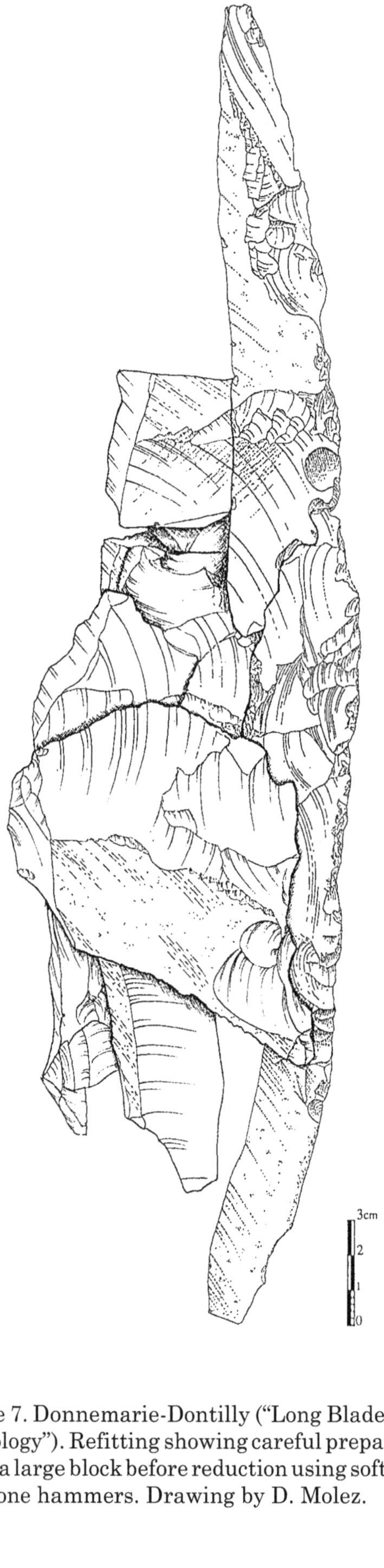

Figure 7. Donnemarie-Dontilly ("Long Blade Technology"). Refitting showing careful preparation of a large block before reduction using soft sandstone hammers. Drawing by D. Molez.

off the site. Bone has not been preserved, but the scarcity of burned materials suggests that consumption activities were not the goal of this occupation. The infrequent retouched tools are primarily scrapers suggesting some processing or manufacture activities on the site, but the most common tools are "bruised" blades and flakes (Figure 8). These large, rough pieces are by-products of laminar production that bear traces of heavy damage that, according to preliminary wear-use studies (Plisson, in Bodu and Valentin 1992), suggest they may have been used by flintknappers to prepare or restore their sandstone hammers. At present, the site of Donnemarie appears to be a very specialized occupation, aimed at production of laminar blanks for later use. This site, apparently an intermediate halting-place between a nearby raw material acquisition locus and other sites where the selected blanks were used, seems to have been occupied for a short time by skillful specialized knappers, and could correspond well to the definition of a lithic workshop.

Donnemarie shares these technological features with about forty sites currently known between the Thames in England and the Loire in central France. The best comparisons can be made with the upper level of Belloy-sur-Somme (France), one of the few sites that have been excavated over a large surface (about 2,000 square meters). The general trends observed at Donnemarie are confirmed, on a larger scale, at Belloy, where it is important to note that no combustion features were found (Fagnart 1992b, 1993). It is therefore probable that most of these sites are not dwelling-sites. The proposed relation between these sites and southern sites of the Ahrensburgian tradition, suggested by the rare presence of Ahrensburgian points in some English sites and at Belloy, is not yet confirmed since dwelling sites of the Ahrensburgian tradition are still not known south of the Belgian Ardennes (Gob 1991).

The most characteristic feature of these Long Blade Technology sites is a specific economic behavior: namely, evidence for large time-and-space breaks in the operational sequences of blank production and transformation, denoting a strong emphasis on anticipation of raw material needs. This pattern of temporal and spatial discontinuity in lithic production is much more apparent than in the Magdalenian, and certainly contrasts with the Federmesser groups of Allerød age. At the present state of research, it is possible, though still difficult, to argue that this specific strategy is related to splitting of social groups during the course of production and transformation processes. The technical homogeneity of the knapping production on the sites could provide some support for this hypothesis. These economic and probably social transformations are the more important because they present a supra-regional dimension. It might be pointed out, for instance, that the same general trend seems to appear at the end of the late glacial in Poland (Swiderian) and probably in northern Germany (Ahrensburgian). At present, these considerations can only suggest a hypothesis of transmission of ideas: technical and economic concepts shared by different populations.

References Cited

Alix, P., A. Averbouh, L. Binter, P. Bodu, A. Boguszewski, C. Cochin, V. Deloze, P. Gouge, V. Krier, C. Leroyer, D. Mordant, M. Philippe, J.-L. Rieu, P. Rodriguez and B. Valentin

1993 Nouvelles recherches sur le peuplement magdalénien de l'interfluve Seine-Yonne: Le Grand Canton et Le Tureau des Gardes à Marolles-sur-Seine (Seine-et-Marne). *Bulletin de la Société Préhistorique Française* 90(3):196-218.

Audouze, F.

1992 L'occupation magdalenienne du Bassin parisien. In *Le peuplement magdalénien: Paléogéographie physique et humaine,* edited by J.-Ph. Rigaud, H. Laville, and B. Vandermeersch, pp. 345-355. C.T.H.S., Paris.

Audouze, F., C. Karlin, D. Cahen, E. de Croisset, P. Coudret, M. Larriere, P. Masson, M. Mauger, M. Olive, J. Pelegrin, N. Pigeot, H. Plisson, B. Schmider, and Y. Taborin

1988 Taille du silex et finalité du débitage dans le Magdalénien du Bassin Parisien. In *De la Loire à l'Oder. Les civilisations du Paléolithique final dans le Nord-Ouest européen,* edited by M. Otte, Vol. 1, pp. 55-84. British Archaeological Reports, S 444, Etudes et Recherches Archéologiques de l'Université de Liège 25. BAR, Oxford.

Barton, R. N. E.

1989 Long blade technology in Southern Britain. In *The Mesolithic in Europe,* edited by C. Bonsall, pp. 264-271. John Donald, Edinburgh.

1991 Technological innovation and continuity at the end of the Pleistocene in Britain. In *The Late Glacial in North-West Europe: Human Adaptation and Environmental Change at the End of the Pleistocene,* edited by N. Barton, A. J. Roberts, and D. A. Roe, pp. 234-245. CBA Research Report *77.* Council for British Archaeology, London.

Bodu, P.

1993 Analyse typo-technologique du matériel lithique de quelques unités du site magdalénien de Pincevent (Seine-et-Marne). Applications spatiales,

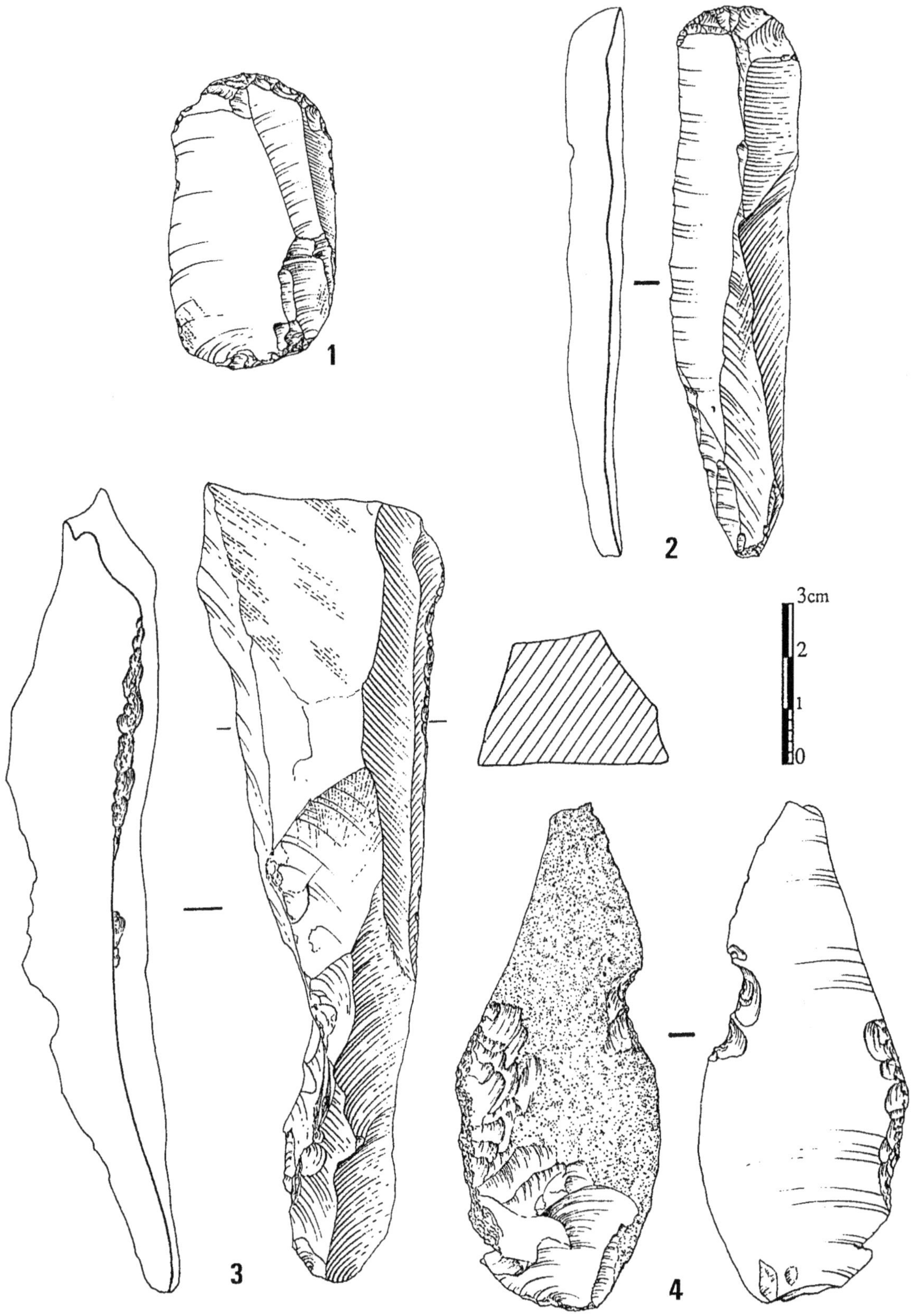

Figure 8. Donnemarie-Dontilly ("Long Blade Technology"). 1: scraper on flake, 2: end-scraper, 3: "bruised" blade, 4: "bruised" flake. Drawings by D. Molez.

économiques et sociales. Thèse de doctorat, Université de Paris I.
1996 Analyse technologique du niveau épipaléolithique III de Pincevent. In *Fouilles de Pincevent II. Le site et ses occupations récentes.* Mémoire de la Société Préhistorique Française 23, edited by G. Gaucher, pp. 69-77. Société Préhistorique Française, Paris.

Bodu, P. and B. Valentin
1992 L'industrie à pièces mâchurées de Donnemarie-Dontilly (Seine-et-Marne, France): Un faciès tardiglaciaire inédit dans le Bassin Parisien. *Préhistoire Européenne* 1:15-34.
1993a Nouvelles recherches sur le peuplement du Bassin Parisien au Paléolithique final. In *Actes des journées archéologiques d'Ile-de-France* (16 et 17 mars 1991), pp. 79-86. Mémoires du groupement archéologique de Seine-et-Marne 1. Association pour la promotion de la recherche archéologique en Ile-de-France, Nemours.
1993b Nouveaux résultats sur le site tardiglaciaire à pièces mâchurées de Donnemarie-Dontilly (Seine et-Marne). *Préhistoire Européenne* 4:85-92.

Bolus, M.
1992 Niederbieber and Andernach: Examples of final Palaeolithic settlement in the Neuwied Basin (Central Rhineland). In *Les bassins du Rhin et du Danube au paléolithique supérieur: Environnement, habitat et systèmes d'échange,* edited by A. Montet-White, pp. 116-133. Etudes et Recherches Archéologiques de l'Université de Liège 43. Université de Liège, Liège.

Bosinski, G.
1987 Upper and final Palaeolithic settlement patterns in the Rheinland Western Germany. In *Upper Pleistocene prehistory of Western Eurasia,* edited by H. L. Dibble and A. Montet-White, pp. 375-386. University Museum Symposium Series 1, University Monograph 54. University of Pennsylvania, Philadelpha.

Bridault, A. and C. Bérnilli
1999 La faune du Grand-Canton à Marolles-sur-Seine. In *Occupations du Paléolithique supérieur dans le sud-est du Bassin parisien,* edited by M. Julien and J. L. Rien. Documents d'Archéologie Française, No. 78. Editions de la Maison des Sciences de l'Homme, Paris.

Burdukiewicz, J. M.
1986 *The Late Pleistocene Shouldered Point Assemblages in Western Europe.* E. J. Brill, Leiden.

David, F. and J. G. Enloe
1992 Chasse saisonnière des Magdaléniens dans le Bassin parisien. *Bulletin et Mémoires de la Société d'Anthropologie de Paris* 4(3-4):167-174.

David, S.
1992 Le peuplement magdalénien dans le nord-est de la France. In *Le peuplement magdalénien: Paléogéographie physique et humaine,* edited by J.-Ph. Rigaud, H. Laville, and B. Vandermeersch, pp. 87-96. C.T.H.S., Paris.

Degros, J., B. Schmider and B. Valentin
1994 Le site du Tilloy à Ville-Saint-Jacques (Seine-et-Marne). In *L'Environnement et l'habitat magdalénien au centre du Bassin parisien,* edited by Y. Taborin, pp. 176-178. Documents d'Archéologie Française 43. Editions de la Maison des Sciences de l'Homme, Paris.

Enloe, J. G. and F. David
1989 Le remontage des os par individus: Le partage du renne chez les Magdaléniens de Pincevent. *Bulletin de la Société Préhistorique Française* 86(9):275-281.

Fagnart, J.-P.
1988 *Les industries lithiques du Paléolithique supérieur dans le Nord de la France.* Revue Archéologique de Picardie, special number. Direction des Antiquités de Picardie, Amiens.
1992a Le Paléolithique final dans le nord de la France : le Magdalénien et les cultures septentrionales. In *Le peuplement magdalénien: Paléogéographie physique et humaine,* edited by J.-Ph. Rigaud, H. Laville, and B. Vandermeersch, pp. 247-258. C.T.H.S., Paris.
1992b Nouvelles observations sur le gisement paléolithique supérieur de Belloy-sur-Somme (Somme). *Gallia Préhistoire* 34:57-83.
1993 *Le Paléolithique supérieur récent et final du Nord de la France dans son cadre paléoclimatique.* Thèse de doctorat, Université de Lille.

Floss, H.
1992 Sur l'approvisionnement des matières premières au Magdalénien et Paléolithique final en Rhénanie (Bassin de Neuwied). In *Les bassins du Rhin et du Danube au Paléolithique supérieur: environnement, habitat et systèmes d'échange,* edited by A. Montet-White, pp. 104-113. Etudes et Recherches Archéologiques de l'Université de Liège 43. Université de Liège, Liège.

Gob, A.
1991 The early Postglacial occupation of the

southern part of the North Sea Basin. In *The Late Glacial in North-West Europe: Human Adaptation and Environmental Change at the End of the Pleistocene,* edited by N. Barton, A. J. Roberts, and D. A. Roe, pp. 227-233. CBA Research Report 77. Council for British Archaeology, London.

Hantai, A.
1993 *Approche technologique et typologique de l'industrie lithique recueillie en 1991 sur le site magdalénien du Tureau des Gardes à Marolles-sur-Seine.* Unpublished Mémoire de Maîtrise, Université de Paris I.

Julien, M.
1989 Activités saisonnières et déplacements des magdaléniens dans le Bassin Parisien. In *Le Magdalénien en Europe: la structuration du Magdalénien,* edited by J.-Ph. Rigaud, pp. 177-189. Etudes et Recherches Archéologiques de l'Université de Liège 38. Université de Liège, Liège.

Karlin, C.
1991 Analyse d'un processus technique: le débitage laminaire des Magdaléniens de Pincevent (Seine et Marne). In *Tecnologia y cadenas opérativas liticas,* edited by R. Mora, X. Terradas, Z. Parpal, and C. Plana, pp. 125-162. Universitat autonoma, Barcelona.

Karlin, C., S. Ploux, P. Bodu and N. Pigeot
1993 Some socio-economic aspects of the knapping process among groups of hunter-gatherers in the Paris Basin area. In *The Use of Tools By Human and Non -human Primates,* edited by A. Berthelet and J. Chavaillon, pp. 318-337. Fyssen Foundation Symposium. Oxford University Press, Oxford.

Lauwers, R.
1988 Le gisement tjongérien de Rekem (Belgique), premier bilan d'une analyse spatiale. In *De la Loire à l'Oder. Les Civilisations du Paléolithique final dans le Nord-Ouest européen,* edited by M. Otte, Vol. 1, pp. 217-234. British Archaeological Reports, S 444, Etudes et Recherches Archéologiques de l'Université de Liège 25. BAR Publishing, Oxford.

Leroi-Gourhan, A. and M. Brezillon
1966 L'habitation magdalénienne n°1 de Pincevent, près de Montereau (Seine-et-Marne). *Gallia préhistoire* 9(2):263-371.
1972 Fouilles de Pincevent: essai d'analyse ethnographique d'un habitat magdalénien (la section 36). *Gallia Préhistoire,* supplément 7, C.N.R.S, Paris.

Mauger, M.
1985 Les matériaux siliceux utilisés au Paléolithique supérieur en Ile de France: occupation du territoire, déplacements et approche des mouvements saisonniers. Thèse de doctorat, Université de Paris I.

Olive, M. and N. Pigeot
1992 Les tailleurs de silex magdaléniens d'Etiolles: Vers l'identification d'une organisation sociale complexe. In *La pierre préhistorique,* edited by M. Menu and P. Walter, pp. 173-185. Ministére de la Culture, Paris.

Pigeot, N.
1987 Magdaléniens d'Etiolles: économie de débitage et organisation sociale. *Gallia-Préhistoire, supplément* XXV. C.N.R.S., Paris.

Ploux, S., C. Karlin and P. Bodu
1991 D'une chaîne l'autre: normes et variations dans le débitage magdalénien. *Techniques et culture* 17-18:81-114.

Rensink, E., J. Kolen, and A. Spieksma
1991 Patterns of raw material distribution in the Upper Pleistocene of northwestern and central Europe. In *Raw Material Economies Among Prehistoric Hunter-Gatherers,* edited by A. Montet-White and S. Holen, pp. 141-159. University of Kansas Publications in Anthropology 19. University of Kansas, Lawrence.

Schmider, B., editor
1993 *Marsangy, un campement des derniers chasseurs magdaléniens sur les bords de l'Yonne.* Etudes et Recherches Archéologiques de l'Université de Liège 55. Université de Liège, Liège.

Valentin, B.
1995 Les groups humains et leurs traditions au Tardiglaciaire dans le Bassin parisien. Thèse de doctorat, Université de Paris I.
1996 Approche typologique et technologique de l'industrie tardiglaciaire recueillie en surface à La Grande Paroisse (Seine-et-Marne) - série U.-J. Guinard. Comparaisons et hypothèses concernant son attribution culturelle. *Bulletin du G.A.S.M.,* n° 32-34:43-55.

Valentin B., P. Bodu, M. Philippe and A. Hantai
n.d. Approche technologique de l'industrie du Grand-Canton à Marolles-sur-Seine. In *Le Paléolithique supérieur sur l'Autoroute A5,* coordinated by M. Julien. Documents d'Archéologie Française. Editions de la Maison des Sciences de l'Homme, Paris.

Van Noten, F.
1978 *Les Chasseurs de Meer.* Dissertationes Archaeologicae Gandenses 18. De Tempel, Bruges.

10. Tardiglacial Lithic Raw Material Utilization in Vasco-Cantabria (Spain) and Gascony (France)

Lawrence Guy Straus

Shifting the geographical focus to southwestern Europe, this chapter offers an exhaustive presentation of the current state of research with respect to the use of lithic raw materials in the late glacial and early postglacial of Vasco-Cantabrian Spain and French Gascony. Based on a thorough review of available data on Upper Magdalenian and Azilian assemblages, the author finds that stone raw materials used in Paleolithic sites in this region closely track local bedrock geology. There is no evidence for long-distance transport of lithic raw materials. Observed variations in raw material use in areas with diverse raw material sources are mostly limited to selection of higher-quality local flints for blade tools, while more accessible and lower-quality materials were used for expedient tools. Those areas in which good flint is more uniformly available show use of flint for all tools. Within this region procurement patterns do not change markedly from the Magdalenian to the Azilian. While the author finds a reduced use of the best-quality flints in the Azilian at some (but not all) sites, it is argued that the use of predominantly local material in all time periods renders the raw material evidence unsuited to drawing broad conclusions about changing sizes of annual territories in this particular region. This consistent use of local stone is due to the nature of the region's complex geology and topography, which makes diverse stone raw materials as well as a wide array of mountain, coastal, and intermontane resources available within a very restricted area. –*Editors.*

Introduction

Fifteen years ago, I published an article entitled "The role of raw materials in lithic assemblage variability" (Straus 1980). At the time there was scant interest in this subject among prehistorians working (mostly under the Bordesian typological paradigm) in Spain, although in the 1920's and 30's J. Carballo had made fairly extensive observations on the prehistoric use of various lithic raw materials and on their geological source locations in the Province of Santander (Carballo 1938, 1957, 1960:39-42). The situation has changed considerably in recent years (e.g., Straus *et al.* 1986; Spanish papers in the V and VI International Flint Symposium in Bordeaux and Madrid respectively [Séronie-Vivien and Lenoir, eds. 1990; Bustillo and Ramos-Millán, eds. 1991] and in Reunión Internacional de Tecnología y Cadenas Operativas Líticas in Barcelona [Mora *et al.* 1992]), although the precise likely geological sources of most Vasco-Cantabrian lithic artifacts are still not usually known.

The general conclusion of my initial study (mainly of Solutrean and early Magdalenian lithics) was that the lithics used in Paleolithic sites closely track local bedrock geology. This overall perspective has been validated recently for the late Magdalenian and Azilian by González Sainz (1989:185-189). In terms of the use of lithic raw materials, Vasco-Cantabrian Spain is fundamentally different from the vast loess- or alluvium-covered North European Plain or Pannonian Basin, where there is abundant evidence of extensive transport of flints and other lithics over long (and even very long, i.e., 100's of kilometers) distances during the terminal Paleolithic (e.g., Bosinski 1988; Floss this volume; Montet-White 1994; Rensink *et al.* 1991; Schild 1996; Sulgostowska this volume). The late Magdalenian Cantabrian lithic economy seems to have involved transport generally even less extensive than that of the Paris Basin during the same period — now argued to have usually involved movements (of people and stones) on the order of <30-50 kilometers and, more rarely, up to 70-80 kilometers (Mauger 1994).

In the adjacent Pyrenean region of southwestern France, interest in Magdalenian use of lithic raw materials extends back at least as far as the work of R. de Saint-Périer (1936) at Isturitz. Over a long period of time, R. Simonnet (1981, 1982, 1985) has, through rigorous fieldwork, established the patterns of flint transport both within this region and into the Pyrenees from the Périgord. Yet, as shown below, though some medium-distance transport and even extra-local procurement of flint <u>did</u> occur in some Magdalenian contexts in the Pyrenean region, this was <u>not</u> the case under all circumstances. The case of the Abri Dufaure shows that under certain geological and residential conditions, lithic procurement in an area of southwestern France could be just as local as it seems to have been more generally in the Cantabrian region.

The Temporal Framework

The focus of this paper is on the Upper Magdalenian and Azilian, both archeologically defined by the presence of antler harpoons (round and flat section, respectively) in many sites of Cantabria and Gascony. The former geographical region, bounded by the Bay of Biscay (Cantabrian Sea) and Cantabrian Cordillera, includes the modern political regions of Asturias, Cantabria (formerly, Santander), and Euskadi (Autonomous Basque Country, including the provinces of Vizcaya, Alava and Guipúzcoa), plus adjacent upland areas of Navarra. In this paper we are concerned with that part of Gascony that borders the westernmost Pyrenees (modern departments of Les Landes and Pyrénées-Atlantiques, which includes the French Basque Country plus Béarn) (Figure 1).

The Magdalenian and, to a somewhat lesser extent, Azilian of both regions are dated by large numbers of radiocarbon dates at dozens of sites. Numerous sites have provided palynological, geological and paleontological data, which complement regional paleoenvironmental evidence from lake and deep sea cores.

Upper Magdalenian technology developed ca. 13,000 BP and was gradually, often unevenly transformed into Azilian technology ca. 11,000-10,500 BP. The Azilian lasted until ca. 9,000 BP, at which time a variety of geometric microlithic and macrolithic industries appear in the regional records. The Upper Magdalenian spanned the environmental fluctuations of the Bølling, Dryas II and Allerød, when some of the first assemblages classifiable as "Azilian" appear in some sites. Otherwise, the bulk of the Azilian assemblages date to the Dryas III and Preboreal. Note that both culture-stratigraphic units occurred under both relatively cold and more temperate conditions, with both more open and more wooded landscapes respectively (Clottes 1989; González Sainz 1989, 1994; Straus 1985, 1991, 1992a).

Relevant Aspects of Regional Lithologies

The bedrock of the Vasco-Cantabrian region is dominated by Mesozoic lithology in the eastern half and Paleozoic in the western half (Figure 2), with the approximate boundary between the two falling midway through the province of Santander (González Sainz 1989:185). While flints of excellent quality (albeit not of large nodule size) are abundant in the widespread Cretaceous limestones of the Basque Country and eastern Santander, the situation becomes much more complex to the west. There are several well-known flint sources in Euskadi (e.g., Jaizkibel, Pedernales), but other primary and secondary sources (e.g., river beds) are fairly ubiquitous. However, in central and western Santander there are exposed bands and pockets of Cretaceous, Carboniferous and Jurassic limestones, Ordovician quartzites, Triassic ophites, quartzes, etc. (Freeman 1991). Some of the limestones are quite hard and were worked, especially in the Middle and early Upper Paleolithic at El Castillo (Breuil 1952; Bernaldo de Quirós and Cabrera 1993). The various Carboniferous flints (from Griotte and Montaña limestones) and Jurassic flints are generally of very poor quality and are found in the form of small river or beach cobbles. The Griotte

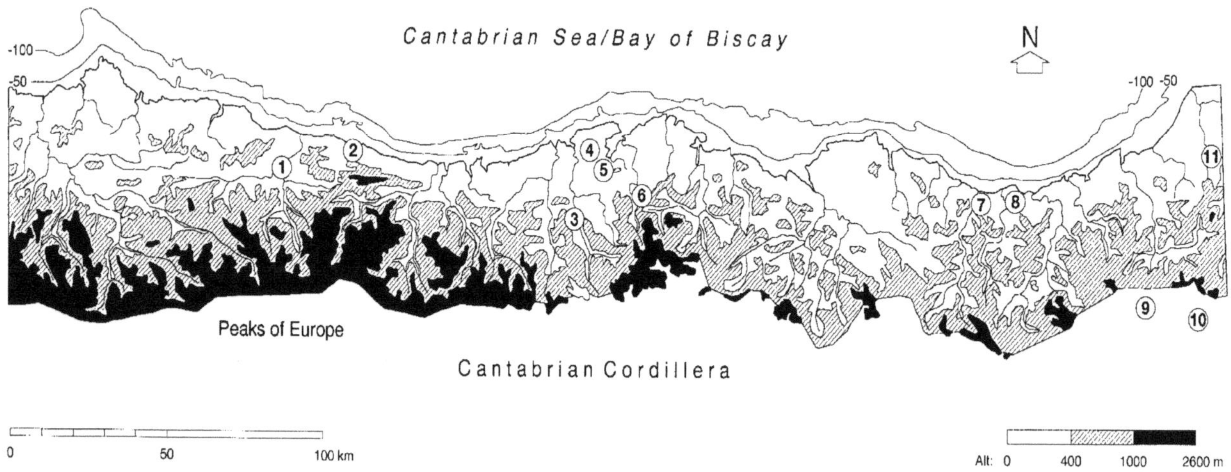

Figure 1. Topographic map of Cantabrian Spain and southwestern Gascony, showing contours above and below present sea level and locations of sites mentioned in the text. 1, Los Azules; 2, La Riera; 3, El Castillo; 4, El Pendo; 5, Cueva Morín; 6, El Rascaño, El Piélago; 7, Ekain; 8, Erralla; 9, Abauntz; 10, Zatoya; 11, Dufaure.

Figure 2. Simplified map of the bedrock lithology of Cantabrian Spain and southwestern Gascony.

flint contains radiolarite micro-fossils. Good-quality Cretaceous flint becomes increasingly rare toward the west, and occurs in limited outcrops such as a band along the present central coast of Santander.

In eastern Asturias, localized bands of flint-bearing limestones are dominated by the poor Carboniferous types, with only very small Cretaceous outcrops (including narrow outcrops along the coast near Llanes), all yielding very small nodules. Bands of quartzites (both fine-grained good-quality types and much coarser grained ones) become increasingly dominant in the lithology of the Hercinian arcs that make up the complex geology of central Asturias. Finally, even Carboniferous limestones end at the Nalón River and Cambrian and Silurian quartzites and slates dominate western Asturias and eastern Galicia (Terán and Sole Sabaris 1968). The bulk of Galicia is a Precambrian granitic shield area — a fact which partly explains its uniformly "crude", archaic-looking lithic industries in <u>all</u> prehistoric and historic periods (Vázquez 1994).

As noted above, the lithic compositions of Paleolithic industries from east to west track the change in regional lithology from dominant Cretaceous limestones to dominant Paleozoic quartzites, although, in Santander especially, it is typical to find higher percentages of quartzites and ophite in Mousterian and Early Upper Paleolithic assemblages than in Late Upper Paleolithic (LUP) assemblages of the same sites (e.g., El Castillo, Cueva Morín, El Pendo) (Straus and Heller 1988; Freeman 1991; Bernaldo de Quirós and Cabrera 1993). Basque Country assemblages are regularly 100% flint or close to that. In the LUP, assemblages from central Santander range from ca. 60-99% flint. Assemblages from eastern Asturias range from ca. 20-80% flint. In central Asturias, flint can make up between ca. 60-97% of LUP lithic assemblages. (All percentages here are by count; percentages of flint by weight in Santander and Asturias would be consistently smaller, since flint artifacts [cores, debitage and tools] are consistently smaller than quartzite ones [Straus 1996]). There can be major differences among sites in terms of the amounts of flint present, probably based on accidents of very local geology, but there also exist differences among levels within sites that are probably related to different uses of the caves at different times by different mixes of human task groups. In most cases in Santander and Asturias, flint was preferentially selected for making certain kinds of (usually small, more delicate) LUP tools and weapons, while quartzite was more often reserved for larger, heavier-duty tools, or was more often left unretouched than flint, in the form of flakes that could have been utilized expediently.

A Tardiglacial Case Study from Asturias: La Riera Cave

To date, the most detailed petrographic and lithic utilization study of a Paleolithic site in northern Spain is the one conducted by the La Riera Paleoecological Project at this cave on the narrow coastal platform of eastern Asturias near Llanes (Straus *et al.* 1986). Descriptions of flints and cherts based on thin-section analysis were done by J. Ordaz, L. Suarez and R. Esbert of the University of Oviedo Department of Petrology. Here I will summarize some observations on the lithics from the Upper Magdalenian, Azilian and Asturian levels (21-30), although the full sequence extends back to the beginning of the Solutrean or slightly earlier, ca. 21,000 BP (Straus and Clark, eds. 1986).

The Magdalenian (both Lower and Upper) levels at La Riera are among the poorest in quartzite artifacts, both in terms of weight and count, with the exception of the highly specialized basal Solutrean levels which represent visits to the cave for ibex-hunting on nearby cliffs, during which lithics were mainly transported to the site in the form of blanks and points. In the early Upper Magdalenian levels 21-23 (late Dryas I or Dryas II, but C^{14} dated to 12,600 ± 300 BP), quartzite makes up 20% of all lithic artifacts by count and 55% by weight, yet 21% of the tools are made on this strictly locally available material (actually two intergrading types both available as stream cobbles). The flints used (79% of both debris and tools, by count) include a locally available material from the Carboniferous Caliza (limestone) de Montaña and two better-quality flint varieties possibly from small, local outcrops of Cretaceous limestone. The poorest quality Griotte flint, otherwise often found at La Riera, is nearly absent in these levels. As is usual in the Magdalenian levels, the flint artifacts are far lighter (hence, smaller) than the quartzite ones (average weights: 7 grams vs. 35 grams).

Interestingly, in comparison with the Lower Magdalenian levels (18-20) which have backed bladelet/backed micropoint indices of 62-74%, levels 21-23 have only 16% of these non-geometric microliths. This may explain why there is so little flint among the retouched pieces in these assemblage — these levels may be the product of occupations involving relatively little hunting activity and more activities for which larger quartzite tools were more suitable (e.g., woodworking, processing and maintenance). This pattern begins to change in the late Magdalenian, harpoon-bearing level 24, dated to 10,900 ± 400 BP (Allerød). In this level, there is an overall increase in quartzite (28% by count, 60% by weight), yet only 5% of the tools are quartzite. There is a clear preference for flint when making formal tools, and the flint raw material consists almost exclusively of locally available Carboniferous types, with very little of the better Cretaceous types. The flint artifacts are small (average weight 8 grams), while the quartzite tools are on average very large (average weight 120 grams). The high percentage of flint tools (95%) corresponds to a notable increase in frequency of backed bladelets and points (54% of the retouched pieces in level 24).

Levels 26-27 can be attributed to either the terminal Magdalenian or Azilian, as the two lithic technologies intergrade significantly in northern Spain and harpoons are absent in these levels (and in the smaller level 25). Level 27 dates to 10,600 ± 100 BP (Dryas III). Use of flint decreases markedly in the overall artifact assemblage (debris and tools), while quartzites become extremely abundant (66-69% by weight and 53-56% by count). Still, 80-92% of the tools are made on flint, almost exclusively the local Carboniferous types, especially the poor-quality Griotte variety. The tools include 39-44% backed bladelets and points, again corresponding to the preference for flint in tool making. Such microliths could be made with the small available local flint nodules, but virtually never with quartzite.

Level 28 (Dryas III) yielded a flat-section Azilian harpoon. Quartzites make up 64% of the assemblage by weight and 48% by count, but 94% of the tools are on flint, mainly the rather poor-quality Griotte plus some Montaña Carboniferous types. Epipaleolithic (i.e., non-geometric) microliths make up a huge 63% of the tools, a fact consonant with the dominance of formal tools on flint. There is a parallel lithic industry of large, mostly unretouched quartzite flakes, many of which may have been expedient, *de facto* tools.

In the Asturian *conchero* level 29 (dated 8,600 - 7,000 BP, Preboreal and Boreal), backed bladelets and use of flint raw materials are nearly absent, and Cretaceous flint disappears entirely. Quartzite is dominant (69% by count, 64% by weight), which makes sense given the high relative frequency of heavy-duty tools (picks, choppers, large sidescrapers, notches and denticulates). There are only a few burins, small endscrapers, perforators, or truncated pieces — the kinds of classic Upper Paleolithic tools for which flint was generally used in the Magdalenian and Azilian levels. No effort seems to have been made to procure even the limited, but locally available, good-quality flints.

The Upper Magdalenian faunal assemblages are dominated by remains of red deer, followed by ibex. Assemblages of both species include many very young animals. Most of the few specimens for which age could be determined were killed between spring and fall. This pattern continues in the transitional and Azilian levels, but with more roe deer, bovines (aurochs?), boar and chamois, and with more evidence of winter as well as warm-season kills (Altuna 1986). This might suggest more frequent human stays on the coastal plain than in the earlier period. The scanty mammalian fauna associated with the Asturian *conchero* level is strongly dominated by red deer (with spring and winter kills only). Oxygen isotope analyses show that mollusks were collected exclusively in winter (Deith and Shackleton 1986).

In short, across the Pleistocene-Holocene transition at La Riera local quartzites and flints (despite generally poor quality and small nodule size) were regularly used and there is no certain evidence for import of extra-local, higher quality materials. There is an apparent trend toward increasing reliance on quartzites for most uses except for microliths, and for almost everything when microliths disappear from assemblages in the Asturian (for whatever reason — depositional/functional or culture-historical [Straus 1979; González Morales 1982]). All of this implies limited band territories (use-areas) and limited mobility, which makes sense in view of the high regional relief. All the resources of the coast, estuaries, coastal plain, coastal mountain range, intermontane valley and slopes of the Peaks of Europe could have been reached within a day's walk of La Riera. It is possible to make an argument for even greater restrictions in territories and mobility in the Azilian than in the Magdalenian, but the data are still very scarce.

Anecdotally, one can mention information provided to González Sainz (1989:188) by J. Fernández Tresguerres, the longtime excavator of Los Azules cave in the intermontane valley inland of La Riera. The Upper Magdalenian levels are said to contain good-quality flint (presumably from the small Cretaceous outcrops along the present coast), while the poor-quality, strictly local, Carboniferous Griotte radiolarite is far more abundant in the Azilian levels. There is an outcrop of the radiolarite in the same hillside where Los Azules is located; in contrast, the present coast is at a distance of ca. 22 kilometers from Los Azules.

The Magdalenian-Azilian Lithic Trajectory in Santander

The classic culture-stratigraphic sequence for Santander and indeed the entire Cantabrian region is El Castillo, a vast cave dominating the Pas River about 20 kilometers upstream of the present coast. Excavated by Hugo Obermaier and Henri Breuil during four campaigns immediately before World War I, the site was not fully published until 70 years later in a remarkable work of "salvage archeology" by Victoria Cabrera (1984). Although the site contains a virtually complete sequence from the Acheulean to the Bronze Age, the levels that interest us here are 4 (Azilian), 6 and 7 (Upper Magdalenian). Level 6 is dated to 12,400 ± 100 BP, with a second date of 10,300 ± 100 BP that is probably too young (Barandiarán 1988:73). The cave is in a mountain formed of Carboniferous limestone (Caliza de Montaña) in a geologically complex area with a wide variety of other outcrops, including Jurassic and Cretaceous flint-bearing limestones, as well as quartzites and volcanic ophite. In short, all the lithic raw materials used at El Castillo (including hard limestone) could have been procured in the immediate surroundings of the cave, in the river below, or within the short distance between the cave and the coast. The wealth and variety of lithic raw materials (including flint) available to and actually used by Paleolithic inhabitants in Santander have recently been documented in preliminary fashion by P. M. Sarabia (1990a,b).

Unfortunately, details on the raw material composition of the Obermaier collections are unavailable. However, it can be observed that while 70% of the total remaining Lower Magdalenian (Level 8) lithic assemblage and 90% of the tools are flint (followed by limestone, quartzites, phtanite and quartz, mainly among the unretouched debris), flint was used for 87% of the formal tools in level 7 and 100% in level 6 — both attributed to the Upper Magdalenian (the latter with harpoons). The same flint dominance is mirrored in the small surviving collections of debitage and cores, and in Breuil and Obermaier's own counts of their much larger original tool collections (63% in level 7 and 88% in level 6) (Cabrera 1984:363-6). The very small Azilian (with flat-section harpoons) lithic assemblage (level 4) recovered by Obermaier and Breuil is 83% flint, and there is a slight increase in quartzite, the only other material utilized in the post-Magdalenian levels (Cabrera 1984:403-4).

Cueva Morín, 12 kilometers to the east of El Castillo and closer to the area dominated by Cretaceous limestone bedrock, was excavated in the 1960's by González Echegaray and Freeman (1971). The percentage of flint increases (from 92 to 93% of the total assemblages) and that of quartzite decreases (from 7 to 3%) between the Upper Magdalenian and Azilian levels. As at El Castillo, these terminal Paleolithic assemblages lack the very large quantities of quartzites and ophite typical of the Mousterian and early Upper Paleolithic assemblages, probably because of the technological trend toward microlithization.

The huge cave of El Pendo is located on the present coastal plain near the city of Santander, not far from local bands of Cretaceous limestone that include good-quality flint. Not surprisingly, therefore, the Mousterian and early Upper Paleolithic levels, though containing numerous quartzite and ophite artifacts, are rich in flint compared to El Castillo and Morín (González Echegaray 1980). The Upper Magdalenian levels (IIc-g, IIa-b and II), ranging in age from 12,500 ± 200 to 10,800 ± 200 BP (Barandiarán 1989), are strongly dominated by flint artifacts (98, 97 and 97% of the tools). The tools include many very well-made classic pieces (including numerous backed bladelets, despite the vicissitudes suffered by the Martínez Santa-Olalla collections between the time of their excavation in the 1950's and their study by González Echegaray in the 1970's). This overwhelming flint dominance does not change in the Azilian: over 98% of the tools are of this excellent (albeit small nodule-size) locally procured raw material.

One other site — this one rediscovered by me in 1973 and re-excavated by González Echegaray and Barandiarán (1981) — that yields information on raw material utilization across the Magdalenian-Azilian transition in Santander Province is El Rascaño. This is a specialized ibex-hunting site on a mountain gorge-side location, albeit not at very high elevation (ca. 275 meters above sea level), about 25 kilometers up the Río Miera from the present coast and about 20 kilometers southeast of Cueva Morín.

The Upper Magdalenian levels (2.3 and 2.1) are dated to 12,900 ± 100 and 12,500 ± 200 BP respectively, and the Azilian level (1) is dated by two determinations to 10,500 BP. In all three levels virtually all lithic artifacts are made on flint: 98, 99 and 99%. The exact source of the material is of course unknown, but certainly it could have been procured locally, either in the uplands that surround the site or on the adjacent coastal plain, notably in river gravels. Rascaño is situated in the middle of Santander Province at the western edge of the Cretaceous-dominated bedrock of eastern Vasco-Cantabria. The flint artifacts are generally of fairly restricted size, reflecting small nodules. The site was visited at various seasons (cold and warm) of the year for ibex procurement (Altuna 1981). It is easily accessible from base camps presumably on the coastal plain. The distances any transported flint may have been carried when parties moved to Rascaño were certainly never great (probably one day's walk at most). The Upper Magdalenian and Azilian lithic assemblages are not characterized by large numbers of backed bladelets (despite fine water-screening), perhaps because the weapon tips used were antler sagaies which abound in El Rascaño. Nonetheless, since flint was easily available, all tools (not just burins, perforators, etc.) were made on this raw material.

Data are scarce and imprecise on the lithic raw materials from the adjacent sites of El Piélago I and II, just a few hundred meters downstream of El Rascaño (Garcia Guinea 1985). The basal cultural levels of Piélago II (6 and 5b), although called "Protoazilian" by Garcia Guinea and undated, are attributed sedimentologically to the cold oscillation of Dryas II by Butzer (1985) and tentatively (but reasonably) to the Final Magdalenian by González Sainz (1989:87-88). They lack any type of harpoon.

There is a clayey silt layer (5a) that may have been deposited during the Allerød. The overlying levels (levels 4 through 1, all with Azilian harpoons) fall sedimentologically within the Dryas III cold phase, as confirmed by radiocarbon dates of 10,700 ± 100 and 10,300 ± 100 BP for levels 4 and 1 respectively (González Sainz 1989:295). Piélago II is another site dominated by remains of ibex (followed by chamois) hunting on the edge of the steep, rocky, northern cliff face of the Cantabrian Cordillera (López-Berges and Valle 1985).

Garcia Guinea (1985:68-69) indicates that the basal ("Protoazilian") levels in Piélago II have a wide variety of flint raw material types, that all the artifacts are on flint, and that the artifacts are on average larger than those of the upper (Azilian) levels. Those later levels contain limited numbers of quartzite pieces and a narrower range of flint raw material types. Piélago I yielded relative frequencies of flint varieties different from those of Piélago II, despite the supposed parallelism between the stratigraphies of the two caves. In any event, flint makes up 100% of the artifacts in most levels, except for level 4 in the early Azilian (Garcia Guinea 1985:96-97). González Sainz (1989:189) believes the flint raw materials to be local types, although Garcia Guinea (1985) thinks that different sources were visited in this lithologically complex area as people perhaps alternated in their use of one cave or the other.

None of the data from Santander really suggest much difference in lithic collection behavior between the Magdalenian and Azilian (*pace* González Sainz [1989, 1992]; González Sainz and González Morales 1986], who claims to see generalized evidence for a reduction in lithic collection areas and hence territories in the Azilian vis à vis the Magdalenian). All lithics were obtainable either immediately proximate to the sites or in the local area (i.e., within a day's walk or much less). What may be suggested in these imperfect (and hard to compare) data is that Magdalenian and Azilian assemblages in the eastern sites (Rascaño, Piélago) seem to be even more dominated by flint (ca. 100%) than the more western sites within the province, reflecting the location of the former sites at the edge of Cretaceous limestone bedrock that covers eastern Santander and much of the Basque Country. If true, this is another indication of the highly local character of lithic procurement, probably embedded in other foraging activities or the fruit of casual prospection in the neighborhood of each site (especially in stream beds and gravel bars and on steep, soliflucted or gullied slopes).

The Basque Country: More Flint, Some Transport

Recent excavations at a number of sites in Euskadi and Navarra have focused on sites with Magdalenian and Azilian components dating to the Pleistocene-Holocene transition. These include Ekain and Erralla in Guipúzcoa Province and Abauntz and Zatoya in northeastern Navarra, not far from the continental divide but within the Ebro-Mediterranean watershed, unlike all the sites of Cantabrian Spain.

At Ekain, the basal part of level VI dates to 12,100 ± 200 BP, and is attributed to Dryas II, while the upper part probably pertains to Allerød and possibly early Dryas III (Altuna and Merino 1984). The lithic industry is attributable to the Upper Magdalenian and is all flint. The flints include a variety of types and patinas; there are both poor and excellent-quality flints, the latter being of a distinctive shiny, black unpatinated variety often found in the Basque Country (Merino 1984). All are probably found within short distances of the site (which is formed in Cretaceous limestone), possibly including the area around the penecontemporaneous base camp site of Urtiaga, located at a two-hour walk from Ekain near the present coast where there are known outcrops. Ekain was a warm-season site, whereas Urtiaga was used at all seasons of the year, according to dental eruption studies of ibex and red deer (Altuna 1984:348).

Levels V-II are assigned to the Azilian, but the uppermost level has Mesolithic-type geometric microliths, whereas levels V-III have an Azilian harpoon and two C^{14} dates of ca. 9,500 BP. The lithic assemblages continue to be absolutely dominated (ca. 100%) by a variety of flints, but with less of the finest quality black flint. Level IV has a burin on a chocolate-colored flint said to be of "allocthonous origin" (Merino 1984:79). Level II has virtually none of the best flint, and is dominated by a rather poor-quality, fissured, fractured material such as appears 40 kilometers to the east along the coast at the Franco-Spanish border between San Sebastián and Hendaye (Monte Jaizkibel); however, this material was not necessarily obtained from sources so far away (Merino 1984:69-70). This level has one perforator on a schist flake and a reddish Triassic sandstone chopper, whose source could be no closer than 20 kilometers from Ekain (Altuna 1984:348). Again there is no clear evidence of a change in the size of the lithic procurement catchment area between the Magdalenian and Azilian, though there is an (unquantified) appearance of a trend toward lessened use of the best-quality, homogeneous black flint throughout the Tardiglacial and early Postglacial.

The nearby and also recently excavated cave site of Erralla produced a series of Upper Magdalenian levels (III-I), all attributed to the Allerød, the lowest one of which is dated to 12,300 ± 200 BP (Altuna *et al.* 1985). The lithic assemblages are said to show signs of "azilianization". All lithic artifacts are on flint, mainly of one light grey type. Most of the tools seem to have come from just a few nodules, although there are a few items on an unusual flint that may have been manufactured elsewhere and brought to the cave. All the lithics are believed to have been available within a distance of at most 20 kilometers of Erralla; indeed, the most common flint type was probably collected on the Cretaceous limestone slopes immediately surrounding the cave or in the stream right below it (Baldeón 1985:127-8,175).

Abauntz is a cave located 700 meters above sea level on the south face of the Cantabrian Cordillera, near its junction with the Pyrenees in Navarra, close to a low (847 meters) pass between the Ebro (Mediterranean) and Bidasoa (Atlantic) drainages. In recent excavations by P. Utrilla (1982), it yielded a Lower Magdalenian (15,800 ± 300 BP) and an Azilian (9,500 ± 300 BP) level. Both levels have tooth eruption data for exclusively warm season (summer and fall) human occupation, primarily for hunting chamois, plus some ibex and red deer (Altuna and Mariezkurrena 1982). Artifacts in both levels are made on three types of raw materials: good quality flint, calcite and quartz crystal. The flint is said to be diverse in kind and apparently not local, having been brought to Abauntz during the human visits to the site mainly in the form of finished (generally small) tools (Utrilla 1982:267). Flint debris is said to be relatively scarce and is dominated by microdebitage. The few flint cores are exhausted and all were reworked into scrapers or burins. The sources of the flints are unknown, but it is possible that they come from Guipuzcoa. The present coast at Fuenterrabía is at distance of ca. 52 kilometers via the Puerto de Velate and Río Bidasoa. The calcite was used to make large (expedient?) tools, despite its relative softness. Quartz crystal was used to produce small numbers of flakes, some with a little bit of (use?) retouch. The latter two raw materials are locally available.

It is interesting to note that the same raw material patterns pertain in both Lower Magdalenian and Azilian levels, separated by some 6000 years. Abauntz is one of the few documented (albeit imprecisely) cases of medium-distance transport of flint in northern Spain. It is a specialized, warm season, relatively high altitude, hunting camp site, not a coastal area residential site (like La Riera, El Pendo or Cueva Morín) or a near-coastal residence or hunting camp in the foot-hill ranges (like Los Azules, El Castillo, El Rascaño, El Piélago, Ekain or Erralla).

Another such interior site in Navarra, Zatoya, is even higher (ca. 1000 meters above sea level) on the southern flank of the Pyrenees. Zatoya Cave is near the headwaters of the Río Salazar (an Ebro tributary), not far from the Pyrenean crest and the 1550 meter divide with the Gave de Mauléon (a tributary of the Gave d'Oloron and hence of the Adour). Via that route, Zatoya is ca. 90 kilometers from the Pastou sites (Duruthy, Dufaure, etc.) in Gascony. The coast at Irún-Hendaye is about the same distance away via south Pyrenean valleys and the Río Bidasoa. Zatoya was recently excavated by Barandiarán and Cava (1989), who uncovered a series of terminal Magdalenian (IIb and II) and Epipaleolithic (Ib) levels. Level II is C^{14} dated to >10,900, 11,800 ± 200, 11,600 ± 400, and 11,500 ± 300 BP, and level Ib to 8,300 ± 600 and 8,200 ± 200 BP. The Magdalenian levels may have been deposited during Dryas I (an age confirmed by the presence of reindeer in level IIb) and Allerød respectively, and level lb in Preboreal or early Boreal.

Faunal assemblages from all three levels are dominated by red deer, but with substantial numbers of remains of ibex and chamois and, in the upper two (warmer climate) levels, boar (Mariezkurrena and Altuna 1989). The level IIb occupations took place strictly in late spring-early summer, based on dental eruption evidence from several red deer and a horse. The evidence of summer occupation is even more abundant for level II, again with no late fall or winter kills. The number of animals for which age can be determined is somewhat smaller in Epipaleolithic level lb, but still most (5 deer and 1 boar) were killed in summer, while one boar may have been killed in fall. Humans were not at this altitude in winter.

A. Cava (1989) distinguishes three categories of lithic raw materials utilized at Zatoya: excellent quality flint of varying colors, black flint of very poor quality, and a quartzite-like coarse-grained rock. The poor quality, fissured flint is of strictly local origin, whereas the good flint is non-local (exact source unknown, but I presume it could include Cretaceous limestone in the area from which the summer visitors to Zatoya came — either north of the Pyrenees or near the Basque coast, where there are other sites of these ages). The quartzite is probably local, but was very little used. The non-local flint was strongly favored for the manufacture of smaller, hunting-related tools (bladelets, microliths), whereas the local flint was used to make larger, maintenance-type tools that are often crude-looking or "nucleiform". However, the relative frequency of tools made on the non-local flint decreases from 86% to 70% between the Magdalenian and Epipaleolithic, as the local flint increases from 12% to 27%. Among the debris, in level IIb 91% of the flint is of the good variety, 83% in level II and only 57% in level Ib, while the poor flint increases commensurably (Fernández Eraso 1989). In sum, the representation of good flint clearly declined over time, suggesting an argument that the humans had lessened access to its sources, perhaps through restriction or changes in patterns of annual mobility and/or social contacts.

Gascon Crossroads: Very Local Flints at Dufaure

The region of the confluence of the Gaves de Pau and Oloron, Gaves Réunis and Adour in southern Landes and northern Pays Basque (Chalosse, Labourd, Basse-Navarre) has abundant and diverse sources of good-quality flint. These include well-known outcrops around Tercis, Pouillon, Saint-Palais, Bidache and Salies-de-Béarn, some of which were exploited during the Magdalenian of Isturitz, located 20-40 kilometers to the south in the heart of the French Basque Country (Saint-Périer 1936:28). The Tercis and other sources in southern Landes around Dax are of Campanian/Maastrichtian (Upper Cretaceous) age, the "Dalle de Bidache" calcareous flysch band running parallel to the Adour-Gaves axis dates to the Coniacian/Turonian/ Maastrichtian (Middle-Upper Cretaceous), and the highly tectonized sources around Salies on the interfluve between the Gaves de Pau and Oloron also date to the Upper Cretaceous (Séronie-Vivien 1994, 1995; see also Debourle and Deloffre 1976). The first and last produce nodular flint of superb quality in relatively large nodules, whereas the second has veins of tabular flint also of good quality, but perhaps somewhat more orthogonal in its fracture pattern.

In the center of this wealth of flints is the cluster of Magdalenian and Azilian sites at the base of the Pastou Cliff near the confluence of the Gaves and 13 kilometers upstream of the confluence of these tributaries with the Adour. The sites are Duruthy, Dufaure, Grand Pastou and Petit Pastou. Detailed information on lithic raw material utilization is available (and is published) for Dufaure (Straus *et al.* 1988; Straus 1992b, 1996; Séronie-Vivien 1994, 1995; Doggett *et al.* 1995). I will only briefly summarize some of the main conclusions of our study. I would also note that, based on my informal examination of the much larger collections from Duruthy, the same general patterns across the Final Magdalenian-Azilian transition probably obtained at that adjacent, penecontemporaneous site as well (as impressionistically observed by R. Arambourou [1978:26] in his description of a shift in the representation of flint types).

The Final Magdalenian of Dufaure has been attributed to the Allerød chronozone and is dated by four stratigraphically coherent C^{14} dates between 12,300 and 10,900 BP. The Azilian was formed during Dryas III and Preboreal and is dated by 4 determinations to between 10,300 and 9,600 BP. The Final Magdalenian level at Dufaure (like those at Duruthy and Grand Pastou) have evidence of only cold season occupations. There is no indication of summer kills among the dental cementum or eruption sequences of reindeer (the dominant game species) or other ungulates at Dufaure (Spiess 1995, Altuna and Mariezkurrena 1995). The birds and moles also give evidence of only cold season human activity at the site (Eastham 1995a,b).

Flint overwhelmingly dominates the lithic assemblages of both the Final Magdalenian (level 4) and Azilian (level 3): 97% in both. In level 4, 2% of the assemblage is chalcedony, consisting of two nodules knapped *in situ* in an area of about one square meter on the terrace in front of the rockshelter, and whose source may be a known locus 25 kilometers to the east. The interesting thing about the flint raw materials at Dufaure is that they are of two sorts: nodular flint from the Salies area (currently available at no more than 6 kilometers from the site) and tabular flint from the Bidache area (no more than 9 kilometers from Dufaure). Although we also sampled the well-known flints from Tercis and Bénesse-lès-Dax, petrographic analysis by Séronie-Vivien revealed no such flint at Dufaure, although these sources are only 19 and 12 kilometers north of Dufaure. Nor do there seem to be any flints from the distinctive Audignon source, about 40 kilometers to the northeast in Les Landes. Once people settled (for the season) at Dufaure, they moved and procured their flints within a very limited radius around the site, exploiting game migrating along and passing the winter in the Gave d'Oloron corridor between the southern Aquitaine Basin and the Pyrenees.

Yet there is a notable difference between the Final Magdalenian and Azilian occupations in terms of choice of flints. In the former, as in the underlying Middle Magdalenian levels, an overwhelming percentage (75% of cores and chunks, 86% of debitage, and 88% of tools) of the flint is of the excellent nodular chalk cortex variety and much less is of the tabular variety. In the Azilian, a lesser relative frequency (56% of cores and chunks, 71% of debitage, and 84% of tools) of the flint is nodular, whereas the tabular flint increases in frequency, especially among cores and debitage. Although the nodular flint continued to be the variety of choice for making formal tools in the Azilian, most *in situ* knapping was of the tabular flint. There is only scanty evidence (two red deer) for seasonality (winter-spring) from the Azilian level, and there are no winter guest bird species. We cannot definitively explain the change in flint procurement, especially as the earlier Magdalenian assemblages had been very similar to the Final Magdalenian one. It is clear, however, that the nature and scale of the occupation of Dufaure changed between the Magdalenian and Azilian (vast, artifact- and structure-rich, reindeer-dominated versus spatially restricted, poor, red deer *cum* roe deer and boar-dominated) and that there were significant changes in vegetation and geomorphological processes that may have also affected the availability of flints and processes of their procurement. So, although essentially all flint acquisition was local, a shift occurred that begs explanation in behavioral and environmental terms, as procurement was probably still embedded in other foraging activities within about a two hour radius of the site.

Conclusions

In this brief *tour d'horizon* I have presented most of the data (some quite impressionistic and imprecise, others more empirically controlled and precise) that exist at present in the archeological record on the use of lithic raw materials in the Tardiglacial and early Postglacial of Vasco-Cantabrian Spain and French Gascony. The main conclusion is the same one that I observed more than fifteen years ago: human use of stone for tool-making closely tracks local bedrock lithology in this stone-rich corner of Europe. In areas of particularly mixed lithology, flints (or the best flints) may have been chosen for making classic leptolithic tools, while other, more easily accessible stones (e.g., quartzites, calcite, poor quality flints or cherts) were used for making simpler, larger, more expedient tools (or used without deliberate retouching). In areas with more uniformly Cretaceous bedrock and more or less ubiquitous flint, flint was used for every category of tool. There is no evidence for truly long-distance transport, although there are clear suggestions that at high-altitude summer sites in Navarra, good-quality flints were brought in from medium distances, perhaps simply during the seasonal moves of human bands.

There is scant evidence for a *generalized* trend toward use of even more local resources between the Magdalenian and the Azilian. At certain sites such a pattern may exist, but this may reflect on individual site-use histories rather than on some general "evolutionary" tendency. Certainly there is evidence for changes between Magdalenian and Azilian occupations in *some* cases, but they often involve frequency shifts tending to favor one local material over another more or less equally local material.

Even in the case of a site that was almost certainly a part of a much larger annual territory that involved large-scale human movements (i.e., between Dufaure and the summer reindeer pastures of the western Pyrenees during the terminal Magdalenian, when one of the last *Rangifer* herds left in France seems to have survived by means of seasonal altitudinal movements), once people arrived at the site for their habitual winter stay, they made use of only local flints. This was because they knew from past experience that they could find abundant, good flint at their winter base, and thus could travel light. (The reverse, of course, was true, in the case of the site of Les Eglises in the high, flint-poor Central Pyrenees, to which ibex and lagomorph hunters and salmon fishers brought flints from the flint-rich lowlands during logistical visits [Simonnet 1985].) Thus the images of extensive procurement catchments in the Magdalenian (via widespread movements and/or social contacts) and of shrunken "lithic territories" in the Azilian that we have received from other regions of Europe do not seem to apply in Vasco-Cantabrian Spain. This is because of the nature of the region's lithology, orography, topography and limited, spatially constricted scale. Nor do they work very well in the special seasonal, residential circumstances of the Dufaure Magdalenian. Even if the Dufaure Azilian annual territory were smaller than that of its predecessor, it is not the flints that can tell us so. Neither the archeological concept of the Magdalenian nor that of the Azilian is monolithic (pardon the pun) — at least not as far as their patterns of lithic procurement throughout the geologically complex sub-continent of western Europe are concerned.

References Cited

Altuna, J.
1981 Restos óseos del yacimiento prehistórico del Rascaño. In *El Paleolítico Superior de la Cueva del Rascaño*, edited by J. González Echegaray and I. Barandiarán. Santander, pp. 223-269. Centro de Investigación y Museo de Altamira Monografías 3.
1984 Resumen y conclusiones. In *El yacimiento prehistórico de Ekain*, edited by J. Altuna and J. Merino, pp. 347-349. Sociedad de Estudios Vascos, San Sebastián.
1986 The mammalian faunas from the prehistoric site of La Riera. In *La Riera Cave*, edited by L. Straus and G. Clark, pp. 237-274. Anthropological Research Papers 36. Arizona State University, Tempe.

Altuna, J. and K. Mariezkurrena
1982 Restos óseos del yacimiento prehistórico de Abauntz. *Trabajos de Arqueología Navarra* 3:347-353.
1995 Les restes osseux de macromammifères. In *Les derniers chasseurs de renne du monde pyrénéen. L'Abri Dufaure: un gisement tardiglaciaire en Gascogne*, edited by L. Straus. Mémoires de la Société Préhistorique Française, Tome 22, Paris.

Altuna, J., A. Baldeón and K. Mariezkurrena
1985 Cazadores Magdalenienses en la Cueva de Erralla. *Munibe* 37:1-206.

Altuna, J. and J. Merino
1984 *El yacimiento prehistórico de Ekain.*

Sociedad de Estudios Vascos, San Sebastián.

Arambourou, R.
1978 *Le gisement préhistorique de Duruthy à Sorde-l'Abbaye.* Mémoires de la Société Préhistorique Française 13. Paris.

Baldeón, A.
1985 Estudio de las industries lítica y ósea de Erralla. In *Cazadores Magdalenienses en la Cueva de Erralla*, edited by J. Altuna, A. Baldeón and K. Mariezkurrena. *Munibe* 37:123185.

Barandiarán, I.
1988 Datation C14 de l'art mobilier magdalénien cantabrique. *Bulletin de la Société Préhistorique de l'Ariège* 43:63-84 .
1989 Precisión cronológica del Magdaleniense del Pendo. In *Cien Años Después de Sautuola*, edited by M. González Morales, pp. 97-114. Diputación Regional de Cantabria, Santander.

Barandiarán, I. and A. Cava
1989 El yacimiento prehistórico de Zatoya. *Trabajos de Arqueología Navarra* 8:1-354.

Bernaldo de Quirós, F. and V. Cabrera
1993 Early Upper Paleolithic industries of Cantabrian Spain. In *Before Lascaux: The Complex Record of the Early Upper Paleolithic*, edited by H. Knecht, A. Pike-Tay and R. White, pp. 57-69. CRC Press, Boca Raton.

Bosinski, G.
1988 Upper and Final Paleolithic settlement patterns in the Rhineland, West Germany. In *Upper Pleistocene Prehistory of Western Eurasia*, edited by H. Dibble and A. Montet-White, pp. 375-386. University Museum Monograph 54. University of Pennsylvania, Philadelphia.

Breuil, H.
1952 Note sur l'outillage en calcaire taillé du Magdalénien du Castillo. *Bulletin de la Société Préhistorique Française* 49:23-24.

Bustillo, M. and A. Ramos-Millán, editors
1991 *VI International Flint Symposium.* International Copy, Madrid.

Butzer, K.
1985 Observaciones sobre la geologia de la Cueva de El Piélago. *Sautuola* 4:19-24.

Cabrera, V.
1984 *El yacimiento de la Cueva de El Castillo.* Bibliotheca Praehistorica Hispana 22. Madrid.

Carballo, J.
1938 Materias primas de las industries prehistóricas. Unpublished manuscript, Museo Prehistórico Regional, Santander.
1957 La cristalografía y la técnica del hombre prehistórico. In *Investigaciones Prehistóricas* I, pp. 51-59. Diputación Provincial, Santander.
1960 Excavaciones en la caverna de El Pendo. In *Investigaciones Prehistóricas* II, pp. 17-108. Diputación Provincial, Santander.

Cava, A.
1989 La industria lítica: los utensilios. In *El yacimiento prehistórico de zatoya*, edited by I. Barandiarán and A. Cava. *Trabajos de Arqueología Navarra* 8:37-136.

Clottes, J.
1989 Le Magdalénien des Pyrénées. In *Le Magdalénien en Europe*, edited by J.-P. Rigaud, pp. 281-360. ERAUL 38, Liège.

Debourle, A. and R. Deloffre
1976 Guides géologiques réqionaux: Pyrénées Occidentales, Béarn, Pays Basque. Masson, Paris.

Deith, M. and N. Shackleton
1986 Seasonal exploitation of marine mollusks. In *La Riera Cave*, edited by L.Straus and G. Clark, pp. 299-314. Anthropological Research Papers 36. Arizona State University, Tempe.

Doggett, S., K. Kramer and L. Straus
1995 La signification des matiéres premiéres lithiques. In *Les derniers chasseurs de renne du monde pyrénéen. L'Abri Dufaure: un gisement tardiglaciaire en Gascogne*, edited by L. Straus. Mémoires de la Société Préhistorique Française, Tome 22, Paris.

Eastham, A.
1995a L'écologie avienne. In *Les derniers chasseurs de renne du monde pyrénéen. L'Abri Dufaure: un gisement tardiglaciaire en Gascogne*, edited by L. Straus. Mémoires de la Société Préhistorique Française, Tome 22, Paris.
1995b La microfaune. In *Les derniers chasseurs de renne du monde pyrénéen. L'Abri Dufaure: un gisement tardiglaciaire en Gascogne*, edited by L. Straus. Mémoires de la Société Préhistorique Française, Tome 22, Paris.

Fernández Eraso, J.
1989 Los residuos de la tecnología de la piedra tallada. In *El yacimiento prehistórico de Zatoya*, edited by I. Barandiarán and A. Cava. *Trabajos de Arcrueologia Navarra* 8: 137-180.

Freeman, L.
1991 What mean these stones? Remarks on raw material use in the Spanish Paleolithic. In *Raw Material Economies Among Prehistoric Hunter-Gatherers*, edited by A. Montet-White and S. Holen, pp. 73-125. Publications in Anthropology 19. University of Kansas, Lawrence.

Garcia Guinea, M.
1985 Las Cuevas Azilienses de El Piélago y sus Excavaciones de 1967-1969. *Sautuola* 4:13-154.

González Echegaray, J.
1980 *El yacimiento de la Cueva de El Pendo.* Bibliotheca Praehistorica Hispana 17. Madrid.

González Echegaray, J. and I. Barandiarán
1981 *El Paleolítico Superior de la Cueva del Rascaño.* Centro de Investigación y Museo de Altamira Monografías 3. Santander.

González Echegaray, J. and L. Freeman
1971 *Cueva Morín: Excavaciones 1966-1968.* Patronato de las Cuevas Prehistóricas, Santander.

González Morales, M.
1982 *El Asturiense v Otras Culturas Locales.* Centro de Investigación y Museo de Altamira Monografias 7. Santander.

González Sainz, C.
1989 El Magdaleniense Superior-Final de la Región Cantábrica. Tantin, Santander.
1992 Algunas reflexiones sobre las materias primas líticas y la variabilidad técnica y tipología, al termino de Paleolítico Superior de la región cantábrica. In *Tecnología y Cadenas Operativas Líticas,* edited by R. Mora, X. Terradas, A. Parpal and C. Plana, pp.57-72. Treballs d'Arqueología 1. Barcelona.
1994 Sobre la cronoestratigrafía del Magdaleniense y Aziliense en la región cantábrica. *Munibe* 46:53-68.

González Sainz, C. and M. González Morales
1986 *La Prehistoria en Cantabria.* Tantin, Santander.

López-Berges, M. and M. Valle
1985 Estudio osteológico de la Cueva de Piélago II. *Sautuola* 4:113-121.

Mariezkurrena, K. and J. Altuna
1989 Análisis arqueozoológico de los macromamíferos. In *El yacimiento prehistórico de Zatoya,* edited by I. Barandiarán and A. Cava. *Trabajos de Arqueología Navarra* 4:237-266.

Mauger, M.
1994 L'approvisionnement en matériaux silicieux au Paléolithique supérieur. In *Environnements et habitats magdaléniens dans le centre du Bassin Parisien,* edited by Y. Taborin, pp. 78-93. Documents d'Archéologie Française 43, Paris.

Merino, J.
1984 Industria lítica del yacimiento de Ekain. In *El yacimiento prehistórico de Ekain,* edited by J. Altuna and J. Merino, pp. 65-175. Sociedad de Estudios Vascos, San Sebastián.

Montet-White, A.
1994 Alternative interpretations of the Late Upper Paleolithic in Central Europe. *Annual Reviews in Anthropology* 23:483-508.

Mora, R., X. Terradas, A. Parpal, and C. Plana
1992 *Tecnología v Cadenas operatives Líticas.* Treballs d'Arqueología 1. Barcelona.

Rensink, E., J. Kolen, and A. Spieksma
1991 Patterns of raw material distribution in the Upper Pleistocene of Northwestern and Central Europe. In *Raw Material Economies Among Prehistoric Hunter-Gatherers,* edited by A. Montet-White and S. Holen, pp. 141-160. Publications in Anthropology 19. University of Kansas, Lawrence.

Saint-Périer, R. de
1936 *La Grotte d'Isturitz II: Le Magdalénien de la Grande Salle.* Archives de l'Institut de Paléontologie Humaine, Mémoire 17. Paris.

Sarabia, P.
1990a Approche de l'étude de la distribution stratigraphique du silex de la province de Cantabria. In *Le silex de sa genèse à l'outil,* edited by M.-R. Séronie-Vivien and M. Lenoir, pp. 141-147. Cahiers du Quaternaire 17. Talence.
1990b L'utilisation du silex dans les industries du Paléolithique de Cantabria. In *Le silex de sa genèse à l'outil,* edited by M.-R. Séronie-Vivien and M. Lenoir, pp. 443-447. Cahiers du Quaternaire 17. Talence.

Schild, R.
1996 The North European Plain and eastern Sub-Balticum between 12,700 and 8,000 BP. In *Humans at the End of the Ice Age: The Archaeology of the Pleistocene-Holocene Transition,* edited by L. G. Straus, B. V. Eriksen, J. M. Erlandson, and D. R. Yesner, pp. 129-157. Plenum Press, New York.

Séronie-Vivien, M.-R.
1994 Pétrographie des silex préhistoriques du gisement paléolithique de l'abri Dufaure. *Bulletin de la Société Linéenne de Bordeaux* 22(l):23-29.
1995 Pétrographie des principaux types de silex. In *Les derniers chasseurs de renne du monde pyrénéen. L'Abri Dufaure: un gisement tardiglaciaire en Gascogne,* edited by L. Straus. Mémoires de la Société Préhistorique Française, Tome 22, Paris.

Séronie-Vivien, M.-R. and M. Lenoir, editors
1990 *Le silex de sa genese à l'outil.* Cahiers du Quaternaire 17. Talence.

Simonnet, R.
1981 Carte des gites à silex des Pré-Pyrénées.

XXI Conqrés Préhistorique de France, vol.1, pp.308-323.
1982 Grandes lames de silex dans le Paléolithique supérieur des Pyrénées centrales. *Bulletin de la Société Préhistorique de l'Arièqe* 37:61-106.
1985 Le silex du Magdalénien final de la Grotte des Eglises dans le Bassin de Tarascon-sur-Ariège. *Bulletin de la Société Préhistorique de l'Ariège* 40:71-98.

Spiess, A.
1995 Etude de la saison d'habitation au moyen du cément dentaire. In *Les derniers chasseurs de renne du monde pyrénéen. L'Abri Dufaure: un gisement tardiglaciaire en Gascogne*, edited by L. Straus. Mémoires de la Société Préhistorique Française, Tome 22, Paris.

Straus, L.
1979 Mesolithic adaptations along the coast of northern Spain. *Quaternaria* 21:305-327.
1980 The role of raw materials in lithic assemblage variability. *Lithic Technology* 9:68-42.
1985 Chronostratigraphy of the Pleistocene-Holocene transition: the Azilian problem in the Franco-Cantabrian region. *Palaeohistoria* 27:89-122.
1991 The role of raw materials in Upper Paleolithic and Mesolithic stone artifact assemblage variability in Soutwest Europe. In *Raw Material Economies Among Prehistoric Hunter-Gatherers*, edited by A. Montet-White and S. Holen, pp. 169-185. Publications in Anthropology 19. University of Kansas, Lawrence.
1992a *Iberia before the Iberians: The Stone Age Prehistory of Cantabrian Spain*. University of New Mexico Press, Albuquerque.
1992b L'Abri Dufaure et la falaise du Pastou dans le systéme adaptatif régional des Pyrénées au Magdalénien. In_*Le peuplement magdalénien*, edited by J.-P. Rigaud, H. Laville and B. Vandermeersch, pp. 335-344. CTHS, Paris.
1996 The use of quartzite in the Upper Paleolithic of Cantabrian Spain. In *Non-Flint Stone Tools and the Paleolithic Occupation of the Iberian Paleolithic*, edited by N. Moloney, L. Raposo and M. Santonja, pp. 37-41. British Archaeological Reports, International Series 649. BAR Publishing, Oxford.

Straus, L., K. Akoshima, M. Petraglia, and M. Séronie-Vivien
1988 Terminal Pleistocene adaptations in Pyrenean France: the nature and role of the Abri Dufaure site. *World Archaeology* 19:328-348.

Straus, L. and G. Clark, editors
1986 *La Riera Cave*. Anthropological Research Papers 36. Arizona State University, Tempe.

Straus, L., G. Clark, J. Ordaz, L. Suarez, and R. Esbert
1986 Patterns of lithic raw material variation at La Riera. In *La Riera Cave*, edited by L. Straus and G. Clark, pp. 189-208. Anthropological Research Papers 36. Tempe.

Straus, L. and C. Heller
1988 Explorations of the Twilight Zone: the Early Upper Paleolithic of Cantabria and Gascony. In *The Early Upper Paleolithic*, edited by J. Hoffecker and C. Wolf, pp. 97-133. British Archaeological Reports S-437. BAR Publishing, Oxford.

Terán, M. de and L. Sole Sabaris
1968 *Geografía Regional de España*. Ariel, Barcelona.

Utrilla, P.
1982 El yacimiento de la Cueva de Abaunta. *Trabajos de Arqueología Navarra* 3:203-345.

Vázquez, J. M.
1994 El Mesolitico costero de Galicia: una panorama actual. Unpublished paper presented at "Old People and the Sea: International Conference on the Mesolithic of the Atlantic Facade," Santander, July 1994.

11. Residential Mobility and Lithic Economizing Behavior: Explaining Technological Change in the Portuguese Upper Paleolithic

Paul Thacker

This paper emphasizes that lithic economy can be fully explained only within the broader context of hunter-gatherer subsistence systems. Data from intensive surface survey and excavation in the Upper Rio Maior, Ribeira da Pa, and Penegral valleys in the eastern Portuguese Estremadura are used to examine lithic raw material use in the Gravettian, Early Magdalenian, and Late Magdalenian. The author use a landscape approach and total coverage survey within a limited area to compare site densities and locations through time and to assess the availability of local raw materials. Patterns of raw material use are examined for different types of local raw material (quartzites and high-quality flint) and several broad tool and blank classes (formal, informal, and multiple tools; bladelets and other blanks) over time. Trends in raw material use are interpreted in light of raw material economizing behavior and changing technological constraints. Results strongly argue against oversimplification of lithic reduction strategies by condensing assemblages into single technological classes without accounting for variation in raw material choice and availability. Examination of the role of a number of different technological objectives (maintainability, transportability, flexibility, reliability) in each prehistoric context allows the author to develop hypotheses about changing settlement organization. Though all time periods show evidence of organized transport of raw materials ("curation"), a trend towards less use of local quartzites, more reliance on bladelet tools, and conservation of raw material is argued to indicate increasing residential mobility (and therefore increased anticipated mobility) throughout the Upper Paleolithic. *–Editors.*

Raw material source areas have influenced prehistoric hunter-gatherer economies in a wide variety of ways. In certain cases/environments, raw material availability affected technological strategies far more than settlement and subsistence decisions (Bamforth 1986, 1990). On the opposite end of the spectrum, some groups organized lithic technology predominantly around subsistence or settlement mobility needs (Bar-Yosef 1991; Demars 1982; Marks 1988; Montet-White 1988, 1991). Proponents of direct correlations between settlement/subsistence and lithic organization are becoming scarcer, as more exceptions to popular middle-range expectations (Binford 1980; Bettinger 1991) are published. This paper will examine the changing lithic economy of Upper Paleolithic hunter-gatherers in Portugal. Based on intensive survey and excavation results utilizing a landscape approach, the results presented here argue against oversimplification of lithic reduction strategies by condensing assemblages into single technological classes without accounting for variation in both raw material choice and availability. The study area, the Upper Rio Maior, Ribeira da Pa, and Penegral valleys, provide an unusual backdrop for lithic studies, as high quality flint, quartz, and quartzites are available within ten kilometers of any site. Yet by analyzing technological organization across tool classes and raw material types, and through time, patterns emerge that can be used to develop hypotheses of group mobility and explanations of technological change.

Overview of Upper Paleolithic Central Estremadura

A six year project co-directed by Anthony Marks and Joao Zilhão has established the basic chronostratigraphic framework for Portuguese Estremadura. The Gravettian has been absolute dated to between 23,500 and 21,000 BP, while the Magdalenian falls into at least two divisions: Early Magdalenian (16,500 - 14,500 BP) and Late Magdalenian (13,000 - 10,000 BP) (Marks *et al.* 1994). Although well documented in other valleys in Estremadura, the Solutrean occupation in this paper's study area was ephemeral or geologically invisible to survey, as only one possible Solutrean occupation was located. Likewise the Aurignacian, the earliest Upper Paleolithic industry of Portugal, is only represented by five sites in the Rio Maior area. This paper excludes the Aurignacian due to high assemblage variability.

Technological differences between and within Upper Paleolithic assemblages are quite complex (see Marks *et al.* 1994; Bicho 1992, 1993; Thacker

1996; Zilhão 1987, 1990a,b, 1995), and cannot receive full justice here. Gravettian single platform blade and bladelet cores fall along a continuum, as the blade/bladelet distinction is, in this situation, arbitrary (Marks *et al.* 1994:61). Multiple platform bladelet cores are more frequent in the Magdalenian periods, and blade technology diminishes in importance. Backed bladelet tools are important in all three periods, but bladelet assemblages are most important in the Magdalenian (for example, geometrics occur with a microburin technique). Some retouched tool typological differences between periods are significant, but many tool assemblages have only been subjected to preliminary analysis.

Paleoenvironmental reconstruction places Central Estremadura on the Late Pleistocene transition between cold and wet northern Portugal and rather temperate southern Portugal (Marks *et al.* 1994). At the glacial maximum a variety of niches would have been compressed into the topographic relief of the area. The high Serra Dos Candeeiros (ca. 350-400 meters above sea level) was probably exposed/open or covered in light scrub, while arboreal species remained in the lower (and sheltered) part of the Rio Maior drainage (Zilhão 1995). The middle section of the three surveyed valleys (hereafter referred to as "middle valley") is the location of nearly all residential campsites and was probably open parkland with some trees present even at the pleniglacial.

Three raw material sources are located within the study region. A high quality red flint occurs mainly on the ridge that separates the Rio Maior and Penegral drainages. The distribution of the secondary flint deposit was determined by total coverage survey, and is shown in Figures 1-3. Flint cobbles occur in sands throughout the shaded area, but were exposed mainly in spring-fed seasonal stream channels near the top of the ridge (Thacker 1996). Cobble sizes vary, but most fall in the 10-40 centimeter range. In contrast to the flint, small cobbles of quartz and quartzite (ca. 5-25 centimeters) are found throughout the valley in Early or Middle Pleistocene gravels. Thus every site discussed below is within a hundred meters of a quartz and/or quartzite source of varying quality. Both the flint and the quartz/quartzite sources were exposed throughout the Upper Paleolithic, with only minor fluctuations occurring as stream channels meandered.

Raw Material Use and Settlement Patterns

Intensive total surface survey was conducted in the study area during the 1991 and 1992 field seasons. The distribution of Gravettian, Early Magdalenian, and Late Magdalenian sites shown in Figures 1-3 is representative of prehistoric patterning, as the valleys have experienced only minor geological events since the Middle Pleistocene. The Gravettian pattern is skewed in that middle valley occupations were occasionally located up to two meters below modern ground surface due to slopewash and local aeolian deposition. The middle valley probably contains more buried Gravettian sites, but the sites known are representative of patterning, based on limited subsurface sampling and geomorphological reconstructions. Deep plowing undertaken for eucalyptus forestry aided survey by exposing deep sites and increasing subsurface exposures. Early Magdalenian and Late Magdalenian sites are rarely deeply buried, and are often partially exposed by agricultural plowing. Lower valley landforms have been continuously exposed throughout the Late Pleistocene, and sampling visibility was extremely good. The lack of Early Magdalenian and Late Magdalenian sites in the lower valley and around the flint source is evidence for absence (or at least scarcity) rather than absence of evidence (Thacker 1996).

Gravettian sites (Figure 1) are located throughout the valley system. A series of lithic workshop sites are found in the lower valley, while several residential sites were excavated in the middle valley. Gravettian lithic workshops, all located within a kilometer from the flint source, are distinguished from residential sites by lower tool diversity and evenness among classes, lower core and tool frequencies, truncated reduction sequences (overrepresentation of large core fashioning without evidence for prolonged tool use and resharpening), and lack of features or other evidence for habitation. Residential sites of all periods display high tool diversity, occasional hearths and other features, and either a complete reduction sequence or only the finished-tool end of manufacturing. The basic distinction between workshop and residential sites was evident to excavators of the sites prior to detailed assemblage analysis (Marks, personal communication).

Assemblage raw material frequencies display unexpected variability given the presence of Gravettian workshop sites (Figure 4). The flint component of Magdalenian assemblages in the middle valleys is near 90 percent, while the Gravettian residential sites contain only 70 - 80 percent flint. Local quartz and quartzites (and occasional rock crystal) make up the difference in all periods, as to date no exotic lithic materials have been identified. While Gravettian groups expended the energy for special-purpose flint extraction workshops, the raw material is exhausted rapidly as distance from the source reaches even a few kilometers (Marks *et al.* 1991; Thacker 1996). Explaining this contradiction to least effort principles requires an understanding of the organizational role of both flint and quartz.

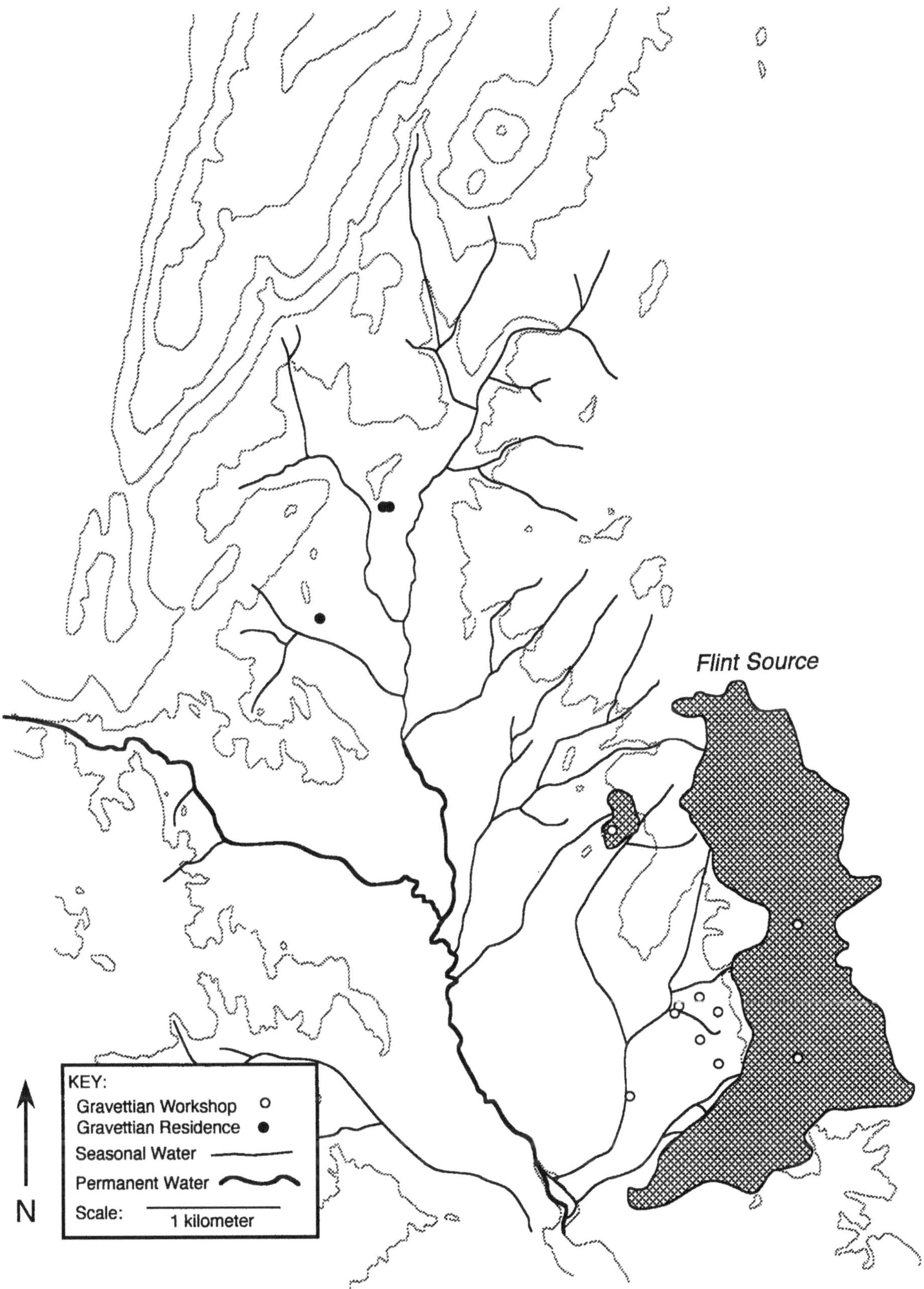

Figure 1. Location of Gravettian sites relative to topographic features, drainages, and flint source.

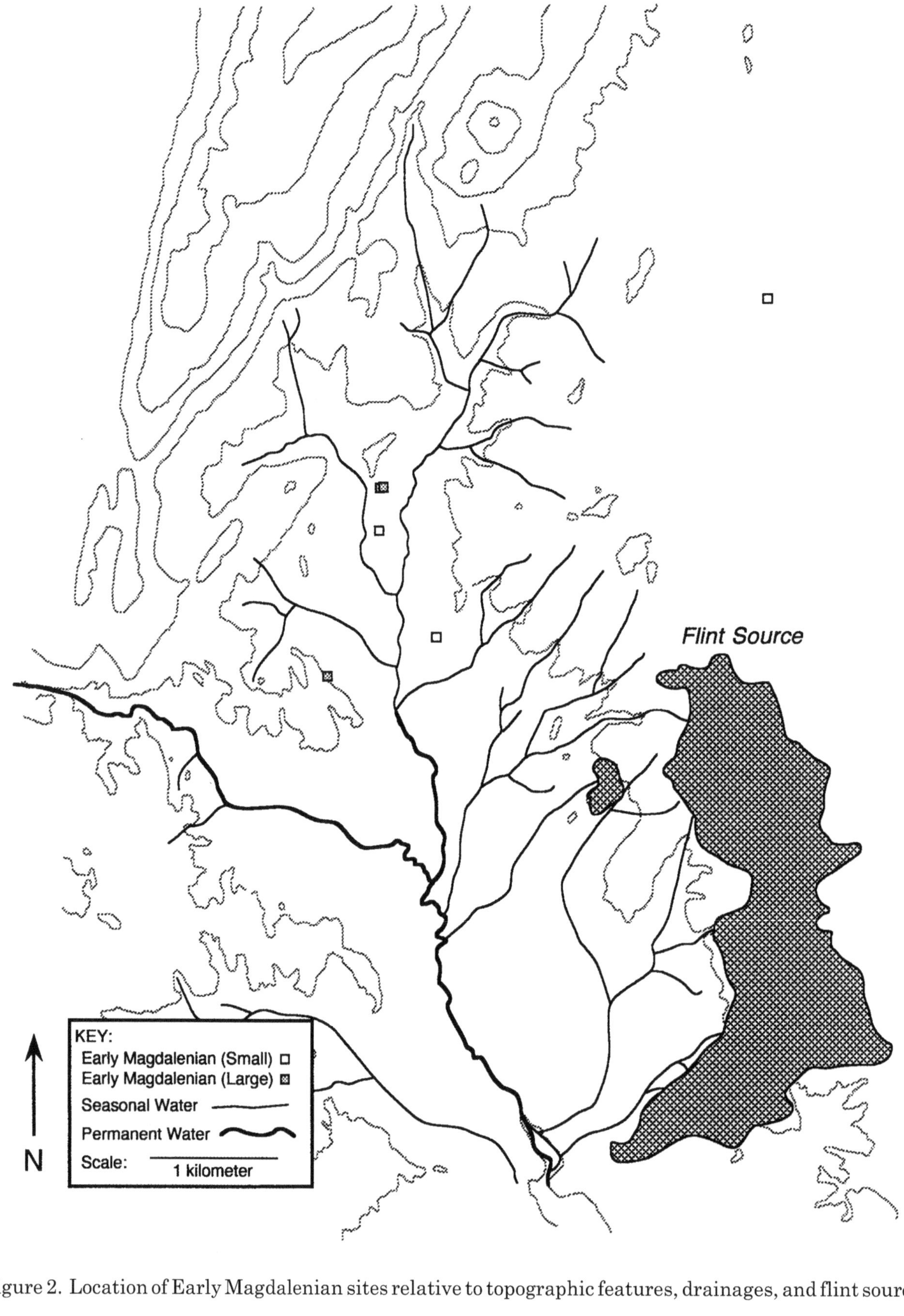

Figure 2. Location of Early Magdalenian sites relative to topographic features, drainages, and flint source.

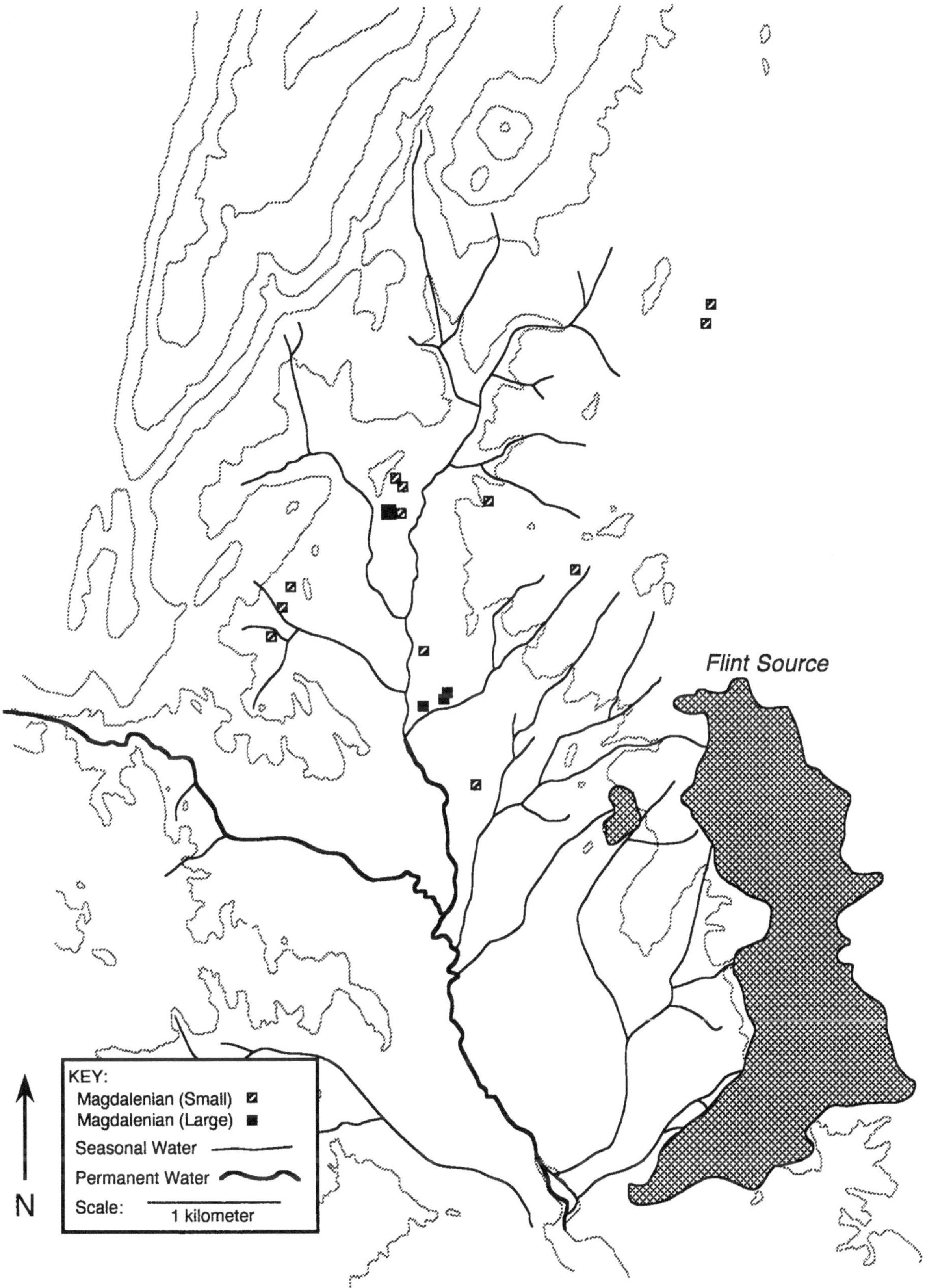

Figure 3. Location of Late Magdalenian sites relative to topographic features, drainages, and flint source. (Fig. does not include Early Magdalenian sites.)

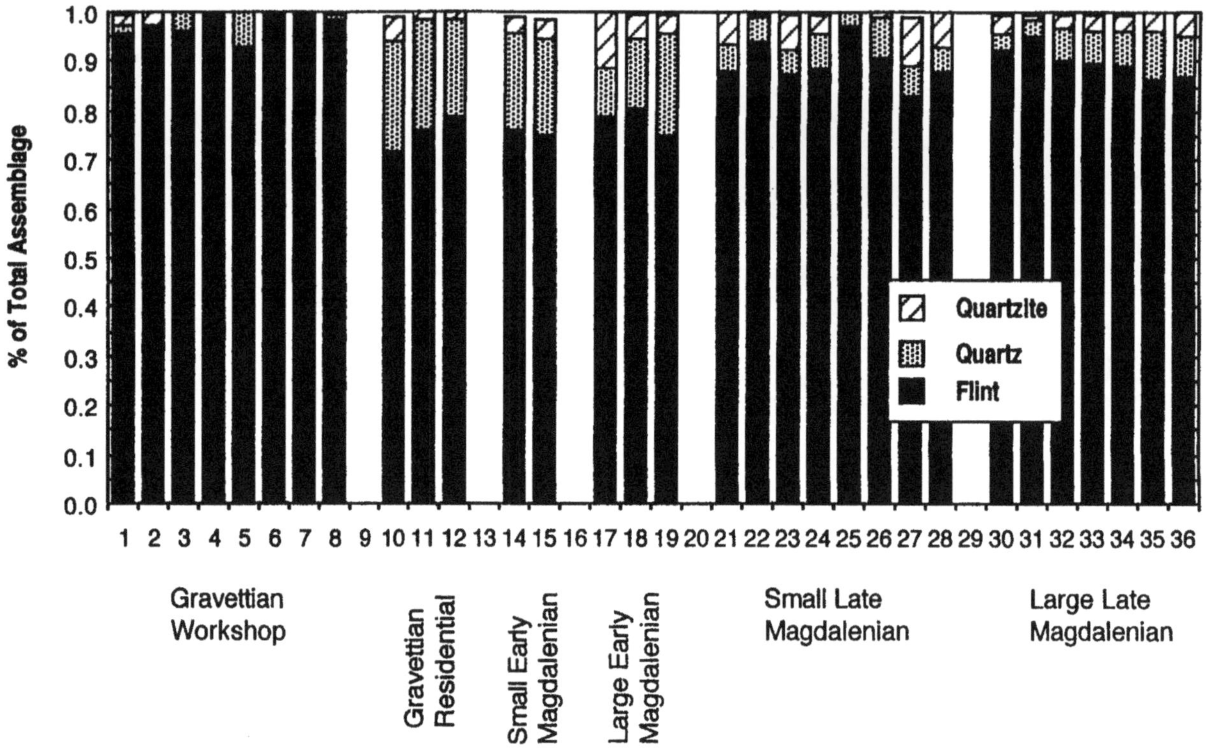

Figure 4. Raw material frequencies within Portuguese Upper Paleolithic assemblages from the survey region.

Quartz and Quartzite Raw Materials: An Alternative to Flint?

In order to gauge the assemblage context of quartz and quartzite use, all tools were grouped into a series of classes based on edge form. These classes, because they condense minor tool type divisions (for example, burin on straight truncation and burin on convex truncation), are more appropriate indicators of tool class diversity relative to assemblage size. For this paper, the classes were further grouped into informal, formal, and multiple tools. Informal tool classes include notches, heavy duty scrapers, retouched flakes, blades, and bladelets, pièces esquillées, and varia. Multiple tools exhibit two complete tool types on a single piece, while all other tools were considered "formal".

Figure 5 displays Gravettian tool class relative frequency by raw material type. The workshop sites contained virtually all flint tools, with about 70 percent of tools being formal. In contrast, the residential sites contained a small but significant number of quartz tools that were distributed across formal and informal classes just as the flint component was. With the exception of bladelet tool classes that will be discussed later, quartz was a commonly used and sufficient substitute for higher quality flint. As far as large tools such as endscrapers and burins were concerned, flint was not a required medium.

Sites from the Early and Late Magdalenian were separated into large and small sites based on a combination of site area and density of artifacts. Figure 6 displays two important trends in raw material use. First, some difference in site function may be present, as informal tools in both flint and quartz outnumber formal ones at small Early Magdalenian sites. This hypothesis is further supported by significant differences in Early Magdalenian site location between small and large sites (Thacker n.d.). Second, quartz again accounts for a small but important component of both formal and informal tool classes.

Late Magdalenian distributions (Figure 7) are different from the two previous periods. Virtually no quartz or quartzite is used for either formal or informal tools. There is also little difference between small and large site lithic assemblages, despite evidence for different paleoenvironmental settings of small and large site locations (Thacker n.d.).

These coarse-grained trends indicate that the cost of transporting flint to residential sites and manufacturing formal tools was increasingly chosen over either using locally available raw material or making informal, less energy demanding tools. Clearly flint was a preferred medium during the Late Magdalenian, while quartz and quartzite supplemented flint in earlier periods. This change in raw material preference could be related to technological needs, or to mobility/settlement shifts. Distinguishing between these two possible factors requires a closer examination of flint use within the assemblage.

The Changing Role of Flint

Of the three available raw materials in the Rio Maior area, the distinctive red flint has the best fracture and flaking qualities. Analyses of prehistoric lithic technologies in both the New and Old World have often found that the best raw material sources (often non-local) were used for manufacturing the most formalized elements of a toolkit (Ellis and Lothrop 1989; Goodyear 1989; Owen 1988). For example, Upper Paleolithic assemblages in Spain exhibit flint-dominated bladelet production in spite of locally available quartzites (Straus 1991, 1992).

The hypothesis of flint use for bladelet tool production was tested by separating the bladelet tool classes from other formalized tools. Table 1 displays the relative frequencies of bladelet tools within the entire tool assemblage, and the percentage of the bladelet tools made on flint. Bladelet tools comprise an increasingly larger proportion of tools through time, and are made on flint over 99% of the time in all periods. The need for bladelet tools

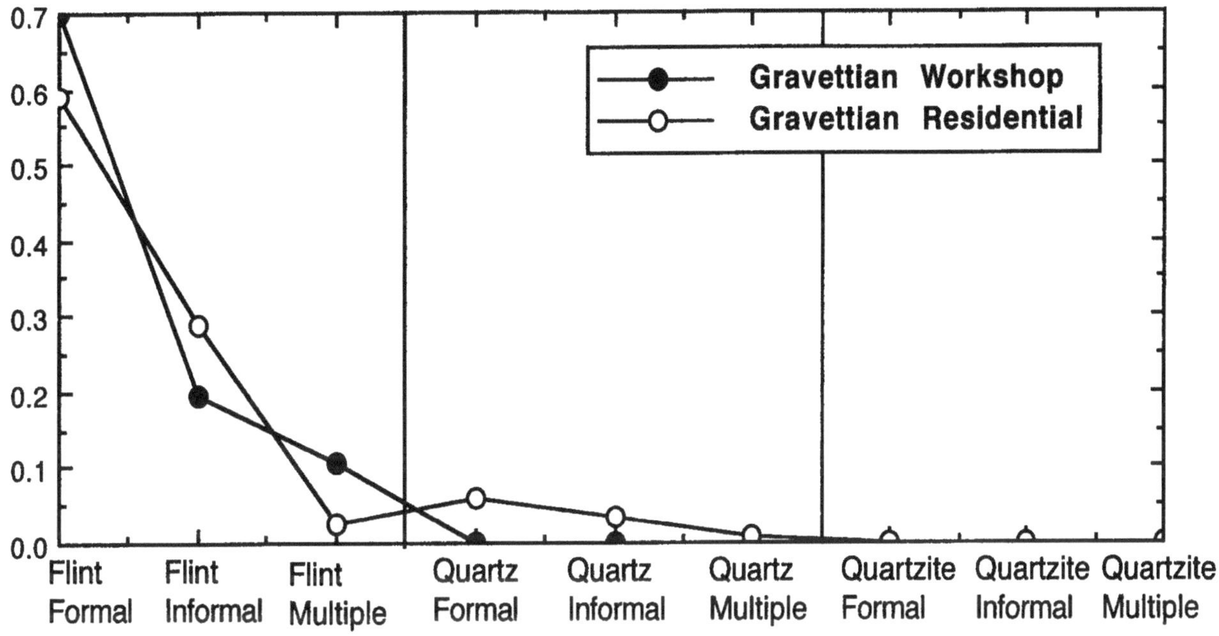

Figure 5. Gravettian distribution of formal, informal, and multiple tools across raw material types.

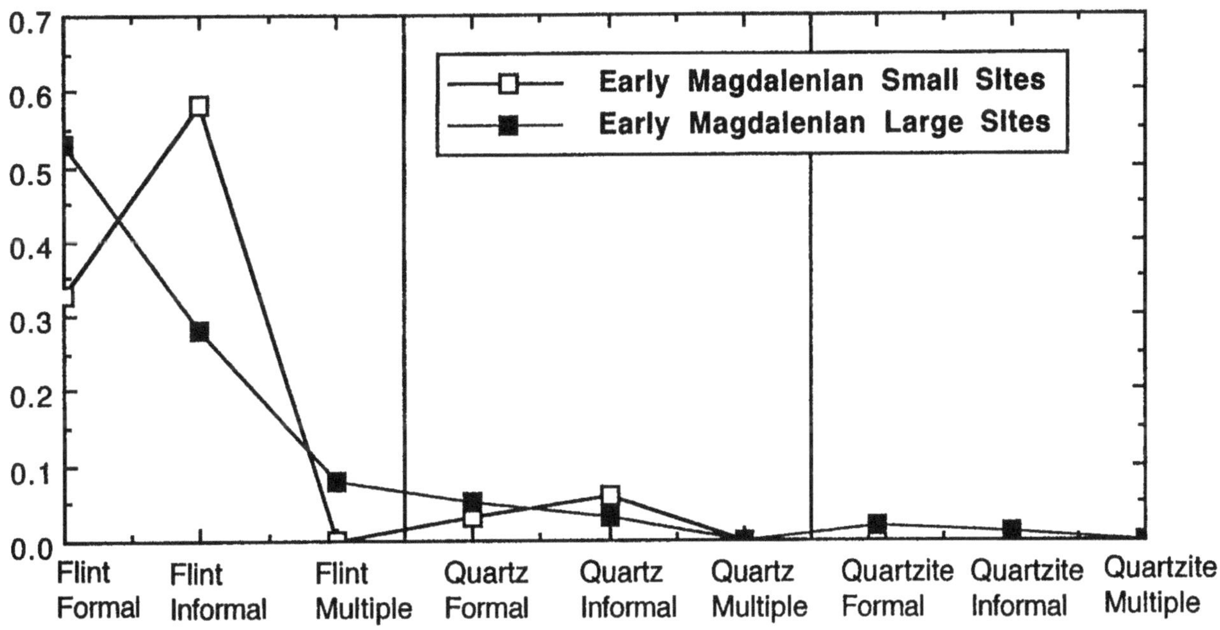

Figure 6. Early Magdalenian distribution of formal, informal, and multiple tools across raw material types.

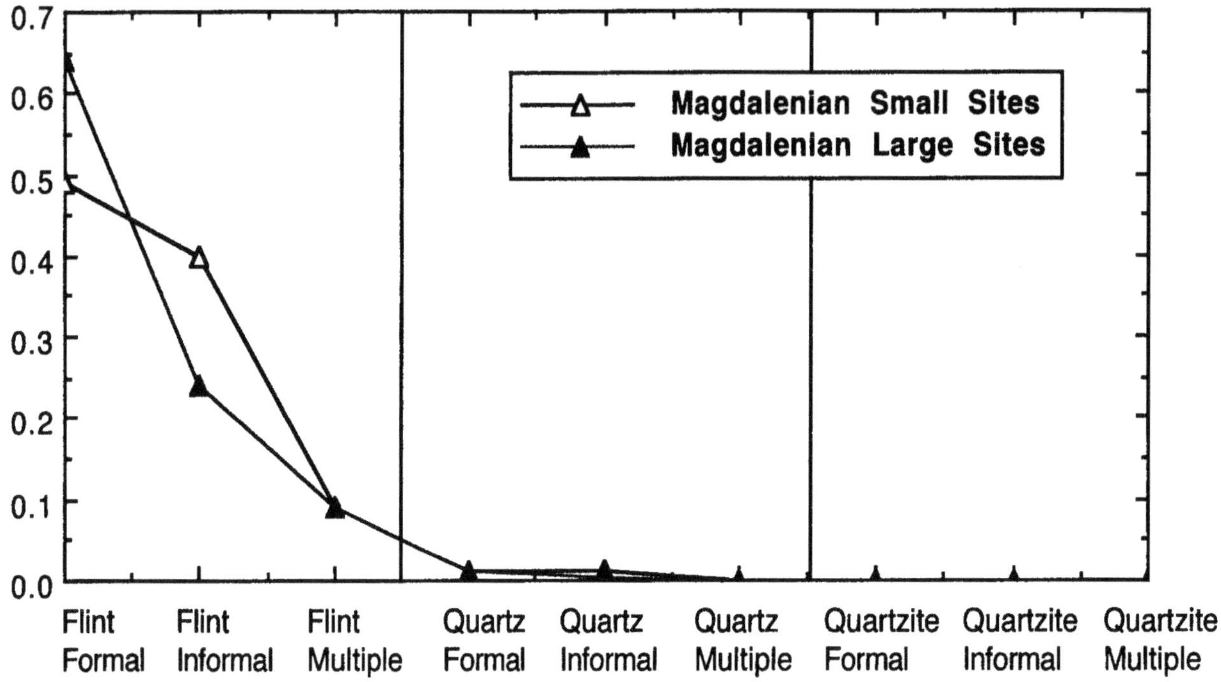

Figure 7. Late Magdalenian distribution of formal, informal, and multiple tools across raw material types.

Table 1. Bladelet Tool Relative Frequencies and Raw Materials

	Bladelet Tool Relative Frequency (among all tools)	% of Bladelet Tools on Flint
Gravettian Workshop	.055	100
Gravettian Residential	.158	100
Early Magdalenian Small Sites	.036	100
Early Magdalenian Large Sites	.131	97.3
Late Magdalenian Small Sites	.086	100
Late Magdalenian Large Sites	.182	99.0

Note: Averages were not weighted based on assemblage size. Preliminary weighted averages produced lower bladelet tool relative frequencies, but only the Gravettian Residential set is lowered significantly.

is part of the explanation for high frequencies of flint in the Magdalenian. During the Magdalenian, bladelet tools are much rarer in small sites than in corresponding large occupations. This observation is not a function of sampling biases, and may be indirect evidence that retooling and "gearing up" activities involving bladelet tools took place at large camps, while the smaller sites were either short duration or functionally distinct occupations.

These results were confirmed by an analysis of core types in the Gravettian. Large blade cores were transported from the lithic workshop sites to the middle valley residential sites. At the residential sites, the blade cores were reduced (transformed) into bladelet cores, as bladelet tools were produced (Thacker 1996). All other formal tool categories occur on quartz or quartzite, indicating a flexibility of raw material choice. Gravettian lithic economy capitalized on the nearby flint for all tool classes, but only the need for flint bladelet cores and tools mandated specialized extraction/workshop sites. As distance to the flint source increases, only bladelet cores and bladelet tool classes retain high flint frequencies. But if flint for bladelet tools was worth the energy expenditure to Gravettian groups, why are there no workshop sites for the Magdalenian, when flint bladelet tools were even more important in the toolkit? This dilemma can only be resolved by assessing economizing behavior of the different groups and technologies.

Hunter Gatherer Economizing Behavior

The Rio Maior data illustrate problems with the concept of curation as employed by archaeologists. Numerous factors can create patterns that have been labeled "curation," causing the term to lose its specificity (Nash 1996). In the Portuguese case, Gravettian groups extracted and worked flint at workshop sites, transported cores to residential sites, and maintained a small flint element in their assemblages for bladelet tools. This behavior can certainly be called curation in a broad sense (Nelson 1991). Yet throughout the Magdalenian, entire cobbles of flint were collected in daily activities or through specialized procurement trips (taking only several hours) and brought to the middle valleys (Thacker 1996). More flint moved greater distances during the Late Upper Paleolithic, and particularly during the Late Magdalenian, quartz and quartzite are not used extensively to supplement it. Both the Early and Late Magdalenian, then, seem to be "curated" technologies. Thus, to say that Portuguese Upper Paleolithic industries displays curation is meaningless, and in fact obscures variability. Variations on this problem have led many archaeologists to abandon the term altogether (Odell 1996).

Lithic economizing behavior aims to increase tool/core efficiency and use life, usually by conserving raw material (Odell 1989, 1996). Weight of tools and cores can provide a general comparison of economizing behavior between periods given representative samples. Figure 8 displays a decrease in flint core weights and non-bladelet flint tool weights through the Upper Paleolithic. Important to this analysis is the finding that original cobble sizes did not vary between periods (Thacker 1996), indicating either more intensive reduction of each core or, more likely, manufacture of multiple cores from a single cobble. The decrease in core and tool weights, and increase in conservation, underscores the importance of flint in the Magdalenian. Not only does flint make up more of the Magdalenian assemblages, it is reduced more efficiently. In contrast, except in the case of the bladelet tool component, Gravettian groups exhausted flint materials rapidly. These diachronic differences cannot be explained by distance to raw materials, differential access to raw materials, flaking qualities, or other external variables. Only change in anticipated residential mobility adequately explains these shifts in raw material use.

Residential Mobility and Lithic Systems

The failure of middle range models to encompass lithic variability stems in part from over-simplification. Proponents of the popular forager/collector model (Binford 1977, 1979, 1980) have

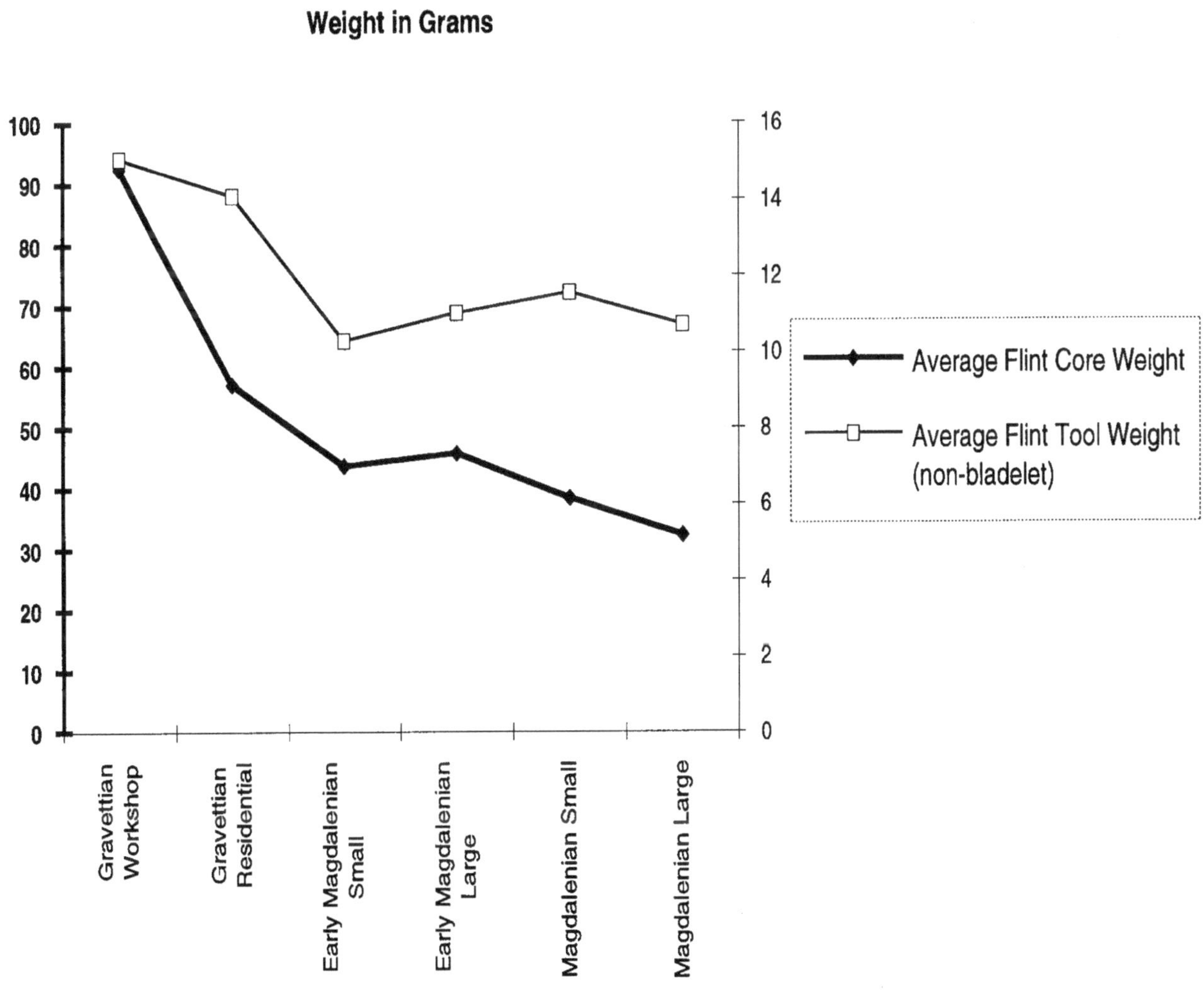

Figure 8. Average flint core weight and average flint tool weight by period and site type. Note that bladelet tools were excluded from the sample.

successfully expanded a perceived dichotomy into a spectrum of hunter gatherer subsistence and mobility organization. Unfortunately the range and complexity of lithic behavior incorporated into the model has not expanded accordingly (Ellis and Lothrop 1993). As illustrated above, the curated/expedient dichotomy should not be applied to an entire assemblage containing numerous raw material types (Nelson 1991).

By understanding technological constraints and economizing behavior, several observations on Upper Paleolithic toolkits can be made. Transportability (Nelson 1991; Parry and Kelly 1987; Shott 1986) is a minimal factor for Gravettian tool manufacture relative to the Early Magdalenian or Late Magdalenian. Gravettian tools are heavier and made on locally available materials. As such, the Gravettian toolkit (bladelet tools excluded) is more flexible than later periods in terms of raw material needs. In contrast, both conservation of raw material and transportability are important factors during the Magdalenian, reflected in smaller tool and core weights, and the retention of high flint frequencies.

All three periods display reliable toolkits (*sensu* Bleed 1986; Nelson 1991), but reliability indicators culminate in the Late Magdalenian. Bladelet tool forms (particularly backed bladelets and geometrics) are good examples of standardized replacement parts, and their increase should correlate with increasing reliability. This reliability of the tool assemblage is achieved at an increased cost of raw material transport, as locally available gravels were apparently considered a substandard medium for bladelet manufacture.

This project's landscape approach and resulting spatial control allows a final distinction between residential mobility models from lithic assemblage data. The more reliable toolkits that emerge in the Early Magdalenian and peak in the Late Magdalenian could result from either *increased* residential mobility (that is *decreasing* duration of occupations at large residential sites), or the *opposite*: a shift to a more "collector"-oriented subsistence strategy (Binford 1980; Bettinger 1991). If a shift in subsistence strategy did occur, residential sites would have been occupied for *longer* durations, but technological organization would still have emphasized reliability (Nelson 1991; Parry and Kelly 1987). This ambiguity in predictions for mobility is resolved through an examination of variability in raw material choice and availability within the survey area.

Throughout all periods, residential sites were located very near quartz and quartzite gravel beds. Just as a landscape approach coupled with geomorphological information was able to argue for minimal Late Upper Paleolithic occupation in the lower valleys, the minimal use of locally available cobbles in the Magdalenian is essential to explaining flint use leading up to the Holocene. Technological organization and toolkits recovered from small extraction sites within a collector strategy are expected to be extremely reliable (Bleed 1986; Bettinger 1991) and in the Portuguese case, meet the corresponding expectation of high frequency of flint (or better quality) raw material. But if residential camps were occupied for longer durations, as would be the case in a collector strategy, some elements of the toolkit or assemblage should have been fashioned on cobbles from the most easily obtained raw material source in order to minimize transport costs (Andrefsky 1994). This expectation stems from both the varying nature of "downtime" activities (Torrence, ed. 1989), and energy expense during raw material selection (Bamforth 1986). Late Upper Paleolithic groups did not significantly exploit quartz and quartzite deposits for tool manufacture, despite residing within a hundred meters of them. The relative paucity of non-flint raw materials in all tool classes of Late Magdalenian assemblages thus argues against an explanation based on shifting subsistence strategy, and argues for increased residential mobility within the same settlement strategy parameters as in the Early Magdalenian.

This point can be illustrated by drawing a comparison with Epipaleolithic assemblages from the Rio Maior. These Early Holocene assemblages display a shift in organization and raw material choice that probably does mark a change in subsistence strategy. At residential sites dating to the Early Holocene, large scraping and chopping tools are often made on local quartzes and quartzites, while flint is used for bladelet (or small flake) tools and cores, as well as during geometric (microlithic) tool production. This assemblage configuration/composition is markedly different from the Late Magdalenian, and co-occurs with the first solid evidence for an expanding resource base (e.g. shell fragments found in Epipaleolithic levels at Bocas rockshelter). In light of these patterns, the Late Upper Paleolithic choice not to fashion tools on local gravel cobbles provides important clarification: Late Magdalenian occupation of residential sites was shorter in duration (rather than longer) than in the Early Magdalenian or Gravettian.

Reliable tool technologies tend to occur in time-stressed systems (Torrence 1983, 1989), as the cost of raw material transport is balanced by assured availability when needed. When the Portuguese data set is examined from this perspective, several patterns converge. Greater flint use, increased bladelet tool manufacture, and technological conservation of raw materials fit into a model of increasing anticipated mobility throughout the Upper Paleolithic. The entire set of behaviors is interrelated, and reinforces itself. Gravettian groups were less mobile, and procurement of quartzes and quartzites was sufficient for all needs except technologically demanding bladelet tool production.

Conclusions

By examining lithic assemblage components across several raw material classes and assessing lithic economizing behavior, models of settlement mobility can be developed without the equifinality of large scale survey patterns. That is, information about the distances over which particular high quality local raw material was transported must be understood in the light of availability and use of local raw materials. The survey reported here almost certainly does not correspond to a prehistoric range, even if temporal control was sufficient to consider the possibility of relative contemporaneity. However, by examining space, in conjunction with technological organization and raw material diversity, this paper demonstrates that models can be assessed using evidence of choices that include *not* occupying an environmental niche or location, or *not* using quartz and quartzite for reduction.

In the case of the Portuguese Upper Paleolithic, increasing anticipated mobility was identified within periods that all display some aspect of organized raw material transport ("curation"). Methodologically, this paper demonstrates the complexity of lithic organizational variables and some possible interrelationships between them. Middle range constructs integrating lithic data with settlement/subsistence organization should incorporate a more sophisticated understanding of lithic raw material differences, technological constraints, and conservation practices. All raw materials are not equal, nor are they distributed evenly. Pigeon-holing lithic assemblages into single theoretical categories (be they "curated/expedient" or "forager/collector") without regard to raw material availability and choice leads to an archaeological record that, besides displaying little variation, contradicts the very model that was built to explain it.

Acknowledgments

Much of the work for this paper was carried out under a National Science Foundation Dissertation Improvement Grant, an Institute for the Study of Earth and Man (SMU) seed grant, and a Weber Fellowship (SMU). Anthony Marks, Joao Zilhão, and Nuno Bicho provided data and insight throughout my work. I also thank Joao Ladeira, Chris Ellis, Frederic Sellet, and George Odell for discussions that produced many of the ideas in this paper. Finally, I thank Lynn Fisher and Berit Eriksen for inviting me to participate in their symposium and their patient feedback on this paper.

References Cited

Andrefsky, W.
1994 Raw-material availability and the organization of technology. *American Antiquity* 59: 21-35.

Bamforth, D. B.
1986 Technological efficiency and tool curation. *American Antiquity* 51: 38-50.
1990 Settlement, raw material, and lithic procurement in the Central Mohave Desert. *Journal of Anthropological Archaeology* 9: 70-104.

Bar-Yosef, 0.
1991 Raw material exploitation in the levantine Epi-paleolithic. In *Raw Material Economies Among Prehistoric Hunter-Gatherers*, edited by A. Montet-White and S. Holen, pp. 235-250. Publications in Anthropology 19. University of Kansas, Lawrence.

Bettinger, R. L.
1991 Hunter-Gatherers: Archaeological and Evolutionary Theory. Plenum Press, New York.

Bicho, N. F.
1992 *Technological Change in the Final Upper Paleolithic of Rio Maior, Portuguese Estremadura.* Ph.D. Dissertation, Southern Methodist University, Dallas. University Microfilms, Ann Arbor.
1993 Late Glacial prehistory of central and southern Portugal. *Antiquity* 67:761-775.

Binford, L. R.
1977 Forty-seven trips: a case study in the character of archaeological formation processes. In *Stone Tools as Cultural Markers: Change, Evolution and Complexity*, edited by R. V. S. Wright, pp. 24-36. Australian Institute of Aboriginal Studies, Canberra.
1979 Organization and formation processes: looking at curated technologies. *Journal of Anthropological Research* 35:255-273.
1980 Willow-smoke and dog's tails: hunter-gatherer settlement systems and archaeological site formation. *American Antiquity* 45(1):4-20.

Bleed, P.
1986 The optimal design of hunting weapons: maintainability or reliability. *American Antiquity* 51(4):737-747.

Demars, P-Y.
1982 *L'utilisation du silex au paleolithique supérieur: choix, approvisionnement,*

circulation. Cahiers du Quaternaire 5. CNRS, Paris.

Ellis, C. and J. Lothrop
1989 *Eastern Paleoindian Lithic Resource Use*. Westview Press, Boulder.
1993 Technological change during Paleoindian and Archaic periods in eastern North America. Paper presented at the 58th Annual Meeting of the Society for American Archaeology, St. Louis.

Goodyear, A. C.
1989 A hypothesis for the use of cryptocrystalline raw materials among Paleoindian groups of North America. In *Eastern Paleoindian Lithic Resource Use*, edited by C. Ellis and J. Lothrop, pp. 1-9. Westview Press, Boulder.

Marks, A. E.
1988 The curation of stone tools during the Upper Pleistocene: a view from the Central Negev, Israel. In *Upper Pleistocene Prehistory of Western Eurasia*, edited by H. L. Dibble and A. Montet-White, pp. 275-286. University Museum Monograph 54. University of Pennsylvania, Philadelphia.

Marks, A. E., N. Bicho, J. Zilhão, and R. Ferring
1994 Upper Pleistocene prehistory in Portuguese Estremadura: results of preliminary research. *Journal of Field Archaeology* 21(l):53-68.

Marks, A. E., J. Shokler, and J. Zilhão
1991 Raw material usage in the Paleolithic. The effects of local availability on selection and economy. In *Raw Material Economies Among Prehistoric Hunter-Gatherers,* edited by A. Montet-White and S. Holen, pp. 127-139. Publications in Anthropology 19. University of Kansas, Lawrence.

Montet-White, A.
1988 Raw material economy among medium-sized Late Paleolithic campsites of Central Europe. In *Upper Pleistocene Prehistory of Western Eurasia*, edited by H. L. Dibble and A. Montet-White, pp. 361-374. University Museum Monograph 54. University of Pennsylvania, Philadelphia.
1991 Lithic acquisition, settlements and territory in the Epigravettian of Central Europe. In *Raw Material Economies Among Prehistoric Hunter-Gatherers*, edited by A. Montet-White and S. Holen, pp. 205-220. Publications in Anthropology 19. University of Kansas, Lawrence.

Nash, S.
1996 Is curation a useful heuristic? In *Stone Tools: Theoretical Insights into Human Prehistory,* edited by G. Odell, pp. 81-100. Plenum Press, New York.

Nelson, M. C.
1991 The study of technological organization. *Archaeological Method and Theory* 3: 57-100.

Odell, G. H.
1989 Fitting analytical techniques to prehistoric problems with lithic data. In *Alternative Approaches to Lithic Analysis*, edited by D. Henry and G. Odell, pp. 159-182. Archaeological Papers of the American Anthropological Association, Washington.
1996 Economizing behavior and the concept of curation. In *Stone Tools: Theoretical Insights into Human Prehistory*, edited by G. Odell. Plenum Press, New York.

Owen, L. R.
1988 Blade and Microblade Technology: Selected Assemblages from the North American Arctic and the Upper Paleolithic of Southwest Germany. British Archaeological Reports, International Series 441. BAR Publishing, Oxford.

Parry, W. and R. Kelly
1987 Expedient core technology and sedentism. In *The Organization of Core Technology*, edited by J. Johnson and C. A. Morrow, pp. 285-304. Westview Press, Boulder.

Shott, M.
1986 Technological organization and settlement mobility: an ethnographic examination. *Journal of Anthropological Research* 42(l) 15-51.

Straus, L. G.
1991 The role of raw materials in Upper Paleolithic and Mesolithic stone artifact assemblage variability in southwest Europe. In *Raw Material Economies Among Prehistoric Hunter-Gatherers*, edited by A. Montet-White and S. Holen, pp. 169-186. Publications in Anthropology 19. University of Kansas, Lawrence.
1992 Iberia Before the Iberians: The Stone Age Prehistory of Cantabrian Spain. University of New Mexico Press, Albuquerque.

Thacker, P.
1995 A Landscape Perspective on Upper Paleolithic Settlement in Portuguese Estremadura. Unpublished Ph.D. dissertation, Southern Methodist University, Dallas.
1996 Hunter-gatherer lithic economy and settlement systems: understanding regional assemblage variability in the Upper Paleolithic of Portuguese Estremadura. In *Stone Tools: Theoretical Insights into Human Prehistory*, edited by G. Odell, pp. 101-128. Plenum Press, New York.

2000 The relevance of regional analysis for Upper Paleolithic archaeology: A case study from Portugal. In *Regional Approaches to Adaptation in Late Pleistocene Western Europe*, edited by Gail Peterkin and Heather Price. BAR Publishing, Oxford.

Torrence, R.
1983 Time budgeting and hunter-gatherer technology. In *Hunter-Gatherer Economy in Prehistory: A European Perspective,* edited by G. Bailey, pp. 11-22. Cambridge University Press, Cambridge.

Torrence, R., editor
1989 *Time, Energy, and Stone Tools.* Cambridge University Press, Cambridge.

Zilhão, J.
1987 *O Solutrense da Estremadura portuguesa. Una proposta de interpretação paleoantropológica.* Trabalhos de Arqueologia 4. Departamento de Arqueologia do Instituto Português do Património Cultural, Lisbon.
1990a The Portuguese Estremadura at 18,000 BP: the Solutrean. In *The World at 18,000 BP (Volume 1: High Latitudes)*, edited by 0. Soffer and C. Gamble, pp. 109-125. Unwin Hyman, Boston.
1990b Le solutréen du Portugal: environnement, chronologies industries, peuplement, origines. In *Les industries a pointes foliacées du paleolithique supérieur*, edited by J. Kozlowski, pp. 486-501. Etudes et Recherches Archéologiques de l'Université de Liège 42, Liège.
1995 O Paleolitico Superior da Estremadura Portuguesa. Ph.D. Dissertation, University of Lisbon.

12. Lithic Raw Material Economy and Hunter-Gatherer Mobility in the Late Glacial and Early Postglacial in Portuguese Prehistory

Nuno Ferreira Bicho

The Portuguese Estremadura is also the geographical focus of the following paper where use of local and non-local raw materials are examined at two levels of analysis: the regional level, with emphasis on mobility patterns in Estremadura; and the local level, with emphasis on diachronic change in the Rio Maior area between 16,000 and 8,500 BP. The paper provides a summary of regional evidence for subsistence and mobility in the late glacial and early postglacial, focusing on the increased importance of maritime resources as indicated by shell midden sites beginning in the Preboreal (ca. 10,000 BP) and an apparent shift towards more logistical mobility at about 10,500 BP. Both are linked to what the author refers to as a "demographic explosion" occurring between 11,000 and 10,000 BP. The bulk of the paper is devoted to an analysis of raw material use at two multi-component sites that have been the subject of detailed studies (Cabeço do Porto Marinho and Carneira, both in the Rio Maior), with comparative material from several additional sites. The author discusses differences in core technology and tool production based on several different types of raw material, and argues that the better-quality cherts were used for careful, standardized production of blanks used in formal tools, while quartz and quartzite were reduced by a more expedient core technology and rarely retouched into formal tools. Results also document a trend toward increased use of the higher-quality cherts over time, associated with higher frequencies of projectile points and microliths. It is further argued that the increase in chert use was a response to higher demand for chert in the context of periodic mass production of projectile points in preparation for logistical forays. *–Editors.*

Introduction

The main goal of most Paleolithic archaeologists is the study of Paleolithic human behavior, which, as Binford (1983) and Torrence (1983) have stated, cannot be observed directly from the archaeological record. Thus, archaeologists are forced to rely on methods and data external to the archaeological record. Examples can then be found, among others, in the use of ethnographic analogy (e.g., Binford 1978; Gould 1969), replication of stone tools (e.g., Pelegrin and Chautat 1993) or hunting techniques (e.g., Frison 1989), refitting (e.g., Marks and Volkman 1983), and taphonomic studies (e.g., Behrensmeyer 1984; Brain 1982; Hill 1980; Shipman and Rose 1983; Russell 1987; Stiner 1991), all of which have been grouped into larger research areas such as ethnoarchaeology or experimental archaeology (Gould, ed. 1979). In any case, the use of these sub-disciplines of anthropology has certainly expanded exponentially during the last decade due to new theoretical perspectives among which the study of formation of the archaeological record seems to be the main focus. While all of these are certainly very important to the study of past human behavior, all require a leap of faith when integrated into the interpretation of archaeological materials (cf. Yellen 1977).

The study of prehistoric hunter-gatherers has recently focused on several categories of archaeological materials, including faunal and floral remains. These are considered now, as Torrence said (1983), central for hunter-gatherer research. Unfortunately, in sites dated to Paleolithic times most artifacts belong to a single category: lithic artifacts. The study of lithic artifacts has traditionally focused on typology for understanding Paleolithic hunter-gatherer behavior as well as for defining cultural (i.e., ethnic) groups within both temporal and spatial dimensions. This traditional perspective is based on time-consuming analysis and long descriptive narratives detailing changes in percentages of stone tools within and across sites, and in some cases lengthy descriptions of the color, size and shape of every artifact found at a site (e.g., Breuil *et al.* 1943). Some of these studies, however, have provided important information on stone tool use (e.g., Bordes 1950; Sonneville-Bordes and Perrot 1954) and were the basis for historical debates such as that known as the Bordes-Binford debate (e.g., Bordes 1968; Bordes and Sonneville-Bordes 1970; Binford and Binford 1966; Binford 1973; Mellars 1973; Rolland and Dibble 1990; White 1982), presently in a dormant status but very likely not yet dead. As Jeske (1989:34) has stated, however, the traditional studies have not explained why people behave differently at different times in different places. As a response to this critique, archaeologists working with lithic materials have developed a series of techniques and methodologies that include, among others, microwear, residue analysis, refitting, replication and, finally, a change from typological to technological studies of lithic materials.

During the last decade a new element of research has been integrated into the study of hunter-gatherer subsistence systems. This new element is lithic raw material economy, which includes a large set of human behavioral operations such as raw material acquisition, reduction sequences, and the production, resharpening, curation, and discard of tools. Although the results of these operations were the focus of traditional typological studies, they were not seen as part of an active system that was interlocked with particular aspects of the prehistoric subsistence system. The connection between raw material economy and technology, time budgeting, activity failure, mobility patterns and settlement system only very recently has been the focus of lithic studies, as is demonstrated by the increasing number of papers, edited volumes and symposia (e.g., Moloney *et al.*, eds. 1996; Montet-White and Holen, ed. 1991) such as this one.

Raw material economy studies tend to focus on the importance of exotic raw materials and long distance acquisition of lithic raw materials, underestimating the importance of local raw materials in the general economic system of prehistoric groups. This paper will focus on the importance of both local and non-local raw materials at two levels of analysis: a) the regional level, with emphasis on mobility patterns in the Estremadura (central Portugal); and b) the local level, with emphasis on diachronic change using materials from the Rio Maior area (Estremadura) between 16,000 and 8,500 BP.

Background: Lithic Technology in Central Portugal after the Last Glacial Maximum

During the Last Glacial Maximum, Portugal saw the development of Solutrean bifacial technology. This bifacial technology, characterized by the production of foliates, shouldered and tanged projectile points, disappeared around 17,000 BP with the beginning of the Magdalenian (Table 1). Magdalenian lithic assemblages were characterized by a wide variety of weapon tip forms, lighter and smaller than those seen in the Solutrean (Bicho 1993a, 1994). These new projectile points suggest not only a different lithic raw material economy, but also different types of mobility patterns and hunting techniques. This aspect will be discussed in detail in the following section.

Magdalenian projectile points included a wide range of backed and retouched pointed bladelets that tended to decrease in size through time. This size reduction was probably due to the introduction of new technological elements: the bow and arrow, and composite tools. There were a number of different types of backed projectile points (Figure 1) including Sauveterre, Malaurie, microgravettes, simple pointed and arched backed bladelets, and denticulated and backed bladelets as well as geometrics (Bicho 1993a). Another type of projectile point was the Dufour bladelet, very common at some of the Estremadura sites.

Two different types of reduction sequences were responsible for the production of these two broad types of weaponry (i.e., backed vs. Dufour). The most common reduction sequence used in the production of backed microliths involved removal of parallel-sided, flat, often pointed bladelets from pyramidal and prismatic cores with single or opposed platforms. The microburin technique was often used for the production of geometrics. A second, technically simpler type of reduction sequence, using flakes as cores, produced very small, thin, twisted and pointed bladelets used in the production of Dufour bladelets. These blanks were produced from carinated cores commonly called carinated endscrapers or burins. These two reduction sequences are characteristic of different assemblages, and probably correspond to two different technological facies: one named "Rio Maior," characterized by backed points, geometrics and microburin technique, and the other named "Carinated" (Bicho 1992, 1993a, 1993b).

The Rio Maior facies was probably the basis for the Mesolithic reduction sequence, responsible for massive production of geometrics and microburins after 8,500 BP. Both facies were also characterized by the presence of various types of endscrapers and burins, as well as by high frequencies of notches and denticulated pieces. Splintered pieces were also common. These assemblages, unlike those of the traditional French and Spanish Magdalenian, were mostly flake assemblages, and all retouched tools except for the microliths were made on flakes. Blades were rarely produced. Their occurrence corresponds to the upper tail of a unimodal size distribution of bladelets. Bladelets always make up less than 30% of the debitage, and average about 20%. Formal bladelet tools make up a similar percentage of tool assemblages.

In summary, the lithic assemblages from Portuguese Estremadura represent two different facies. One, "Rio Maior", dating to between 16,000 BP and 9,000 BP, is characterized by retouched tools made on flake blanks, backed bladelets used as projectile points, geometrics, and use of the microburin technique. This facies can be divided into three phases, the first dated to between ca.

Table 1. Late Upper Paleolithic and Epipaleolithic Radiocarbon Dates from Estremadura

Site	Level	Phase	Facies	Date BP	Lab. N°	Material
CPM*	area I/Lower	Lower Mag.	Rio Maior	16,340±420	SMU-2015	charcoal
				15,820±400	ICEN-542	charcoal
	area VI/Lower	Lower Mag.	Rio Maior	15,420±180	SMU-2634	charcoal
	area II/Middle	Lower Mag.	Rio Maior	15,410±195	SMU-2476	charcoal
	area IIIS/QLower	Middle Mag.	Rio Maior	14,050±850	SMU-2668	charcoal
Caldeirão	Eb base	Magdalenian		14,450±890	ICEN-70	bone
	Eb top			10,700±380	ICEN-72	bone
CPM	area I/Upper	Late Mag.	Rio Maior	12,220±110	ICEN-687	charcoal
				11,680±60	SMU-2011	charcoal
	area IIIS/Upper	Late Mag.	Rio Maior	11,810±110	ICEN-689	charcoal
	area III/Upper	Late Mag.	Rio Maior	11,160±280	ICEN-545	charcoal
	area II/Upper	Late Mag.	Rio Maior	11,110±130	SMU-2637	charcoal
	area III Trench	Late Mag.	Rio Maior	10,940±210	ICEN-690	charcoal
Carneira	Pinhal	Late Mag.	Carinated	10,880±90	SMU-2635	charcoal
CPM	area VI/Upper	Late Mag.	Rio Maior	10,160±130	SMU-2637	charcoal
Bocas I	Fundo	Terminal Mag.	Rio Maior	10,110±90	ICEN-901	bone
	0	Terminal Mag.	Rio Maior	9,880±220	ICEN-900	bone
	1/2	Terminal Mag.	Rio Maior	9,900±70	ICEN-903	shell**
Casal Papagaio	base	Epipaleolithic		9,710±70	ICEN-369	charcoal
				9,270±90	ICEN-372	shell
				8,870±105	Hv-1351	shell
Magoito		Epipaleolithic		9,970±70	ICEN-80	shell
				9,790±120	ICEN-81	shell
				9,580±100	GrN-11229	charcoal
				9,490±60	ICEN-52	charcoal
CPM	IIISW/Upper	Terminal Mag.	Rio Maior ?	9,270±170	SMU-2666	charcoal
CPM	area V/Lower	Terminal Mag.	Rio Maior	9,100±160	ICEN-688	charcoal
S. Julião	area A	Epipaleolithic		8,800-7,500	?	?
	area B	Epipaleolithic		8,250-7,900	?	?
AR III	Test VIb	Epipaleolithic	Carinated	8,860±80	ICEN-547	charcoal
	area 1/hearth 2			8,850±50	ICEN-494	charcoal
	area 1/hearth 1			8,570±130	ICEN-546	charcoal
	area 2			8,380±90	ICEN-688	charcoal
Ponta da Vigia	hearth	Epipaleolithic		8,730±110	ICEN-51	charcoal
Curral Velho		Epipaleolithic		8,040±70	ICEN-270	shell**

* CPM, Cabeço do Porto Marinho.

**Dates corrected by the subtraction of 360±35 years, apparent age of estuarine shells accumulated in the Sado shell middens, according to Soares (1989) and Soares and Cabral (1989).

16,000 and 14,000 BP, the second dated to between ca. 12,000 and 10,500 BP, and the last dated to between ca. 10,500 and 8,500 BP. No sites are known dated to between 14,000 and 12,000 BP. The second facies ("Carinated"), dating to between ca. 11,000 BP and 8,500 BP, was characterized by a lack of geometrics and of the microburin technique, rare backed points, and high frequencies of carinated elements and Dufour bladelets.

Subsistence, Mobility Patterns and Raw Material Economy in Central Portugal

Subsistence and Mobility Patterns

From the end of the Last Glacial Maximum (ca. 18,000 BP) through Boreal times, the subsistence of hunter-gatherer groups in Estremadura was certainly composed of a wide range of foods, although the main diet consisted of terrestrial mammals. The most important prey species were red deer, aurochs, wild boar and horse. The relative frequency of these species at each site was probably related to the local vegetation density and proximity to river valleys (Bicho 1993a, 1994; Straus 1995). Rabbit and hare were also common prey species, as inferred from Caldeirão cave (Zilhão 1987, 1992b) (see Figure 2). Other prey species, whose frequency depended both on climatic oscillations (e.g., Dryas III) and topography, were ibex and chamois (Bicho 1994). Nuts, berries, and fruits were available, and were certainly used, as suggested by the presence of grinding stones and charcoal from pine, olive and wild strawberry trees at some of these sites (Bicho 1993a, 1994; Zilhão *et al.* 1995).

There is some evidence of use of maritime

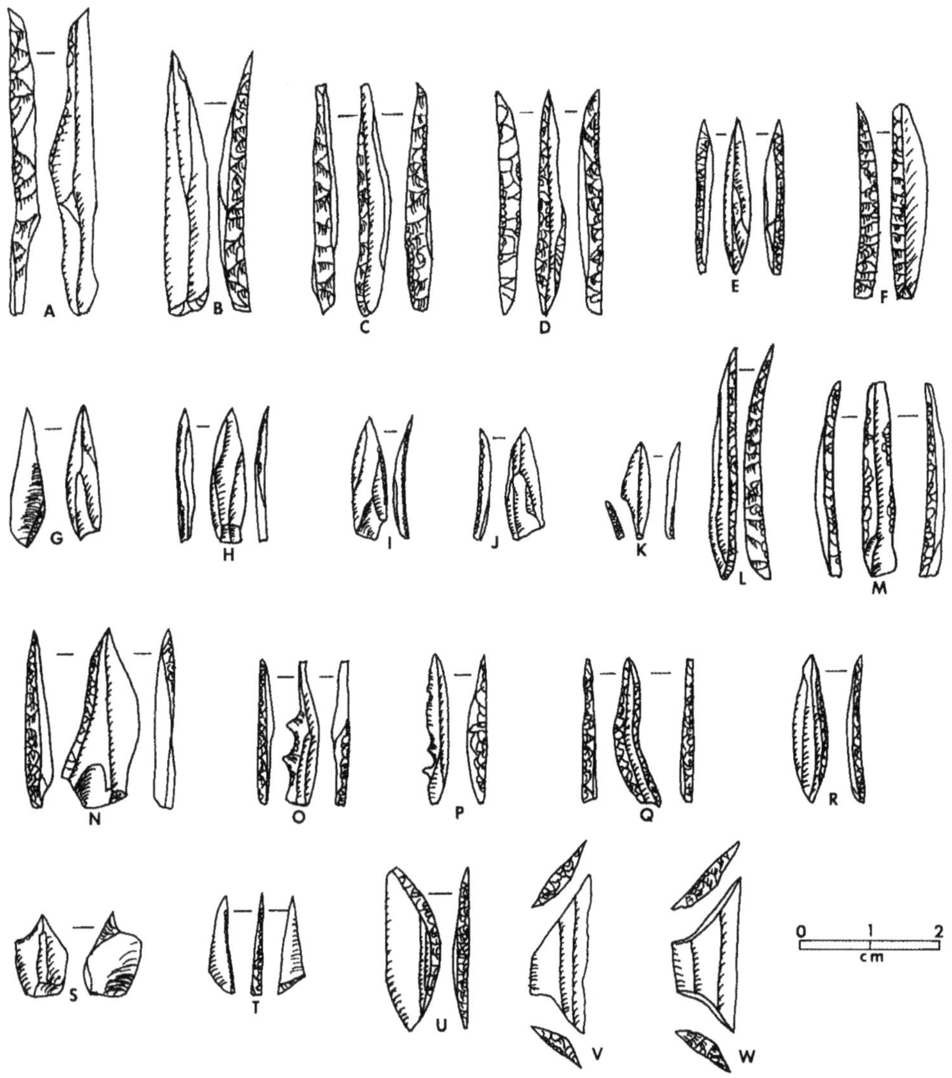

Figure 1. Microliths from Rio Maior sites. Carneira II: S-W; Pinhal da Carneira: E, G-J; Cabeço do Porto Marinho IIIS: A-D, F, M, N, O; Cabeço do Porto Marinho IIIT: K, L-P, Q, R.

resources before 11,000 BP in Portugal at the inland cave of Caldeirão (Zilhão 1992a). It is, however, at the beginning of Preboreal times (ca. 10,000 BP) that these resources became important (Bicho 1994). Fish and shellfish remains were present at Suão, an inland cave (Roche 1964, 1979, 1982), and shell middens dated to ca. 10,000 BP are present at two inland sites, Bocas and Casal Papagaio (Arnaud and Bento 1988; Bicho 1993a, 1994), as well as in one coastal site, Magoito (Arnaud 1986, 1993; Morales and Arnaud 1990). After 9,000 BP, shellfish was present in the coastal sites of Magoito, S. Julião, Curral Velho, and Toledo (Araújo 1994; Arnaud 1986; Bicho 1993a, 1994; Morales and Arnaud 1990). The main marine species present at these coastal sites were the common cockle (*Cerastoderma edule*), oyster (*Ostrea edule*), clams (*Ruditapes decussata*, *Venerupis decussata*, and *Scrobicularia plana*), mussel (*Mytilus*), limpets (*Patella*), periwinkles (*Littorina littorea*) and top shells (*Monodonta lineata*), indicating use of varied aquatic resources from estuarine habitats as well as sandy and rocky shores.

In summary, the scarce data seem to point to a general trend towards subsistence intensification through both specialization (aurochs and red deer) and diversification (Bicho 1993a, 1994). The diversification is seen in increased use of maritime resources such as fish and shellfish. This trend culminated during Mesolithic times (after 8,000 BP) with a very wide range of animal and plant species present in the extensive shell middens from the Tagus and Sado Valleys. Faunal assemblages in these Mesolithic shell midden sites include large game (e.g., red deer, aurochs), small species such as rabbit, hare and birds, and aquatic resources such as estuarine and open-sea fish, sea mammals, crustaceans, urchins, and shellfish.

The land use system in Portugal also points to a different pattern before and after ca. 10,500 BP, in what can be called, respectively, Final Upper Paleolithic and Epipaleolithic (Bicho 1994). During the Final Upper Paleolithic (before ca. 10,500 BP), there were two major areas of human occupation in Portugal, one located inland around the town of Rio Maior and the other on the coast near Torres Vedras (Figure 2). All the sites found in these areas are open-air localities. Although in Rio Maior there are small surface scatters of artifacts probably dated to late glacial times, most of the known lithic assemblages come from two sites (Cabeço do Porto Marinho and Carneira) with a total of more than 10 archaeological levels. Both sites probably correspond to a number of different human occupations, since some, if not all, of the levels are dense palimpsests. The sites from Torres Vedras (i.e., Rossio do Cabo,

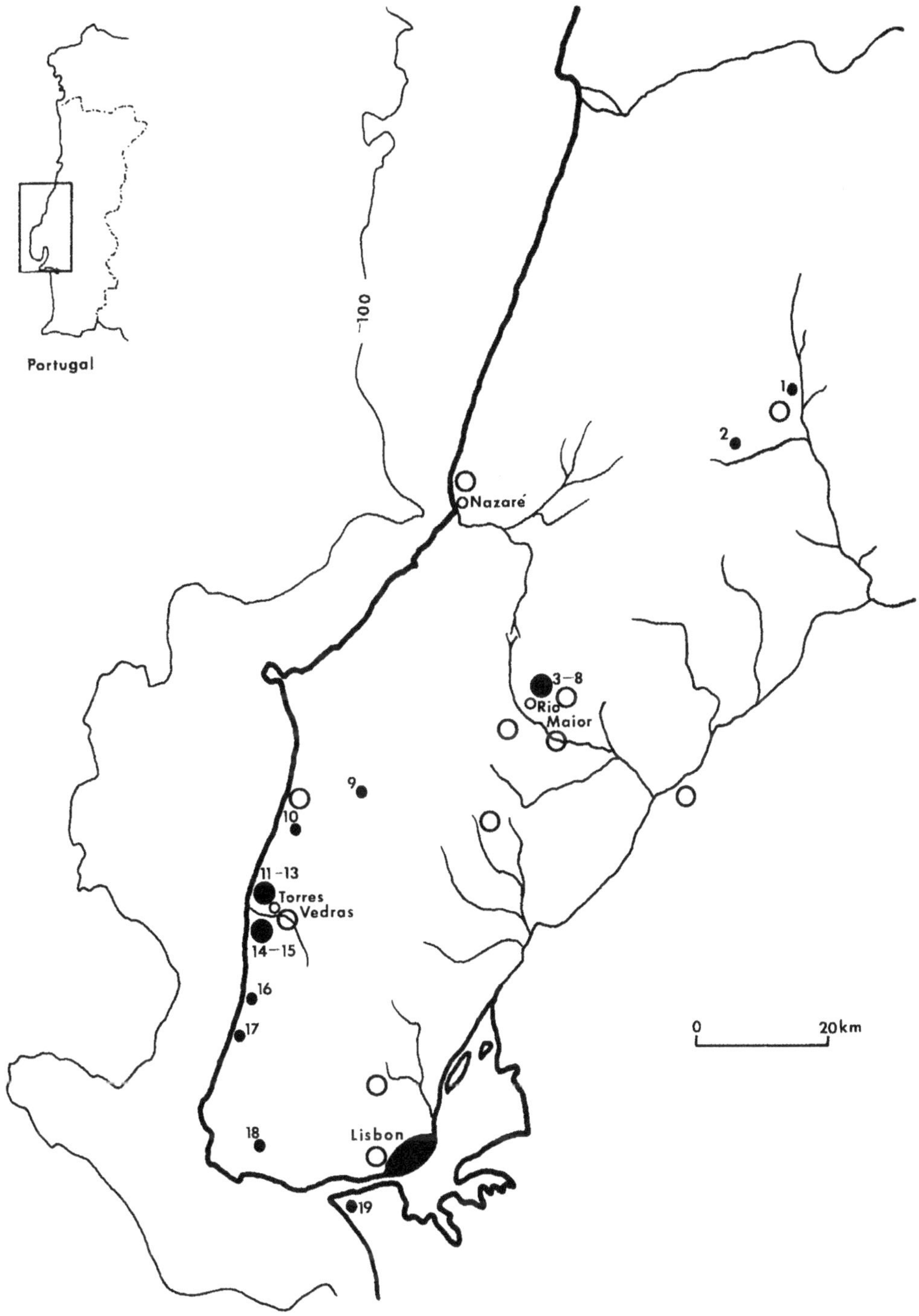

Figure 2. Map of Portuguese Estremadura showing location of archaeological sites discussed. Open circles represent chert sources. 1, Caldeirão Cave; 2, Casal Papagaio Cave; 3, Cabeço do Porto Marinho (CPM); 4, Areeiro I; 5, Areeiro III; 6, Carneira; 7, Vascas; 8, Bocas rockshelter; 9, Suão Cave; 10, Toledo; 11, Rossio do Cabo; 12, Cerrado Novo; 13, Ponta da Vigia; 14, Vale da Mata; 15, Cabeço do Curral Velho; 16, S. Julião; 17, Magoito; 18, Penha Verde; 19, Ponta do Cabedelo.

Cerrado Novo and Vale da Mata), on the other hand, tend to be more dispersed, as well as smaller. The only isolated site is Caldeirão Cave, located further north and inland from Rio Maior, in an area of higher elevation and more irregular terrain.

After ca. 10,500 BP, the land use system showed some changes. These changes were probably due to a demographic explosion. The number of sites as well as the number of occupied areas of Estremadura increased. In contrast to the period before 10,500 BP, human occupation was not limited to Rio Maior and Torres Vedras, but extended to other areas of central Portugal as well as the Alentejo and Algarve coasts (southern Portugal). Nearly 30 sites are known from this period. Unlike earlier times, cave sites such as Suão, Bocas, and Casal Papagaio, located inland in hilly limestone country, were occupied. At Rio Maior, human occupations still tended to concentrate in specific areas, including those of Cabeço do Porto Marinho, Carneira, and Areeiro III (Bicho 1994).

Although it is impossible to connect specific inland sites with coastal sites, it is evident that there was frequent movement, possibly seasonal, between the inland and coastal regions. Evidence of coastal-inland movement includes the presence of marine materials at inland sites, such as the large quantities of estuarine and marine shell at Bocas rock shelter and Casal do Papagaio cave, as well as some fish remains and marine shell ornaments at Caldeirão and Suão caves. With the exception of Suão cave, located some 20 kilometers from the coast, all the cave sites lie between 30 and 50 kilometers from the modern coastline. All dates from these sites cluster around 10,000 BP, corresponding to the end of Dryas III, when sea level was as much as 60 meters below present sea level (Dias 1985). Thus, the late glacial coast could have been up to 10 kilometers further away from these sites than the modern coastline. This distance would correspond to a minimum of one day's walk, more likely two or more days, if carrying a fairly large pack of shellfish as well as the "common" tool kit used during excursions or large scale movements.

Based on currently available data, it is difficult to determine whether the mobility pattern in Estremadura was residential (i.e., seasonal migration of the hunter-gatherer groups from inland to coastal areas) or logistical (i.e., semi-permanent base camps from which small excursions were used to exploit non-local resources) (Binford 1980; Jeske 1989). The presence of two technological facies at both inland and coastal sites (Table 1 and Figure 2) and the presence of large shell middens at inland sites seems, however, to point to two different hunter-gatherer groups, both with a logistical mobility pattern and a similar range of resource acquisition but using different regions for their base camps: one inland and the other coastal. The presence of inland shell middens also suggests, using Jeske's words (1989:36), that these hunter-gatherers did not schedule their movements so as to be in the right place at the right time. More likely, they arranged to be in the wrong place at the right time, or in other words, rather than living where the resources were they brought the resources to their residential sites. Therefore, there was a need for small excursions to gather marine resources and bring them to the base camps during the right season.

The association of lithic assemblages characterized by the Rio Maior facies (the most common in the Rio Maior region) with the inland shell midden at Bocas suggests that the hunter-gatherer group responsible for this facies established residential sites in the inland region and used the coastal area for seasonal/episodic exploitation of specific resources. Since the lithic assemblages from the other shell middens, both inland and coastal, have not yet been published, the data are still far too scarce to support or refute this hypothesis.

Raw Material Economy

Consideration of lithic raw material economy has so far been left out of the land use patterns described above. How important was the role of lithic raw materials such as chert in the scheduling of activities, and in the inland-coastal excursion routes of these hunter-gatherers?

The lithic assemblages from Estremadura dated to between ca. 16,000 BP and 8,000 BP were composed mostly of chert (70-95%), quartz (1-25%), and quartzite (1-10%). Other raw materials such as rock crystal, dolerite, basalt, and sandstone are often present in percentages totaling less than 1% of the assemblages. With the exception of chert and the igneous rocks, all of these raw materials are found in gravel deposits of Miocene to Pleistocene age. These gravel deposits are ubiquitous in Estremadura, and are usually present within a few meters of every archaeological site.

Estremadura is characterized by numerous chert sources (Figure 2). There are both primary and secondary sources. Primary sources are located in Eocene and Cretaceous limestone formations in the coastal areas of Lisbon, Runa (Torres Vedras) and Nazaré. The secondary sources are Miocene and Pliocene gravel deposits located near primary sources or concentrated in certain areas across the landscape, such as Rio Maior. As a result, any chert source is never further than a day's walk from any point in Estremadura. Not surprisingly, chert is by far the mostly commonly used raw material in the Upper Paleolithic of central Portugal. The relative frequency of chert, however, changes through time, probably as a consequence of cultural and technological traits and not due to economy (Bicho 1993a,

1996; Marks *et al.* 1991). This aspect will be discussed later in the paper.

Non-local raw materials are very rare since the lower quality rocks (e.g., quartz and quartzite) are ubiquitous and chert is abundant across the landscape. Some non-local chert, however, is present and, therefore, has important implications for the mobility patterns of the local prehistoric hunter-gatherers.

The results presented here come from extensive analysis carried out on lithic assemblages from two sites, Cabeço do Porto Marinho (CPM) and Carneira, both in the Rio Maior, and from preliminary analysis of assemblages from Torres Vedras. On the basis of the preliminary examination of Torres Vedras assemblages, it is possible to say that the chert used was mostly local. The assemblages from Rio Maior are predominantly composed of local chert (i.e., materials found within a radius of 10 kilometers). Non-local cherts in these assemblages make up less than 1% of the total assemblage, amounting to only a dozen or so artifacts in any one level (Table 2). Cores, core maintenance products, and primary flakes of non-local chert are rarely represented in these assemblages. The low representation of these artifact classes suggests that the rare "exotic" or non-local pieces brought to the site arrived in an already shaped form. There is, however, one exception, in the case of a core from CPM IIISW L from which 5 retouched tools were produced. The sample of non-local chert, however, is far too small and therefore this pattern is yet to be confirmed. Most non-local chert pieces were recognized macroscopically as coming from the Torres Vedras area and in two cases from the Tomar and Nazaré areas.

The extremely low frequencies of non-local chert and the nearly complete lack of retouched tools of non-local chert suggest that only local raw materials were used both in the residential area and during acquisition excursions. In rare circumstances, non-local chert was carried around and used — thus, the presence of extremely low frequencies of non-local chert in every assemblage. The patterns described above have some important implications. First, the acquisition of local raw material was probably embedded in the daily hunting and gathering activities of the group, as suggested by the presence of non-flaked chert cobbles in every assemblage. Second, though a pattern of logistical mobility was probably used and chert acquisition was likely to have been embedded, the location of the chert sources certainly was a key factor in determining the route taken in resource-gathering excursions (or seasonal migrations if the mobility pattern was residential). This model is supported by the concentration of sites around Torres Vedras and Rio Maior (the main chert sources of Estremadura), and the fact that every site in Estremadura dating later than 16,000 BP falls within ca. 5 kilometers of a chert source. Since there was no reason to go from Rio Maior to coastal areas (i.e., Nazaré and Torres Vedras) just to acquire chert, it is assumed that those locations were used for the gathering of shellfish. Since those locations are not the nearest shore areas to Rio Maior, and those

Table 2. Non-Local Chert Present at the Sites of Cabeço de Porto Marinho and Carneira

	Cabeço do Porto Marinho											Carneira		
	IL*	IIM*	IIISWL	IIISWM*	IU*	IIISU*	IIIU*	IIU*	IIIT	IIISWU	V	CRII	CRIII	PC
Fragments				1										1
Chips			7	2	8	1				2	2		2	10
Flakes														
broken	3		6	4	7	3	3	2	2	1	1	2		9
cortical			1			1								
partially cortical	1		4	1		1			4		4	1		2
non-cortical	8	4	2	6		7	6	1	5		3	2	3	4
Blades														
broken		1												
complete						1						1		
Bladelets														
broken		2								1	1	1		1
complete	1								1					2
Core maintenance products		1	1						1					
Cores				1						1				
Retouched tools			1	5										
TOTAL	13	8	22	20	15	14	9	3	13	5	13	7	5	29
Percentage	.74	.35	1.74	1.48	.32	.13	.42	.27	.61	.27	.23	.17	.16	.48
TOTAL SAMPLE	1766	2260	1262	1346	4703	5179	2148	2393	2114	1807	5635	4162	3165	6074

*Results based on preliminary samples from the excavated assemblages.

activities could not be performed there with a low energy consumption, it is likely that the choice of those particular locations for exploiting coastal resources was due to the presence of chert sources. This mobility pattern would certainly tend to reduce the necessity of carrying a substantial weight in lithic materials in the form of tools or cores. The tool kit used in traveling was probably light, consisting only of personal gear such as antler billets, spear points, and/or bow and arrows. The increasing importance of such a light tool kit is confirmed by a parallel increase in the relative frequency of projectile points in assemblages from the Rio Maior area (Table 3) and in the frequency of non-local maritime resources. Weapons lost while traveling could be replaced on arrival at the coast, since it was known that the new location was rich in chert for the production of stone tools needed. The light tool kit was even more important during the return trip, when the excursion members were laden with shellfish to be brought to the base camps located more than a days' walk away.

In summary, settlement locations as well as routes of travel were likely to have been determined by the presence of chert. Since all economic resources, with the exception of chert and maritime resources, were ubiquitous across the landscape, settling near chert localities would reduce the input of time and energy required for lithic resource acquisition. This resulted in a higher output of maritime resources translated, in this case, into a larger load of resources transported to the inland base camp.

Raw Material Economy, Reduction Sequences and Tool Production in Rio Maior

During the final late glacial in Portuguese prehistory, raw material economy had implications at two levels: one regional, described above, and one local. Archaeological studies of raw materials have tended to focus mostly at the regional level due to their connection to mobility or planning (e.g., Andrefsky 1994; Bamforth 1990; Geneste 1987; Jeske 1989; Kuhn 1989; Roebroeks *et al.* 1988), while the local trends are less frequently investigated in detail (e.g., Bamforth 1991; Bicho 1993c; Freeman 1991; Marks *et al.* 1991; Straus 1991).

"Local" raw material is usually defined as coming from a source within less than 20 kilometers and "little consideration is given to them compared to those sources that are further away" (Marks *et al.* 1991:128). In the case of the Portuguese Estremadura, most sources would fall within the category of "local". A change in perspective is required in order to investigate the use of these sources. Since most sites dated to after the Last Glacial Maximum are between 1 and 5

Table 3. Relative Frequency of Microliths from Upper Paleolithic and Epipaleolithic Sites* in the Rio Maior area

	Geometrics	Backed points	Dufour	All points	Microliths	Sample
16,000-14,000 BP						
AR I.	.49	.49	.49	1.49	4.97	201
CPM IL	1.49	.98		2.48	13.93	201
CPM IIM		1.69	1.69	3.38	16.90	118
CPM IIISWL			2.27	2.27	4.54	44
CPM IIISWM	1.25	1.87	1.25	4.37	6.25	160
12,000-11,000 BP						
CPM IU	.40	1.21	1.01	2.63	9.58	1481
CPM IIISU	1.07	5.60	.80	7.52	15.59	372
CPM IIU	3.20	3.20	1.07	7.48	13.36	187
CPM IIIU	1.90	3.83	.76	6.51	16.47	261
11,000-9,500 BP						
PC		5.00	7.00	12.00	23.5	200
CPM IIIT		4.44	1.11	5.55	15.55	90
Bocas F		8.30	.75	9.05	26.41	265
Bocas 0	.27	13.31	.81	14.40	42.66	368
Bocas 1	.35	6.66	.35	7.36	24.56	285
9,500-8,500 BP						
CPM IIISWU		12.24		12.24	36.73	49
CPM V	3.18	1.91		5.09	25.47	157
CR II	8.18	5.26	.58	14.61	35.08	171
CR III	6.78	5.08		11.86	37.28	59
AR III	.24	.73	8.63	9.60	12.37	1228

* AR I, Areeiro I; AR III, Areeiro III; CPM, Cabeço do Porto Marinho; CR, Carneira; PC, Ponta do Cabedelo.

kilometers from a chert source, any source located 10 kilometers away should be considered non-local since it would have had very little impact on the local economy. Such sources, however, were likely to have been important in terms of the regional economy and in the acquisition of all resources, as was suggested in the previous section of this paper.

This section of the paper will focus on the role of local raw materials in the technology and stone-tool production in two dimensions, space and time. The main focus of the section is on two multicomponent sites, Cabeço do Porto Marinho (CPM) and Carneira (CR) (Figure 3), although other sites from the Rio Maior area, such as Bocas, Areeiro I and III, will be included to illustrate some aspects of the local raw material economy (Bicho 1992, 1993a; Marks *et al.* 1994). Cabeço do Porto Marinho is a large multicomponent site. It was first occupied during Gravettian times, around 23,000 BP, followed by Early and Late Magdalenian, Epipaleolithic, Neolithic, Chalcolithic, and Bronze Age occupations with a total of at least 24 archaeological levels. Thirteen of these have been dated either by C^{14} or stratigraphically to the period between ca. 16,000 BP and 9,000 BP (Table 1). Some of these levels are a palimpsest of occupations that cannot be separated. Not all levels will be used in this study since they are still under analysis.

The archaeological deposits are located in the top strata of a long sequence of sand deposits. The lower units are sterile quartz sands of Miocene/ Pliocene age, followed by Late Pleistocene re-deposited eolian sands mixed with some colluvium. Separating these two deposits is a geologic unit of fluvial terraces marked by clays, sands and gravels. The archaeological materials are present *in situ* in the overlying eolian/colluvial sand formation.

Carneira is also a multicomponent site, marked by the presence of a series of human occupations spanning a period from Solutrean to Chalcolithic times (Marks *et al.* 1994). There were at least five different Magdalenian/Epipaleolithic levels at this site. The total number of levels is not known, since the first excavations (Heleno 1944, 1956), affecting an area as large as 500 square meters, were carried out some 50 years ago without horizontal or vertical control. The site is also characterized by a fluvial terrace (dating to Solutrean times) underlying eolian and colluvial sands where the Magdalenian/ Epipaleolithic occupations took place.

Within a few meters of both sites there is a wide selection of quartz and quartzite gravels. These gravels are present both in natural cuts and on the eroded Miocene/Pliocene surfaces. Chert is present in secondary sources, including Miocene gravels occurring at different distances from the two sites (Table 4). The chert nodules range from small pebble size to large boulders, which tend to have a spongy, thick cortex.

Reduction Sequences and Raw Materials

Chert

Chert is the raw material most common in all assemblages. Chert was carried to the sites in large amounts, but was not used with the same frequency in all assemblages. The nodules brought to the sites were probably relatively small, since most cores as well as primary flakes are fairly small. Based on the debitage assemblages, it seems that chert cores were not prepared at the sources, but were brought intact to CPM and Carneira. Core types are characterized by a wide variety of forms for both flake and bladelet production.

Chert, compared to quartz or quartzite, was an expensive resource. Although chert sources were close to these sites, it is evident that chert cores were fully exploited. Cores suggest that reduction sequences were long and that core shaping and maintenance were usually carried out in such a way that the core produced the maximum possible number of blanks before it was abandoned, frequently due to hinge fractures. Most cores with hinge fractures could have been further re-utilized with the removal of hinged areas, but flaking surfaces would then have become very small. This suggests that there was a specific desired size for flakes. This idea is confirmed by the presence of opposed and multi-platform cores that show a transition from flake production to bladelet production when the flaking surface became too small for flake production (Bicho 1992). Retouched tools made on chert are present in large quantities in every assemblage, showing a very wide range of formal types from endscrapers to geometrics.

Quartzite

In general, quartzite was used in a very simple and expedient manner. Quartzite cores are very homogeneous in form, mostly of the chopper type, apparently with a standardized reduction sequence where decortication was not extensive. Fragments and chips are very rare, suggesting that quartzite shattered very little during flaking, although most flakes produced would break during the knapping process, resulting in few complete flakes.

Quartzite cobbles were only roughly decorticated, and relatively large flakes were produced with cortical platforms and very little cortex on the dorsal surface. This indicates a very simple technology and an expedient reduction sequence in which shaping and maintenance of the core were not performed. Retouched tools had very simple forms. The most frequent tool type made on quartzite was the denticulate, followed by heavy duty tools such as hammerstones, anvils, and large scrapers. There are rare thick endscrapers, notches, and only one

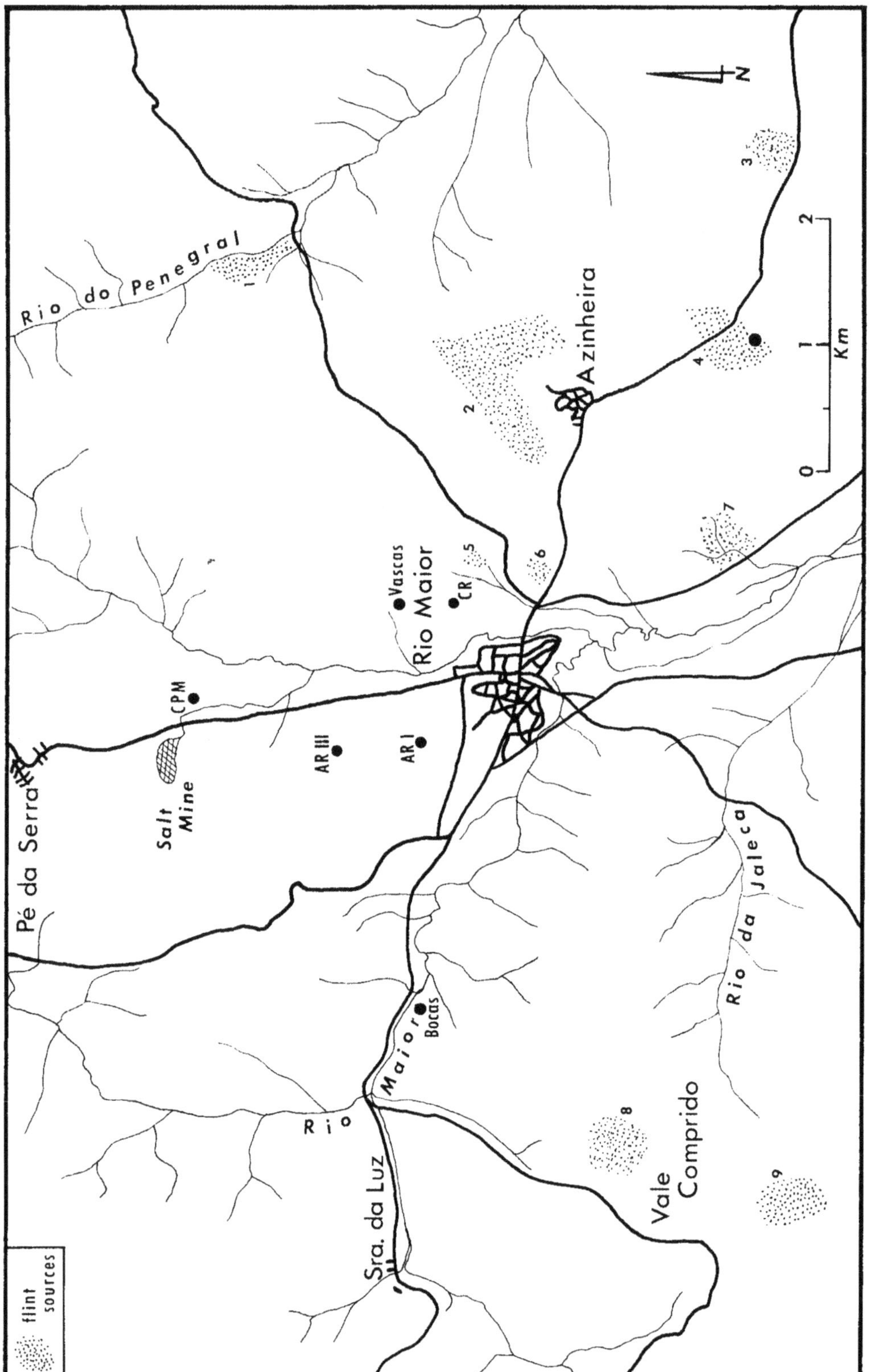

Figure 3. Map of the Rio Maior area, showing location of archaeological sites and chert sources. AR I, Areeiro I; AR III, Areeiro III; CPM, Cabeço de Porto Marinho; CR, Carneira. Numbered chert source areas correspond to sources listed in Table 4.

Table 4. Distance to Chert Sources (km)

Name	Carneira	Cabeço do Porto Marinho
1. Penegral	3.25	3.25
2. Azinheira	2.25	3.25
3. Cabeço do Via Vae	4.25	6.25
4. Estrada da Azinheira	2.75	5.00
5. Minas	.25	2.25
6. Gato Preto	.75	2.75
7. Quinta do Sanguinhal	2.00	4.75
8. Vale Comprido	4.50	4.50
9. Bairradas	5.25	6.00

burin and one retouched piece from all assemblages. There are a total of only 26 quartzite tools in all assemblages from CPM and Carneira, corresponding to less than 1% of all tools. Quartzite tools are very rare in all CPM and Carneira assemblages, with a maximum of less than 3.5% of each assemblage (Table 5). There is a very low ratio of retouched tools per core and, in general, there are more cores than tools in these assemblages. The very low ratio of tools to cores and tools to flakes suggests that quartzite flakes were used mostly without secondary modification, indicating expedient usage of quartzite flakes.

Quartz

Quartz blocks were initially smaller than those of quartzite, although they were brought to the site in two different sizes (Bicho 1992). The larger quartzite blocks were transformed into very simple cores, such as choppers, while smaller blocks were shaped into single platform cores or more rarely into prismatic cores. Opposed platform cores are rare. Flakes, in general, were smaller than those made of quartzite, although of similar thickness.

Quartz exhibits more diversified blanks than does quartzite. Often, these assemblages have some blades and bladelets, but always in extremely low numbers. The breakage of blanks was very common, probably during the reduction process, since there are higher percentages of broken quartz blanks than there are of other raw materials.

Quartz bladelets were produced from the same types of cores as those made on chert, with the exception of single platform and orthogonal cores. The production of both bladelets and flakes from the same core was, in the case of quartz, restricted to prismatic cores. Although there are different core types, and apparently complex ones, such as the multi-platform or pyramidal, core shaping was relatively simple. Decortication was more extensive than on quartzite, but it was restricted to the flaking surface. Although bladelet production was carried out, the shaping of these bladelet cores

Table 5. Relative Frequencies of Raw Material Types for Cores, Flakes and Retouched Tools

	Cabeço do Porto Marinho											Carneira		
	IL*	IIM*	IIISWL	IIISWM*	IU*	IIISU*	IIU*	IIIU*	IIIT	IIISWU	V	PC	CRII	CRIII
Cores														
quartz	25.6	39.0	47.4	37.5	7.8	16.0	14.7	26.7	12.5	10.5	19.4	–	2.2	–
quartzite	11.6	7.3	5.2	12.5	5.1	10.7	2.9	2.2	8.3	5.2	5.6	.4	6.5	–
other non-chert	3.8	–	–	–	2.3	1.3	–	–	–	5.2	–	–	–	–
Total non-chert	41.0	46.3	52.6	50.0	15.2	27.4	17.6	28.9	20.8	20.9	25.0	3.4	8.7	–
chert	55.2	51.3	47.4	50.0	67.7	54.7	76.5	62.2	75.0	58.0	50.0	48.3	73.9	75.0
burnt chert	3.8	2.4	–	–	17.1	18.7	5.9	8.9	4.2	21.1	25.0	48.3	17.4	25.0
Total chert	59.0	53.7	47.4	50.0	84.8	73.4	82.4	71.1	79.2	79.1	75.0	96.6	91.3	100.0
Total	78	41	19	24	217	75	34	90	24	19	36	29	46	24
Flakes														
quartz	18.9	18.9	14.6	15.4	3.9	7.7	8.6	7.4	8.1	9.4	6.3	1.8	2.6	2.7
quartzite	3.3	8.5	5.8	11.3	2.3	6.6	4.9	6.3	5.7	3.8	9.0	5.1	5.9	2.9
other non-chert	.3	.9	–	..3	1.0	.9	1.4	1.3	–	.3	–	.2	.5	–
Total non-chert	22.5	27.4	20.4	27.0	7.2	15.3	14.9	15.0	13.8	13.5	15.3	7.1	9.0	5.5
chert	69.7	66.0	75.0	63.4	74.9	61.0	66.3	68.0	68.7	74.3	71.2	61.6	63.7	73.2
burnt chert	7.8	6.5	4.6	9.6	17.9	23.7	18.8	17.0	17.5	12.2	13.5	31.3	27.3	21.3
Total chert	77.5	72.6	79.6	73.0	92.8	84.7	85.2	85.0	86.2	86.5	84.7	92.9	91.0	94.5
Total	360	445	364	293	685	980	514	447	457	393	921	988	835	489
Retouched Tools														
quartz	8.4	10.8	18.7	1.9	1.1	2.6	4.3	2.6	1.1	1.7	.6	1.0	.6	–
quartzite	3.5	1.7	–	2.6	.5	.3	–	.4	–	–	.6	.5	.6	–
other non-chert	–	–	–	.6	.2	–	–	.4	–	–	–	–	–	–
Total non-chert	11.9	12.5	18.7	5.1	1.8	2.9	4.3	3.4	1.1	1.7	1.2	1.5	1.2	–
chert	79.7	77.5	68.8	83.5	75.3	71.2	80.7	83.2	91.1	88.1	78.8	74.1	67.4	72.9

was very different from those producing chert bladelets. This is indicated by the complete absence of crested blades among quartz debitage, and the absence of either crested or keeled areas on any of the faces of the quartz cores.

In conclusion, quartz seems to have been used in two different ways. Some quartz was reduced in a simple manner without much shaping or decortication, and some quartz was reduced following a different pattern, sometimes showing change in flaking direction, resulting in both opposed and multi-platform cores. Core preparation in the former case was expedient. The latter reduction sequence represents a more complex and expensive technology.

The tools were more diverse than those made on quartzite. The most common tool types were burins (mostly dihedral), thick endscrapers, and notches. Also common were retouched pieces and thin endscrapers. Denticulated pieces and microliths were rarely made of quartz. The total number of quartz tools is 90, or 2.5% of the tool collection. As with quartzite, there are, in general, more cores than tools. Quartz, thus, was rarely used for tool production. This suggests that just as in the case of quartzite, quartz blanks were often the desired end product.

Raw Material Economy

There are a number of different aspects to be considered in the study of raw materials present in an archaeological assemblage. One of these aspects is the difference between the relative frequency of each raw material in three different phases of the tool production process: raw material acquisition, blank production, and tool production. Although one might assume that each raw material should be equally represented in each phase (why carry rocks to the site other than for use?), that is not the case. The difference can probably be attributed to the distance to sources, ultimately reflected in the type of reduction sequences (i.e., expensive vs. expedient) and the general economy of raw materials. Acquisition, blank production and tool production are represented archaeologically by cores, blanks, and retouched tools respectively.

Temporal trends

Temporal trends can best be seen within CPM, since the assemblages from this site span the period from ca. 16,000 BP to ca. 9,000 BP. During the first phase, between 16,000 and 14,000 BP (CPM units IL, IIM, IIISWL and IIISWM), only about half of the core blanks brought to the site were made on chert (Table 5). Quartz was the second most common raw material brought to the site before 14,000 BP, amounting to between 25 and 47% of cores, while quartzite made up only about 10% of the material brought to the site. After 12,000 BP (CPM units IU, IIISU, IIU, IIIU, IIIT and IIISWU), a much higher percentage of chert was brought to the site. In this later period, chert represents between 70 and 85% of the total number of cores, while the frequencies of quartz and quartzite are, respectively, 10 to 27% and less than ca. 10%.

Raw material frequencies among flakes do not correspond to the relative frequency of each raw material as indicated by the number of cores brought to the site. Instead, the relative frequency of each raw material in flake assemblages can be taken to represent the importance of the different raw materials in blank production. During the first phase (16,000 - 14,000 BP), chert flakes make up about 75% of flakes in each assemblage at CPM, while quartz and quartzite amount to 16% and 6% respectively. During the latter phase, after 12,000 BP, more than 85% of flakes are chert, while quartz and quartzite make up the other 15%.

Tool production shows the same temporal trend, with a sharp increase in the production of chert tools from the first phase to the second phase. Formal tools made of quartz and quartzite are much less common in all assemblages. Although there are only complete counts for 4 CPM assemblages (IIISW L, IIIT, IIISWU, and V), the ratio of formal tools to flakes apparently changed through time. During the first phase, one tool was produced for each 10 flakes of quartz, quartzite and chert, while later this ratio increases twofold for chert (1 chert tool per 5 flakes) and decreased significantly in the non-chert raw materials (one retouched tool per ca. 100 flakes).

Spatial variability

Spatial variability in the Rio Maior area has been described in analyses of lithic technology (e.g., Bicho 1992) and settlement pattern (e.g., Bicho 1993a; Zilhão 1992b). With few exceptions (Bicho 1992, 1993c; Marks *et al.* 1991; Straus 1991), however, spatial variability in raw material economy has been ignored in the study of the Paleolithic in Portugal. This section will focus on the data from the site of Carneira and those assemblages from CPM dated to between 12,000 BP and 9,000 BP, corresponding to the time span represented at Carneira. As noted above, CPM is located one to two kilometers farther from chert sources than Carneira (Table 4 and Figure 3).

Chert cores make up more than 90% of the core assemblages at Carneira, while the frequency at CPM is, in general, less than 85% (Table 5 and Figure 4). Similarly, at Carneira chert flakes comprise more than 90% of total flakes (Table 5 and Figure 5). There are two assemblages, CPM V and Carneira II, that are virtually identical temporally,

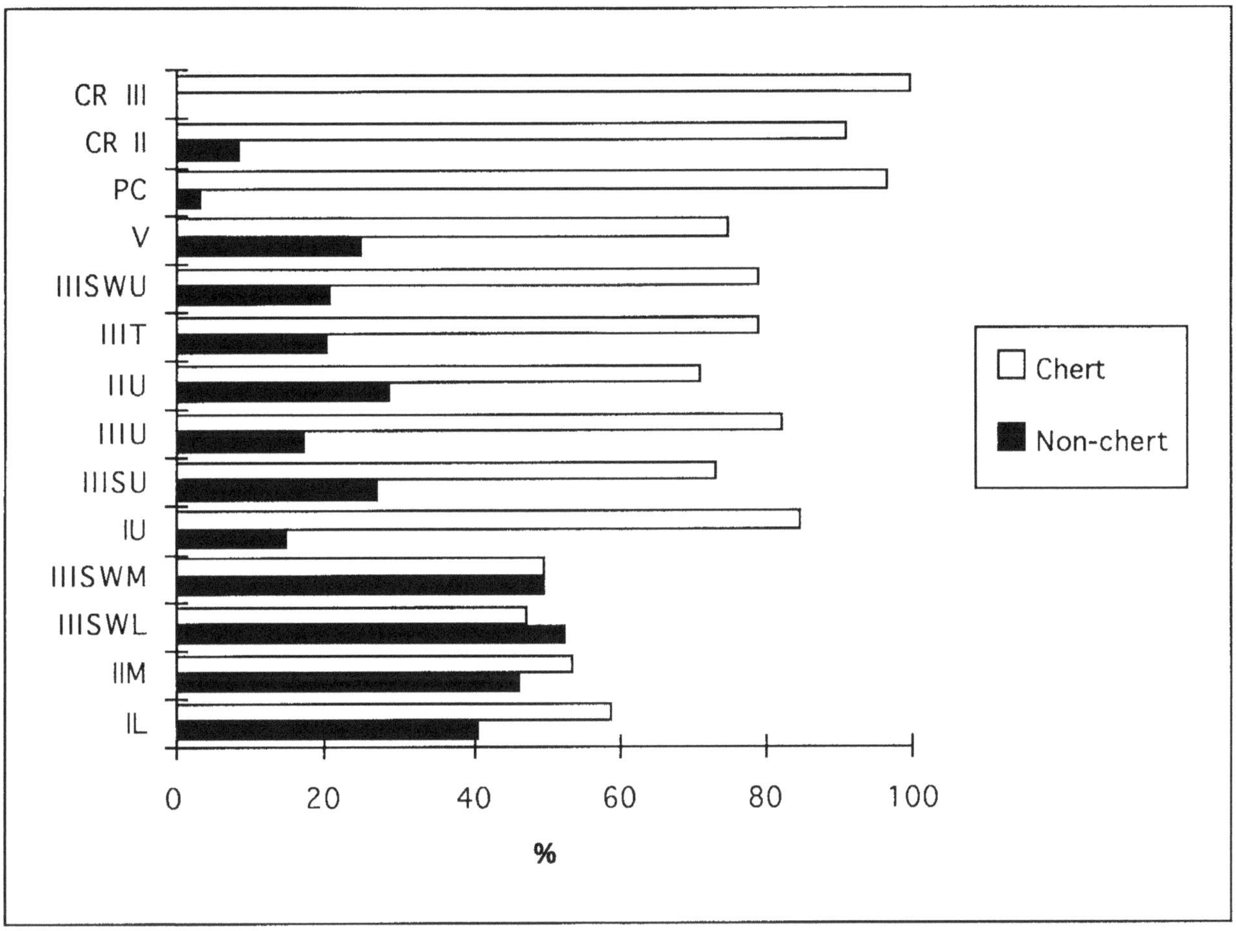

Figure 4. Bar chart of raw material frequencies among cores.

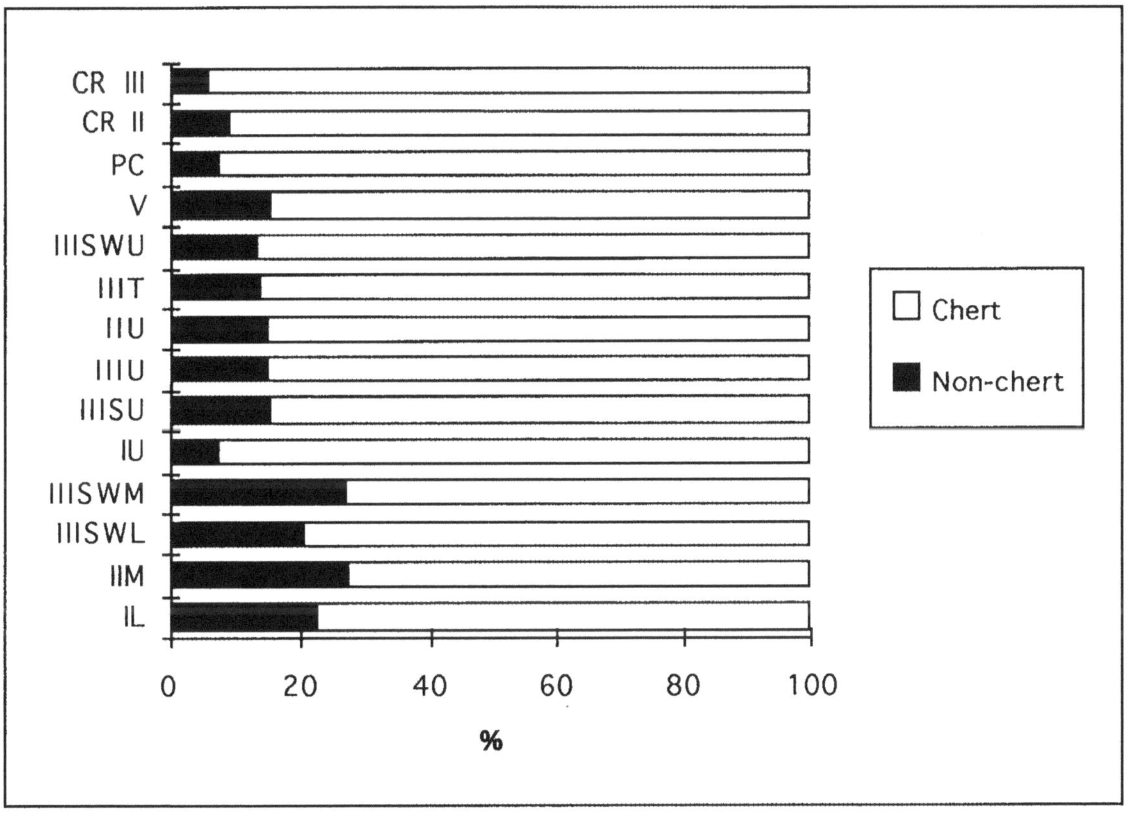

Figure 5. Bar chart of raw material frequencies among flakes.

technologically, morphologically, and typologically (Bicho 1992; 1993b, 1993c). The raw material economy, however, was different in the two assemblages. At Carneira II, chert was brought to the site and used there for blank production in a much higher frequency than at CPM V. Flake and core dimensions are similar across the two assemblages, while bladelet length and thickness are significantly different in the two assemblages: CPM V bladelets are considerably smaller than those from Carneira II. Also, burins, notches and backed bladelets are significantly shorter and thinner at CPM V. The difference in tool and bladelet dimensions between the two sites suggest a more economical exploitation of the chert brought to the site at CPM V than at Carneira II.

Discussion

In Rio Maior, raw material is a useful indicator of synchronic and diachronic variability. Temporal variability seen in Rio Maior lithic assemblages can be explained by either technological changes or cultural preferences, or both. The decrease in use of non-chert raw materials from the first phase (16,000-14,000 BP) to the second phase (12,000-9,000 BP) is evident at all stages of the lithic technological process: acquisition, blank production, and tool production. When all artifacts are included (Table 6) the results indicate the same patterns seen in the cores, flakes, and retouched tools (Table 5). These patterns are clearly shown in the dendrogram presented in Figure 6, obtained through "Cluster Analysis" based on the frequency of chert, burned chert and non-chert raw materials. The method used here uses Euclidean distances as the proximity measure between assemblages, in an agglomerative hierarchical clustering procedure through complete-linkage. Three clusters are evident, representing 1) the assemblages dated to between 16,000 and 14,000 BP, 2) the assemblages dated to between 12,000 and 9,000 BP from CPM, and 3) the assemblages dated to between 11,000 and 9,000 BP from Carneira. If the assemblages composing each cluster are compared as a group, the average frequency of chert is ca. 75% in the early phase of CPM, nearly 90% in the latter phase of CPM, and 94% in the Carneira cluster. An interesting aspect of raw materials in the Rio Maior area is an increase in frequency of burned chert that accompanies the increase in chert use. This parallel trend may be related to an increase in the use of heat treatment, although the lack of clear macroscopic evidence in tools and cores, as well as the differential rate of burning evident in cores, blanks, and retouched tools casts some doubt on this hypothesis. The data from the Rio Maior assemblages clearly indicate the parallel use of two raw material economies: expedient or inexpensive (Binford 1979; Jeske 1989) for low quality raw materials, quartz and quartzite, and expensive (Jeske 1989) for high quality raw materials (chert). The low-quality raw materials had very low procurement costs as well as low energy input during blank and tool production. Since formal tools made of non-chert raw materials are rare and never heavily retouched or standardized, it is likely that they were used in common, simple daily tasks and, thus, activity failure rates were probably quite low. On the other hand, chert had higher procurement costs and certainly very high energy input for blank and formal tool production. The activities in which at least some of the chert

Table 6. Relative Frequencies for all Artifacts by Raw Materials

	NON-CHERT				CHERT				SAMPLE
	Quartz	Quartzite	Other	Total	Local	Burned	Nonlocal	Total	
16,000-14,000 BP at Cabeço do Porto Marinho (CPM)									
CPM IL*	17.9	3.3	.3	**21.5**	62.2	15.6	.7	**78.5**	1,766
CPM IIM*	20.7	3.6	.6	**24.9**	66.5	8.2	.4	**75.1**	2,260
CPM IIISWL	23.5	3.4	.2	**27.1**	64.0	7.2	1.7	**72.9**	1,262
CPM IIISWM*	17.2	5.5	.4	**23.1**	59.7	15.7	1.5	**76.9**	1,346
12,000-9,000 BP at Cabeço do Porto Marinho (CPM)									
CPM IU*	3.4	1.2	.7	**5.3**	73.6	20.8	.3	**94.7**	4,703
CPM IIISU*	6.9	3.3	.7	**10.9**	55.6	33.2	.3	**89.1**	5,179
CPM IIU*	6.4	2.8	.4	**9.6**	61.2	29.1	.1	**90.4**	2,393
CPM IIIU*	6.7	3.2	.5	**10.4**	61.7	27.5	.4	**89.6**	2,148
CPM IIIT	8.3	3.4	.1	**11.8**	62.0	25.6	.6	**88.2**	2,114
CPM IIISWU	7.5	2.5	.3	**10.3**	68.0	21.4	.3	**89.7**	1,807
CPM V	8.2	5.3	.1	**13.6**	54.3	31.9	.2	**86.4**	5,635
11,000-9,000 BP at Carneira									
PC	3.2	3.5	.1	**6.6**	49.3	43.4	.5	**93.2**	6,074
CR II	3.8	4.5	.2	**8.5**	52.5	38.8	.2	**91.5**	4,162
CR III	1.5	1.8	—	**3.3**	48.3	48.2	.2	**96.7**	3,165

*Results based on preliminary samples from the excavated assemblages.

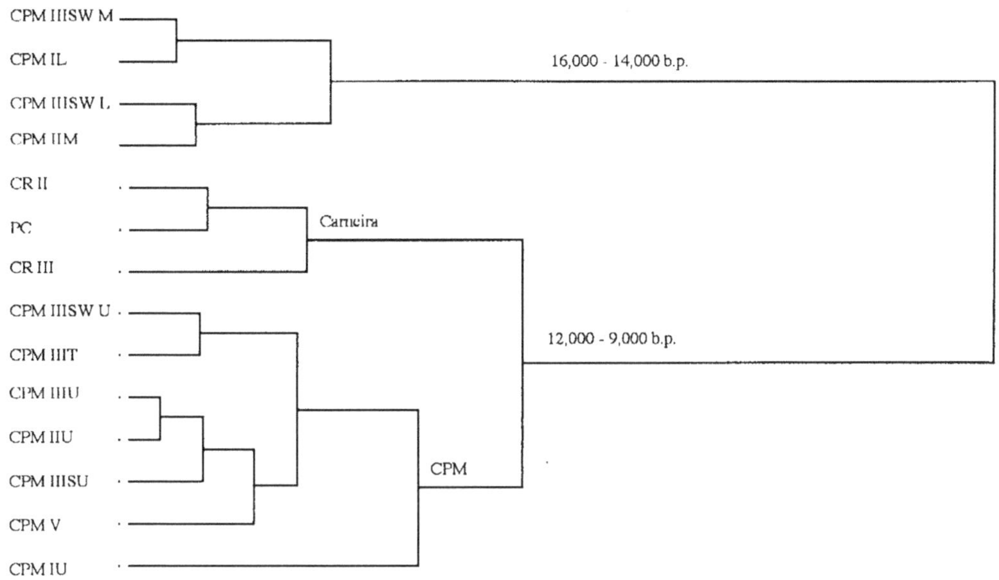

Figure 6. Dendrogram of Cabeço do Porto Marinho (CPM) and Carneira assemblages based on frequency of raw materials (non-chert, chert, and burned chert).

tools were used (i.e., the projectile points) certainly had high activity failure rates.

The increase in the use of chert through time has been explained as a change in cultural preferences (Bicho 1992; Marks *et al.* 1991). Such an explanation seems to be a logical one since the distance to raw material sources did not change through time. There is, however, another possibility. The increase in chert use can be explained as a change in technology in response to a change in mobility patterns. Along with the increase in the frequency of chert, there was a parallel increase in the frequency of projectile points (Table 3). This increase probably corresponds to periodic mass production of these retouched tools, preparing for future expeditions characteristic of a logistical mobility pattern (Kuhn 1989:35). Since only chert was used in the production of weaponry, with the increase in this type of retouched tools there was a higher demand for chert. This trend, however, is not seen in the bladelet production that corresponds to ca. 20% of the blank production in all assemblages. The increases in weaponry, and in chert acquisition and use, were likely the result of the introduction of a new hunting technology, the bow and arrow, as well as of the slow change in land use patterns with a shift from a more residential to a more logistical mobility pattern. This change in land use patterns seems to have been associated with a demographic explosion that occurred between 11,000 and 10,000 BP in the Portuguese Estremadura (Bicho 1994).

The differences in raw material frequencies between sites are certainly the result of distance to the chert sources. A distance of only one or two kilometers is usually seen as having no direct implications for the raw material economy of an area. The data from Rio Maior, however, indicate that in fact even very short distances can have a strong impact not only in the acquisition of raw material but also in the size of formal tools and blanks.

Conclusions

This study has shown that lithic raw materials have played a key role in various aspects of the hunter-gatherer living system. Clearly, lithic raw material availability had direct implications for the organization of both stone tool and hunting technology. Although chert acquisition probably was embedded within the acquisition of other subsistence resources, the high density of chert sources across the landscape had a direct and important role in the routes of logistical excursions, and therefore had direct implications for settlement patterns during the late glacial prehistory of Portugal.

Again, as one would expect, the qualities of the lithic raw materials are reflected in the organization of lithic technology and stone tool production. Low quality raw materials (in Rio Maior, quartz and quartzite) were used in an inexpensive and expedient manner. Formal tools were rarely produced in these raw materials, and when produced resulted in simple forms. In general, little energy was invested in use of quartz and quartzite either during the acquisition or production phases.

Unlike quartz and quartzite, chert was a high quality raw material. It was also an expensive raw material since it had medium to high procurement costs, and very high energy input during the production phase. The planning of activities, including production of formal stone tools as well as detailed knowledge of the location of raw material sources, allowed a low level of stone tool curation in the Portuguese Final Paleolithic. The data from Estremadura indicate thorough knowledge of the availability and scheduling of economic resources present in the region during the year. These resources included different types of large prey species (red deer, wild boar, aurochs, ibex), maritime resources (fish and shellfish), plant species (nuts, berries, and fruits), and, last but not least, chert sources. The organization, scheduling and use of resources were profoundly affected by mobility patterns and hunting techniques. These were closely linked to the availability and high quality of chert in the Estremadura. Therefore, chert played a key role in the whole living system of the Paleolithic hunter-gatherers at the end of the Pleistocene in Estremadura.

Variation in lithic technology has been said to be a multidimensional problem requiring analysts to account for different sources of variation (Andrefsky 1994:31). Different variables such as mobility, settlement pattern, site function, differential attrition rates of stone tools, or quality of raw materials may be combined to explain the organization of specific lithic technologies. Data from the Portuguese Estremadura provide one more example of the wide range of variability in the organization of lithic technology. In Rio Maior, chert had a key role in the organization of lithic technology and stone tool production, as well as implications for the whole web of the economic-subsistence system. The case of Rio Maior and the Portuguese Estremadura is unique, however, due to the large number of raw material sources of both low and high quality.

Acknowledgments

I thank the National Science Foundation (Grant BNS-8803798 to A. Marks and BNS-9107144 to the author), the Portuguese Junta Nacional de Investigação Científica e Tecnológica (Grant BIC-307), the Department of Archaeology of the Instituto Português do Património Arquitectónico e Arqueológico), and the Institute for the Study of Earth and Man, Southern Methodist University (Seed Grant) for funding the Estremadura Research, and Drew University for a research Grant in 1993. Finally, I thank Dr. Maria Masucci for her help in editing this paper.

References Cited

Andrefsky, W.
1994 Raw material availability and the organization of technology. *American Antiquity* 59(1):21-34.

Araújo, A. C.
1994 O concheiro Epipaleolítico do Cabeço do Curral Velho (Cambelas, Torres Vedras). *V Jornadas Arqueológicas* 2:43-51. Associaçao dos Arqueológos Portugueses, Lisbon.

Arnaud, J.
1986 Post-glacial adaptations in southern Portugal. Paper presented at the World Archaeological Congress, Southampton, 1986.
1993 O Mesolítico e a neolitização. Balanço e perspectivas. In *O Quaternário em Portugal: balanços e perspectivas*, edited by G. S. Carvalho, A. B. Ferreira, and J. C. Senna-Martinez, pp. 173-184. Edições Colibri, Lisbon.

Arnaud, J. and J. D. B. Bento.
1988 Caracterização da Ocupação Pré-Histórica da Gruta do Casal Papagaio (Fátima, Vila Nova de Ourém). *Boletim da Sociedade Portuguesa de Espeleologia* 2:27-34.

Bamforth, D. B.
1990 Settlement, raw material, and lithic procurement in the Central Mojave Desert. *Journal of Anthropological Archaeology* 9:70-104.
1991 Technological organization and hunter-gatherer land-use: A California example. *American Antiquity* 56(2):216-234.

Behrensmeyer, A.
1984 Taphonomy and the fossil record. *American Scientist* 72:558-566.

Bicho, N.
1992 *Technological Change in the Final Upper Paleolithic of Rio Maior, Portuguese Estremadura.* Ph.D. Dissertation, Southern Methodist University, Dallas. University Microfilms, Ann Arbor.
1993a Late Glacial prehistory of central and southern Portugal. *Antiquity* 67:761-775.
1993b O Paleolítico Superior Final de Rio Maior: perspectiva tecnológica. *Trabalhos de Antropologia e Etnologia* 33(3-4):15-36.
1993c How two kilometers can make a difference: a raw material study of Epipaleolithic assemblages from Central Portugal. Paper presented at the 58th Annual Meeting of the Society for American Archaeology, St. Louis.
1994 The end of the Paleolithic and the Mesolithic in Portugal. *Current Anthropology* 35(5):664-674.
1996 The role of quartz and quartzite in the Magdalenian of Cabeço de Porto Marinho, Rio Maior, Portugal. In *Non-Flint Stone*

Tools and the Palaeolithic Occupation of the Iberian Peninsula, edited by N. Moloney, L. Raposo and M. Santonja, pp. 175-181. British Archaeological Reports, International Series 649. BAR Publishing, Oxford.

Binford, L.
1973 Interassemblage variability: the Mousterian and the 'functional' argument. In *The Explanation of Culture Change,* edited by C. Renfrew, pp. 227-254. Duckworth, London.
1978 *Nunamiut Ethnoarchaeology.* Academic Press, New York.
1979 Organization and formation processes: Looking at curated technologies. *Journal of Anthropological Research* 35:255-273.
1980 Willow smoke and dogs' tails: hunter-gatherer settlement systems and archaeological site formation. *American Antiquity* 45(1):4-20.
1983 *In Pursuit of the Past.* Thames and Hudson, London.

Binford, L. and S. Binford
1966 A preliminary analysis of functional variability in the Mousterian of Levallois facies. *American Anthropologist* 68:238-295.

Bordes, F.
1950 L'évolution buissonnante des industries en Europe occidentale: considérations théoriques sur le Paléolithique ancien et moyen. *L'Anthropologie* 54:19-34.
1968 *The Old Stone Age.* McGraw Hill, London.

Bordes, F. and D. Sonneville-Bordes
1970 The significance of variability in Paleolithic assemblages. *World Archaeology* 2(1):61-73.

Brain, C.
1982 The Hunters or the Hunted? An Introduction to African Cave Taphonomy. University Chicago Press, Chicago.

Breuil, H., O. Ribeiro and G. Zbyszewski
1943 Les plages quaternaires et les industries préhistoriques du littoral de l'Alentejo entre Sines et Vila Nova de Milfontes. *Communications présentées à la 7ème section du Congrès Luso-Espagnol pour l'Avancement des Sciences*, Volume VIII, pp. 48-62. Porto.

Dias, J. M. A.
1985 Registos de migração da linha de costa nos últimos 18000 anos na plataforma continental portuguesa setentrional. In *Actas da I Reunião do Quaternário Ibérico,* Volume I, pp. 281-295. Lisbon.

Freeman, L. G.
1991 What mean these stones? Remarks on raw material use in the Spanish Paleolithic. In *Raw Material Economies among Prehistoric Hunter-Gatherers*, edited by A. Montet-White and S. Holen, pp. 73-125. Publications on Anthropology, 19. University of Kansas, Lawrence.

Frison, G. C.
1989 Experimental use of Clovis weaponry and tools on African elephants. *American Antiquity* 54(4):766-784.

Geneste, J.-M.
1987 Systèmes d'approvisionnement en matières premières au Paléolithique Moyen et au Paléolithique Supérieur en Aquitaine. In *L'Homme de Neandertal* 8, pp. 61-70. Etudes et recherches archéologiques de l'Université de Liège, 32. Université de Liège, Liège.

Gould, R. A.
1969 *Yiwara, Foragers of the Australian Desert.* Charles Scribner's Sons, New York.

Gould, R. A., editor
1979 *Explorations in Ethnoarchaeology.* University of New Mexico Press, Albuquerque.

Heleno, M.
1944 *O problema Capsence: Contribuição Portuguesa para a sua revisão.* Comunicação ao Instituto de Arqueologia na Sessão de Abril de 1944, Lisbon.
1956 Um quarto de século de investigação arqueológica. *O Arqueólogo Português* II-III:221-237.

Hill, A.
1980 Early postmodern damage to the remains of some contemporary East African mammals. In *Fossils in the Making: Vertebrate Taphonomy and Paleoecology,* edited by A. Behrensmeyer and A. Hill, pp. 131-152. University of Chicago Press, Chicago.

Jeske, R.
1989 Economies in raw material use by prehistoric hunter-gatherers. In *Time, Energy, and Stone Tools,* edited by R. Torrence, pp. 34-45. Cambridge University Press, Cambridge.

Kuhn, S.
1989 Hunter-gatherer foraging organization and strategies of artifact replacement and discard. In *Experiments in Lithic Technology,* edited by D. S. Amick and R. P. Mauldin, pp. 33-47. British Archaeological Reports, International Series 528. BAR Publishing, Oxford.

Marks, A. E. and P. Volkman
1983 Changing core reduction strategies: A technological shift from the Middle to Upper Paleolithic in the Southern Levant.

In *The Mousterian Legacy,* edited by E. Trinkaus, pp. 13-33. British Archaeological Reports, International Series 164. BAR Publishing, Oxford.

Marks, A. E., J. Shokler and J. Zilhão
1991 Raw material usage in the Paleolithic. The effects of local availability on selection and economy. In *Raw Material Economies among Prehistoric Hunter-Gatherers,* edited by A. Montet-White and S. Holen, pp. 127-139. Publications in Anthropology, 19. University of Kansas, Lawrence.

Marks, A. E., N. Bicho, J. Zilhão and R. Ferring.
1994 Upper Paleolithic Prehistory in Portuguese Estremadura: Preliminary results. *Journal of Field Archaeology* 21:53-68.

Mellars, P.
1973 The character of the Middle-Upper Paleolithic transition in Southwest France. In *The Explanation of Culture Change,* edited by C. Renfrew, pp. 255-276. Duckworth, London.

Moloney, N., L. Raposo and M. Santonja, editors
1996 *Non-Flint Stone Tools and the Palaeolithic Occupation of the Iberian Paleolithic.* British Archaeological Reports, International Series 649. BAR Publishing, Oxford.

Montet-White, A. and S. Holen, editors
1991 *Raw Material Economies among Hunter-Gatherers.* Publications in Anthropology, 19. University of Kansas, Lawrence.

Morales, M. G. and J. Arnaud
1990 Recent Research on the Mesolithic in the Iberian Peninsula: Problems and Perspectives. In *Contributions to the Mesolithic in Europe,* edited by P. M. Vermeersch and P. Van Peer, pp. 451-462. Leuven University Press, Leuven.

Pelegrin, J. and C. Chautat
1993 Tecnología y función de las puntas de Paiján: el aporte de la experimentación. *Latin American Antiquity* 4(4):367-382.

Roche, J.
1964 Le Paléolithique supérieur portugais. Bilan de nos connaissances et problemes. *Bulletin de la Société Préhistorique Française* 31:41-47.
1979 Le Magdalénien Portugais. In *La fin des temps glaciaires en Europe, Chronostratigraphie et écologie des cultures du paléolithique final,* edited by D. Sonneville-Bordes, pp. 753-758. Colloques Internationaux du CNRS, 271. CNRS, Paris.
1982 A gruta chamada Lapa do Suão (Bombarral). *Arqueologia* 5:5-18.

Roebroeks, W., J. Kolen and E. Rensink
1988 Planning depth, anticipation and the organization of Middle Paleolithic Technology: The "archaic natives" meet Eve's descendants. *Helinium* 28(1):17-34.

Rolland, N. and H. Dibble
1990 A new synthesis of Middle Paleolithic variability. *American Antiquity* 55(3):480-499.

Russell, M.
1987 Bone breakage in the Krapina hominid collection. *American Journal of Physical Anthropology* 72:371-379.

Shipman, P. and J. Rose
1983 Early hominid hunting, butchering, and carcass-processing behaviors: Approaches to the fossil record. *Journal of Anthropological Archaeology* 2:57-98.

Soares, A. M.
1989 *O efeito de reservatório oceânico nas águas costeiras de Portugal continental.* Instituto de Ciências e Engenharia Nucleares, Lisbon.

Soares, A. M. and J. M. P. Cabral
1989 Idades aparentes das conchas marinhas provenientes da costa portuguesa e a sua variação ao longo do Holoceno. Paper presented to the 2d Conference on the Iberian Quaternary, Madrid, September.

Sonneville-Bordes, D. and J. Perrot
1954 Lexique typologique du paléolithique supérieur: Outillage lithique. *Bulletin de la Société Préhistorique Française* 51 (7):327-333.

Stiner, M.
1991 A taphonomic perspective on the origins of the faunal remains of Grotta Guattari (Latium, Italy). *Current Anthropology* 32(2):103-117.

Straus, L. G.
1991 The role of raw materials in upper Paleolithic and Mesolithic stone artifact assemblage variability in Southwest Europe. In *Raw Material Economies among Prehistoric Hunter-Gatherers,* edited by A. Montet-White and S. Holen, pp. 190-185. Publications in Anthropology, 19. University of Kansas, Lawrence.
1995 Diversity in the face of adversity: Human adaptations to the environmental changes of the Pleistocene-Holocene transition in the Atlantic regions of Aquitaine, Vasco-Cantabria and Portugal. In *Los Últimos Cazadores. Transformaciones culturales y económicas durante el Tardiglaciar y el inicio del Holoceno en el ámbito mediterráneo,* edited by V. Villaverde, pp. 9-22. Ediciones Instituto da Cultura Fundación Gil-Albert, Alicante.

Torrence, R.
1983 Time budgeting and hunter-gatherer technology. In *Hunter-Gatherer Economy in Prehistory*, edited by G. Bailey, pp. 11-22. Cambridge University Press, Cambridge.

White, R.
1982 Rethinking the Middle/Upper Paleolithic transition. *Current Anthropology* 23(2):169-192.

Yellen, J.
1977 *Archaeological Approaches to the Present: Models for Reconstructing the Past.* Academic Press, New York.

Zilhão, J.
1987 *O Solutrense da Estremadura portuguesa. Una proposta de interpretação paleoantropológica.* Trabalhos de Arqueologia 4. Departamento de Arqueologia do Instituto Português do Património Cultural, Lisbon.
1992a *Gruta do Caldeirão, O Neolítico Antigo.* Trabalhos de Arqueologia 6. Instituto do Património Arquitectónico e Arqueológico, Departamento de Arqueologia, Lisbon.
1992b Estratégias de Povoamento e Subsistência no Paleolítico e no Mesolítico de Portugal. In *Elefantes, ciervos y ovicaprinos,* edited by A. M. Romanillo, pp. 149-162. Universidade de Cantabria, Santander.

Zilhão, J., A. Marks, C. Reid Ferring, N. Bicho and I. Figueiral
1995 The Upper Paleolithic of the Rio Maior basin (Portugal). Preliminary results of a 1987-1993 Portuguese-American research project. *Trabalhos de Antropologia Etnologia* 35(4):69-82.

www.ingramcontent.com/pod-product-compliance
Lightning Source LLC
LaVergne TN
LVHW071211210926
848482LV00015B/781